De Novo Quantum Cosmology with Artificial Intelligence

Experiments attempting to recreate the big bang and measurements in deep space point to the tantalizing possibility that our universe may be the relic of something simple, powerful, and highly symmetric. The evidence suggests an entity where matter and energy cannot be told apart and the four fundamental forces are unified into one. Empowered by artificial intelligence, *De Novo Quantum Cosmology with Artificial Intelligence* seeks to unravel the mystery as it searches for an encompassing physical picture where it all falls into place at the aftermath of creation from a quantum void. From the outset, AI reckons that the problem cannot be tackled without proper contextualization, that is, without dealing with other intimately related problems in particle cosmology including the nature of dark matter and dark energy, the hierarchy problem of particle masses, the incommensurably weak coupling strength of gravity, the universe topology, the cosmological constant problem, and the vacuum catastrophe. Accordingly, the book addresses the matter in its full conceptual richness. This monograph addresses a broad readership that includes nonhuman audiences involving AI systems. A background in college-level physics and computer science would be essential. Although informal in the approach, the material is presented with scientific rigor, so that readers gain hands-on experience on the subject. The book is geared at graduate students as well as professional physicists, mathematicians, cosmologists, and big data scientists that seek to venture into some of the core problems in particle cosmology empowered by AI. Notably, the book is also geared at nonhuman audiences, since AI systems may incorporate its fundamental operational tenets and take the matter to unfathomable heights.

Key Features:

- Introduces an artificial intelligence system to tackle core problems in particle cosmology.
- Describes a grand unification scheme to explain the common origin of the fundamental forces.
- Identifies the origin of matter as a phase transition from the quantum vacuum.

Ariel Fernández (born Ariel Fernández Stigliano, April 8, 1957) is an Argentine-American physical chemist and mathematician. He obtained a Ph.D. degree in Chemical Physics from Yale University in record time and held the Karl F. Hasselmann Endowed Chair Professorship in Engineering at Rice University until his retirement. He was also an Adjunct Professor of Computer Science at the University of Chicago. To date, he has published approximately 500 scientific papers in professional journals and has also authored ten books on physical chemistry, molecular medicine, artificial intelligence, cosmology, and mathematical physics. Additionally, he holds several patents on technological innovation. Fernández is a senior member of the National Research Council of Argentina (CONICET) and the CSO of the Daruma Institute for Applied Intelligence, the research arm of AF Innovation, a Consultancy based in Argentina and the USA.

Chapman & Hall/CRC
Artificial Intelligence and Robotics Series
Series Editor: Roman Yampolskiy

Topological Dynamics in Metamodel Discovery with Artificial Intelligence
From Biomedical to Cosmological Technologies
Ariel Fernández

A Robotic Framework for the Mobile Manipulator
Theory and Application
Nguyen Van Toan and Phan Bui Khoi

AI in and for Africa
A Humanist Perspective
Susan Brokensha, Eduan Kotzé, Burgert A. Senekal

Artificial Intelligence on Dark Matter and Dark Energy
Reverse Engineering of the Big Bang
Ariel Fernández

Explainable Agency in Artificial Intelligence
Research and Practice
Silvia Tulli
David W. Aha

An Introduction to Universal Artificial Intelligence
Marcus Hutter, Elliot Catt, and David Quarel

AI: Unpredictable, Unexplainable, Uncontrollable
Roman V. Yampolskiy

De Novo Quantum Cosmology with Artificial Intelligence
Applications of Formal Autoencoders
Ariel Fernández

Transcending Imagination:
Artificial Intelligence and the Future of Creativity
Alexander Manu

Responsible Use of AI in Military Systems
Jan Maarten Schraagen

AI iQ for a Human-Focused Future
Strategy, Talent, and Culture
Seth Dobrin

Federated Learning
Unlocking the Power of Collaborative Intelligence
Edited by *M. Irfan Uddin and Wali Khan Mashwan*

Designing Interactions with Robots
Methods and Perspectives
Edited by *Maria Luce Lupetti, Cristina Zaga, Nazli Cila, Selma Šabanović, and Malte F. Jung*

The Naked Android
Synthetic Socialness and the Human Gaze
Julie Carpenter

For more information about this series please visit: www.routledge.com/Chapman--HallCRC-Artificial-Intelligence-and-Robotics-Series/book-series/ARTILRO

De Novo Quantum Cosmology with Artificial Intelligence

Applications of Formal Autoencoders

Ariel Fernández

CRC Press
Taylor & Francis Group
Boca Raton London New York

CRC Press is an imprint of the
Taylor & Francis Group, an **informa** business
A CHAPMAN & HALL BOOK

Front cover image: Rroselavy/Shutterstock

First edition published 2026
by CRC Press
2385 NW Executive Center Drive, Suite 320, Boca Raton FL 33431

and by CRC Press
4 Park Square, Milton Park, Abingdon, Oxon, OX14 4RN

CRC Press is an imprint of Taylor & Francis Group, LLC

ISBN: 978-1-041-04503-8 (hbk)
ISBN: 978-1-041-05922-6 (pbk)
ISBN: 978-1-003-63292-4 (ebk)

DOI: 10.1201/9781003632924

Typeset in Times
by Newgen Publishing UK

Dedication

In loving memory of Haydée Stigliano, my mother

Contents

1 Propaedeutics of Particle Cosmology in a Quest for the Origin of the Universe

"Φύσις κρύπτεσθαι φιλεῖ".

Heraclitus
(*Nature likes to keep its secrets*,
Translation by the author)

Experimental evidence recreating the big bang and deep-space observation points to the tantalizing possibility that our universe may be the relic of something simple, powerful, and highly symmetric. As to the nature of this primeval entity, physicists' favorite answer involves a fluctuation in the quantum vacuum. The evidence suggests an entity where matter and energy cannot be told apart, the four fundamental forces are unified into one, and the quantum vacuum is somehow endowed with attributes of ur-matter, that is, capable of acting as a precursor to the observable matter.

This book describes an approach to the problem of the origin of the universe empowered by artificial intelligence (AI). As it searches for an encompassing physical picture where it all falls into place at the aftermath of creation from a quantum void, AI reckons that the problem cannot be tackled without proper contextualization, that is, without dealing with other intimately related problems in particle cosmology that bear on the understanding of the primeval forces that shaped in the universe. These problems include the nature of dark matter (DM) and dark energy (DE), the universe topology, the vacuum catastrophe, and the cosmological constant problem. Accordingly, this chapter introduces elements of particle cosmology that enable the pursuit of the problem of the origin of the universe in its full conceptual and relational richness as required by the AI system that empowers the approach.

This preparatory chapter delineates an AI metamodel of the dark universe that is of the invisible matter in all forms, including dark energy, assumed to be present in overwhelming amounts in deep space. By metamodel, we mean an inferential scheme that acquires big evidentiary data and generates a matrix suited for a cogent physically consistent model capable of meaningful prediction. The evidence includes deep-space phenomenology and radiation relics of the big bang and other cosmic reproductive events. By leveraging fairly elementary physical arguments, the chapter first surveys the experimental evidence from deep-space observation that supports the existence of the dark sector. The data are then reconciled with the incorporation of a latent extra spatial dimension, modeled by an AI system named *autoencoder*, that distills the quantum reality from a space–time encoded in five dimensions. In this way, the chapter paves the way to the AI-empowered approach to the origin of the universe described in Chapter 2.

DOI: 10.1201/9781003632924-1

1.1 DARK MATTER AND DARK ENERGY IN THE AFTERMATH OF THE BIRTH OF THE UNIVERSE

This book muses endlessly over the possibility that our universe may be the relic of something simple, powerful and highly symmetric. This tantalizing possibility is surely suggested by the experimental evidence that helped establish the big bang and deep-space observation data. As to the nature of the primeval entity, physicists' favorite answer involves a fluctuation in the quantum vacuum. The evidence suggests an entity where matter and energy cannot be told apart, the four fundamental forces are unified into one, and the quantum vacuum is somehow endowed with attributes of ur-matter, that is, capable of acting as a precursor to the observable matter. The cosmological birth then becomes akin to a first-order phase transition where primeval spatial symmetries are transmuted into gauge symmetries. In this overarching model, the emerging gauge fields with the proper symmetry interact with photons conforming to visible matter, while those fields with incompatible symmetry represent dark matter, the scaffold of the universe in deep space. To complete the picture, dark energy becomes the latent heat released with the first-order phase transition.

That a phase transition may internalize a spatial symmetry as a latent symmetry which is spontaneously broken is by no means an alien concept to physics. However, transferring this picture into a viable *ansatz* for the origin of the universe poses a number of challenges to the theoretician, beginning with the definition of the critical temperature, T_c. Thus, if we are concerned with the origin of mass, T_c should be taken to be $T_c = E_{TV}/k_B$, where E_{TV} is the vacuum expectation value (true vacuum) for the Higgs field and k_B is Boltzmann's constant. An AI system will be implemented to model this crucial phase transition and generate the physics that yields the symmetry breaking, leading to the emergence of mass. In a skeletal representation, the AI system takes the form of a commutative arrow diagram, as shown in Figure 1.1. An extra compact dimension is incorporated into space–time, yielding the ur-manifold W to account for the dramatic geometric dilution of gravity relative to the other three fundamental forces of established quantum origin. The gauge fields thus emerge as supported by a 4D coarse-grained manifold $W/\sim$, the standard space–time, rendered as quotient space via an equivalence relation "$\sim$" that filters out the extra dimension. The horizontal arrows $W \rightarrow W/\sim$ become canonical projections, surjectively assigning the corresponding equivalence class to each point in the 5D manifold, while the vertical arrows represent the symmetry breaking associated with the phase transition, $T \gg T_c \rightarrow T < T_c$, at the two different levels of resolution. This book upholds and advocates for this overarching picture and justifies this position as it approaches the problem of the origin of the universe empowered by AI. It is necessary to understand that the book covers uncharted territory, where AI formalizes quantum cosmology by implementing large language models (LLM) enshrined in formal proof assistants. The AI system is delineated in Chapter 3, while the formalization and mathematical proving processes are described in detail in the Appendix. The AI system, termed *autoencoder*, distills the essential physical picture from a vastly more complicated representation. This is the type of the AI system that can assist the reader coming to grips with quantum gravity, the holy grail in the grand unification of fundamental forces, purportedly holding the key to the origin of the universe.

As it searches for an encompassing physical picture where it all falls into place at the aftermath of creation from a quantum void, AI reckons that the problem cannot be tackled without proper contextualization, that is, without dealing with other intimately related problems in particle cosmology that bear on the understanding of the primeval forces that shaped in the universe. These problems include the nature of dark matter and dark energy, the universe topology, the vacuum catastrophe and the cosmological constant problem. Accordingly, this chapter introduces elements of particle cosmology that enable the pursuit of the problem of the origin of the universe in its full conceptual and relational richness.

Dark matter (DM) and dark energy (DE) become the key relics in the aftermath of the birth of the universe. As shown in Chapter 2, DM along with visible matter emerged from the first-order

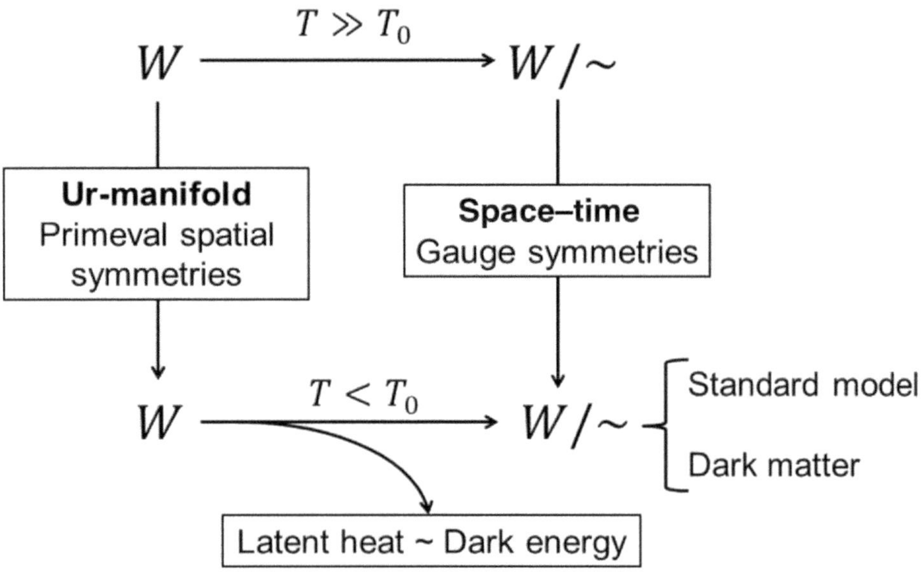

FIGURE 1.1 Blueprint for the birth of the universe as a phase transition internalizing spatial ur-symmetries as gauge symmetries. An AI system will be implemented to model this phase transition and generate the physics that yields the symmetry breaking leading to the emergence of mass. The AI system is represented as a commutative arrow diagram. An extra compact dimension is incorporated to space–time, yielding the ur-manifold W to account for the dramatic geometric dilution of gravity relative to the other three fundamental forces. The gauge fields thus emerge as supported by a 4D coarse-grained manifold $W/\sim$, the standard space–time, rendered as quotient space via an equivalence relation "$\sim$" that filters out the extra dimension. The horizontal arrows $W \to W/\sim$ become canonical projections, while the vertical arrows represent the symmetry breaking associated with the phase transition, $T \gg T_c \to T < T_c$.

phase transition as two different alternative ways of internalizing or "gauging" the primeval symmetry, and since DM is the gravitational scaffold of the universe, the gravitational waves sprouting from the primeval phase transition should be detectable even today. On the other hand, if DE is the latent heat released with the primeval phase transition, how can the DE concentration remain constant, as experimentally verified in a universe subject to accelerated expansion? Quantum vacuum energy generated autocatalytically in an ever-expanding vacuum does not seem to be the straightforward answer, as suggested by the so-called vacuum catastrophe problem. This conundrum is resolved in Chapters 2 and 5, where a model of DE depletion/replenishment dynamics is put forth by the autoencoder to account for the sustainability of the universe topology in the observed runaway mode.

This preparatory chapter delineates an AI metamodel of the dark universe that is of the invisible matter in all forms, including DE, assumed to be present in overwhelming amounts in deep space. By metamodel, we mean an inferential scheme that acquires big evidentiary data and generates a matrix suited for a cogent physically consistent model capable of meaningful prediction. The evidence includes deep-space phenomenology and radiation relics of the big bang and other cosmic reproductive events. By leveraging fairly elementary physical arguments, the chapter first surveys the experimental evidence from deep-space observation that supports the existence of the dark sector. The data are then reconciled with the incorporation of a latent extra spatial dimension, modeled by the autoencoder, that distills the quantum reality from a space–time encoded in five dimensions. In this way, the chapter paves the way to the AI-empowered approach to the origin of the universe described in Chapter 2.

1.2 THE LATENT SPACE OF DARK DIMENSION AND THE NATURE OF THE QUANTUM VACUUM: AI SYSTEM TO ENCODE THE DARK SECTOR

The dark universe constitutes a vast proportion of matter and energy in deep space that cannot be accounted for with extant physical models. To shed light on the problem, AI is leveraged and testable assumptions are made concerning the nature of dark matter and dark energy. As described in the subsequent chapters, cosmological data will be inputted into an appropriate AI system under the assumption that there exists a "dark dimension" such that any particle field concentrated along this dimension will not interact with photons unless there is a spillover or "geometric dilution" of the field into the observable dimensions.

This picture prompts us to tentatively model the universe adopting a generic AI system under the guise of what we have termed an *autoencoder*. In a standard view, autoencoders can be thought of as embodiments of machine learning systems suited to predict the dynamical behavior of a physical system from a time series of a set of observables. This time series can be spliced into a training set and a testing set. As described in Chapter 3, the predictive power of autoencoders hinges on distilling the dynamics into a simplified version named "latent dynamics" which is assumed to be operative in a lower dimensional space (usually a differentiable manifold), known as "latent space". The dimensionality reduction is presumed to capture the essential dynamics, that is, the modes that enslave fast-relaxing degrees of freedom that can be averaged out for predictive purposes.

To illustrate the concept of autoencoder, suppose the universe was a giant molecule, an assemblage of N atomic nuclei and J electrons. In a classical physics view, the state of the system at time t would correspond to a point $\boldsymbol{x} = \boldsymbol{x}(t)$ in the space $W = \mathbb{R}^{3(N+J)}$ and a time series training the AI system would be given as $\left\{\boldsymbol{x}(n\tau), \boldsymbol{x}((n+1)\tau)\right\}_{n=0,1,\ldots,M}$, where the period τ indicates the level of time coarse graining. Then, the prediction would entail an estimation of $(L\tau), L = M+2, M+3, \ldots$. By making physically reasonable assumptions, it is possible to simplify the dynamics with significant dimensionality reduction, thereby enabling the prediction. For example, it is well known that the electron "motion" can be treated "adiabatically", that is, in an averaged way relative to the nuclei. This is so since the electron degrees of freedom are entrained by those of the nuclei. This is the so-called Born–Oppenheimer (B-O) approximation adopted in molecular and atomic physics, as described in Chapter 3, and enables the autoencoder to adopt a latent manifold $\Omega = \mathbb{R}^{3N}$, thus significantly reducing dimensionality by averaging out the 3J electron degrees of freedom.

A more generic way of visualizing the encoding, more akin to particle physics, would be to consider the level of delocalization (uncertainty in the determination of position) of the particles in the assemblage. The delocalization is given by the so-called Compton wavelength defined as $\lambda_C = \frac{1}{2}\frac{h}{mc}$, where h is Planck's constant, c is the speed of light, and m is the particle mass. Thus, it becomes clear that, relative to the nuclei, the electrons may be represented as "clouds" with averaged-out motion, whereas the nuclei, with their far larger mass and hence significantly smaller Compton wavelengths, may be treated as localized objects.

Based on the B-O *ansatz*, we may construct an autoencoder adopting a multi-layered neural network technology (Chapter 3), so that the dynamics in W becomes encoded or simplified at a coarser but fundamental level in Ω and subsequently decoded back to W (Figure 1.2). Essentially, if $F: W \to W$ represents the time evolution map such that $F\boldsymbol{x}(n\tau) = \boldsymbol{x}((n+1)\tau)$, the autoencoder is determined by a tern (γ, K, μ) consisting of the map $K: \Omega \to \Omega$, the projection $\gamma: W \to \Omega$, and the hologram (injection) $\mu: \Omega \to W$, optimized in such a way that the following commutativity relations hold: $K \circ \gamma = \gamma \circ F$ and $F \circ \mu = \mu \circ K$ (Figure 1.1). In other words, the autoencoder is able to predict the output $\boldsymbol{x}((L+1)\tau)$ from input $\boldsymbol{x}(L\tau)$ as $\boldsymbol{x}((L+1)\tau) = (\mu \circ K \circ \gamma)\boldsymbol{x}(L\tau)$.

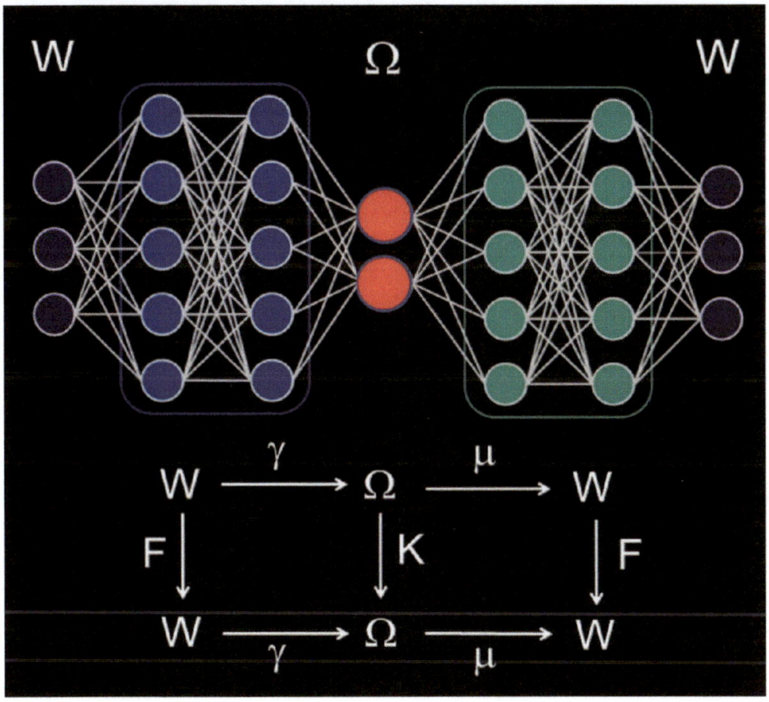

FIGURE 1.2 Observable universe (Ω) modeled within an AI system under the guise of an autoencoder constructed using multi-layered neural network technology and defined by the term (γ, K, μ). The dark universe becomes the portion of a higher dimensional universe *W* that incorporates a dark dimension and excludes the hologram $\mu\Omega$. The autoencoder parameters are optimized so that the diagram at the bottom becomes commutative.

Autoencoders are introduced in different guises throughout the book, from deep learning machines (Chapter 3) to formal language models (Appendix), trained by big data (Chapters 4 and 5) or by formal axiomatic systems (Chapters 2 and 6). The details on the implementations carried out by the author are provided in Chapter 3 and in the Appendix.

Human civilization has been able to physically model the fundamental particle fields that constitute the fabric of the universe, as described by the so-called standard model (SM) of particle physics. However, the model clearly does not encompass the dark portion of the universe, the vast portion now known to be impervious to interaction with photons and other gauge bosons in the SM. Thus, we may regard the extant physical model as actually corresponding to the evolution map on a latent manifold $K_{SM}: \Omega \to \Omega$, but not on *W*. Unlocalizable (undetectable) particles, altogether lacking Compton wavelength, are excluded from the latent physics, requiring (AI postulates) a dark dimension to materialize.

We know that the K_{SM} model is unsatisfactory in deep space because it does not incorporate the bearing of the dark universe on the detectable dynamics. When this influence is incorporated, especially at the large scales of deep space, we end up with a model of broader applicability $K^{\odot}_{SM}: \Omega \to \Omega$ with $K_{SM} \neq K^{\odot}_{SM}$. Under the u niverse-as-autoencoder *ansatz*, determining the physics of the dark universe becomes tantamount to identifying the dark universe as the large portion, $W \setminus \mu\Omega$, of *W* that excludes the hologram $\mu\Omega$ and determining the autoencoder model (γ, F, μ) that satisfies the commutative relations $K^{\odot}_{SM} \circ \gamma = \gamma \circ F$ and $F \circ \mu = \mu \circ K^{\odot}_{SM}$.

The autoencoder scheme will also be used in a new development as an abstract formalization tool embodied as a large language model (LLM) enshrined within a theorem proof assistant (Chapter 2 and Appendix). This AI-empowered tool is aptly named "formal autoencoder" and enables the formalization of a physical theory cast on one representation in order to distill it as a projection into a simpler representation. Chapter 2 will make extensive use of formal autoencoders as it makes general relativity compatible with quantum physics in order to seek the ultimate unification in a theory of quantum gravity. Both theories represent towering achievements in the understanding of the universe at what appeared to be vastly different scales, from cosmological to subatomic. Yet, as we come to grips with cosmological singularities such as black holes, the locality and dimensions under consideration enabled the possibility of a common ground for quantum gravity, purportedly bridging the two towering theories [1]. At the same time, the awareness that a vast proportion (95%) of the universe contains invisible or undetectable forms of matter and energy added further impetus to the pursuit of quantum gravity as a vantage point to understand unyielding aspects of deep space and interpret the phenomenology that arises thereof. The hitherto unknown forms of matter and energy constitute the dark universe and are likely to trigger a paradigm shift [2,3]. This assertion served in good measure as the motivation for this book.

Three pivotal tenets, justified and elaborated in the subsequent chapters, define our approach to understanding the dark universe in the context of the birth of the universe:

I. The topology of space–time is a determinant of the nature of the dark universe.
II. The universe is compact and cannot have boundaries because there is no interface with nothingness, and therefore, space–time is a multiply connected manifold.
III. The universe contains a hologram in the sense that the quantum fabric of space is apparent within a dimension reduction yielding a latent manifold that encodes the relativistic space–time [4].

The last tenet implies that an AI system in the guise of an autoencoder may be ideally suited to study quantum gravity [5], as shown in Chapter 2. Autoencoders may be thought of as learning systems that distill the fundamental physics underlying a process or phenomenology by encoding it in a lower dimensional manifold, known as a latent manifold. Thus, *a hologram becomes styled as an autoencoder*. A physical embodiment of the autoencoder may be tuned with adherence to statistical mechanics laws in such a way as to "realize" quantum gravity [5].

The existence of a dark universe summons a hologram setting in which the phenomenology associated with the dark universe manifests itself in the latent manifold (Ω) only indirectly and through gravitational perturbation. This manifold corresponds to the realm of the detectable, the realm where, in terms of quantum physics, an act of observation collapses the wave function. However, as shown in this book, the dark universe is contained in a higher dimensional space (W), an ur-universe, that admits no observer. Only 5% of matter in all forms in W is encoded by Ω, with the remaining 95% corresponding to the dark universe. We may state that "*the dark universe is the part of W that is not part of the hologram*". In rigorous terms, if $\eta : \Omega \to W$ is the holographic function, the dark universe $\mathcal{W}$ is defined as $\mathcal{W} = W \setminus \eta(\Omega)$.

Let $K^{\odot}{}_{SM} \supset \mathcal{K}_{SM} : \Omega \to \Omega$ define a generic transformation beyond the SM, then an overarching goal is to identify the transformation $\mathcal{F} : W \setminus \mathcal{W} \to W \setminus \mathcal{W}$ jointly with the map $\gamma : W \setminus \mathcal{W} \to \Omega$ such that the diagram in Figure 1.3 is commutative, that is, $\gamma \circ \mathcal{F} = K^{\odot}{}_{SM} \circ \gamma$. The dark universe is thus determined by the compatibility of the observable universe (Ω, η), and the ur-universe (W) guaranteed by the commutativity of the diagram in Figure 1.3, which delineates the metamodel for the dark universe.

This picture is incomplete and in a sense unsatisfactory in so far as the ur-universe remains undefined. To grasp the nature of the problem, we need to incorporate the tenets I and II. The topology of the universe prompts the autoencoder to determine the embedding of the four-dimensional space–time W_4 into the higher dimensional space W. As discussed in the subsequent chapters, the spatial dimensions in W_4 must be compact because they once were, if we grant validity to the big

$$\begin{array}{ccc} W \setminus \mathcal{W} & \xrightarrow{\gamma} & \Omega \\ \mathcal{F} \big\downarrow & & \big\downarrow \mathcal{K}_{SM}^{\odot} \\ W \setminus \mathcal{W} & \xrightarrow{\gamma} & \Omega \end{array}$$

FIGURE 1.3 Metamodel of the dark universe $\mathcal{W} = W \setminus \eta(\Omega)$ understood as an AI system in the form of an autoencoder. The dark universe is identified with the portion of the ur-universe W that excludes the hologram, $\eta(\Omega) \subset W$, encoded by the latent manifold Ω. The dark universe is characterized by the lifting $(\mathcal{F}, \gamma)$ of a "corrected" standard model map $K^{\odot}_{SM}$ that makes the diagram commutative.

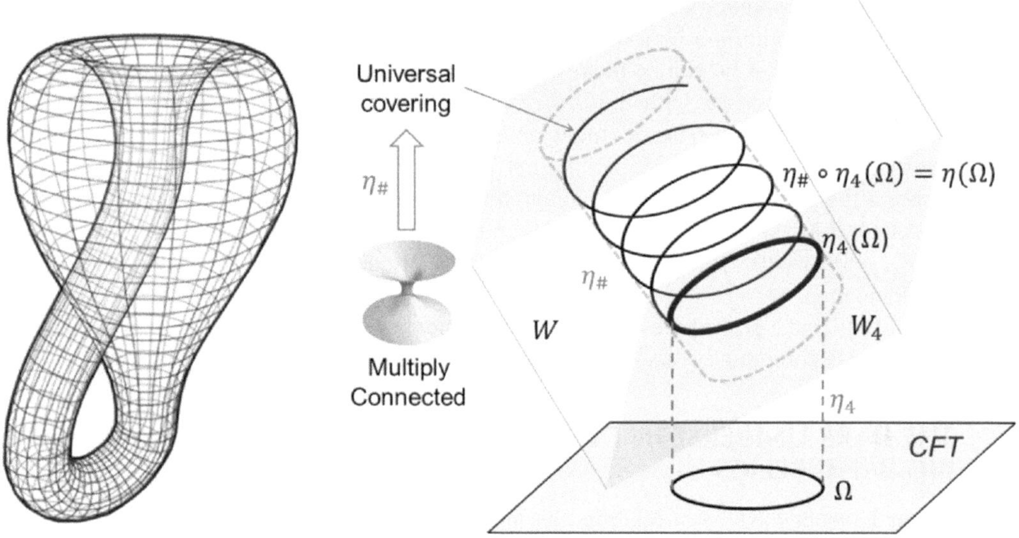

FIGURE 1.4 Artefactual singularities of space–time smoothed out by lifting at the level of the universal covering. (a) The Klein bottle, a nonorientable two-dimensional manifold that self-intersects when represented in a three-dimensional space, but not in four dimensions. (b) The multiverse or ur-universe W topologically represented as the universal covering of a holographic autoencoder.

bang scenario in a relativistic context, and relativistic space–time cannot afford topological change. Furthermore, Ω cannot have a boundary since that would imply interfacing with nothingness, that is, not with vacuum but with the absence of geometry. Such an interface represents a physical and metaphysical impossibility. As a compact manifold without boundaries, Ω must be multiply connected and, consequently, so must be also its hologram $\eta_4(\Omega) \subset W_4$. However, as shown in Chapter 5, to retain rotational symmetry (more specifically, axis permutation), the multiply connected space–time must be endowed with a primeval wormhole. This singularity, however, should be regarded as artefactual of the dimensionality of the space that admits the quantum observer. The situation is akin to that of the Klein bottle, a nonorientable surface that features self-intersection when displayed in the three-dimensional space but not in four dimensions, where it is not amenable to representation (Figure 1.4a). Furthermore, as shown in Chapter 4, W_4 should be embeddable in a space, W_5, with an extra spatial compact dimension required to store dark matter in the form of stationary waves with no geometric dilution into the other spatial dimensions.

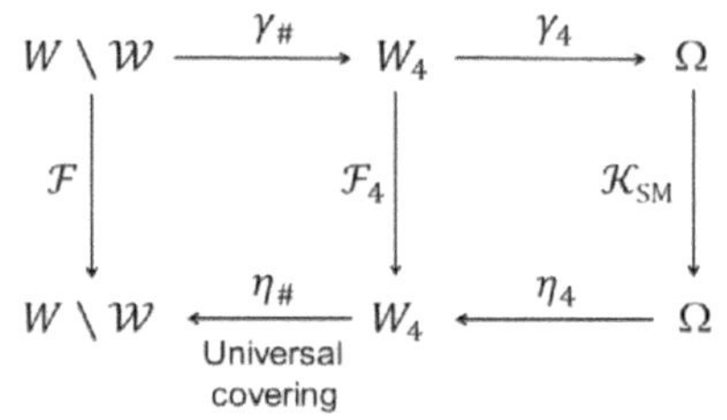

FIGURE 1.5 Characterization of the dark universe through a cogent metamodel of the universe represented as an autoencoder. The compatibility between the holographic and lifted (multiverse) representation of the latent physics enshrined in the standard model is warranted by the commutativity of the diagram.

The ur-universe W becomes then the universal (simply connected) covering of W_5. In this way, the simply connected topology that smooths out the inherent singularities of W_5 is recovered at the expense of incorporating a noncompact spatial dimension (Figure 1.4b). This implies that the holographic map $\eta: \Omega \to W$ can be factorized as $\eta = \eta_\# \circ \eta_4$, through the embedding of the latent space $\eta_4 : \Omega \to W_4$ composed with the lifting $\eta_\# : W_4 \to W$, which is a multi-valued map. This implies that *the universal covering W (together with the lifting* $\eta_\#$*) corresponds to an AI encoding of the many-worlds interpretation of quantum mechanics.* This interpretation is thus validated by AI through the commutativity of the diagram presented in Figure 1.5, whereby the dark universe is identified as $\mathcal{W} = W \setminus (\eta_\# \circ \eta_4)(\Omega)$ and the following relations hold: $\mathcal{F} = \eta_\# \circ \mathcal{F}_4 \circ \gamma_\#$, $\mathcal{F}_4 = \eta_4 \circ \mathcal{K}_{SM} \circ \gamma_4$, where $\gamma_\#, \gamma_4$ are, respectively, the projections associated with the lifting and the holographic map.

One would say that a goal of the book is to determine the maps $\mathcal{F}$, $\gamma = \circ\gamma_4\gamma_\#$, that make the diagram in Figure 1.5 commutative and thereby enable the elucidation of the dark universe.

1.3 THE DARK UNIVERSE IN DEEP-SPACE OBSERVATION AND BIG BANG RELICS

Radiation and matter have been regarded as interrelated concepts since humans pondered about nature and the cosmos, way before the advent of physics as a formal discipline. One of the latest musings over this relationship came with Albert Einstein, who enquired how could the photon, the quantum of light, carry momentum while it has no mass. That perplexity eventually led him to formulate what may well be the most famous equation in all of physics: $E = mc^2$, where E = energy, m = mass at rest, and c = speed of light. The idea of finding the massive particle equivalent to the photon also led him to the idea that gravity may affect radiation, which had no clear precedence in physics.

Rather than writing the momentum of the photon in standard form as $p = mc$ (p = momentum), he chose the more fundamental expression $p = E/c$, where E is the "kinetic energy" of the photon, which, as shown by the physicist Max Planck (1858–1947), is proportional to its frequency f as radiation carrier. The proportionality factor is Planck's constant h, so that we may write $E = hf$. But the crux of Einstein's argument is the question: What would be the mass of a particle that would carry the same momentum as the photon? Reciprocally, we may ask what would be the frequency of the photon "equivalent" (i.e., with the same momentum) to a particle of rest mass m. Based on Einstein's relation – and assuming momentum to be far smaller than mc – the answer at hand is $f = \frac{mc^2}{h}$.

Now, what if we suddenly decide to postulate that only 5% of matter, made up of particles endowed with mass or light quanta equivalents, actually obeys the relation $f = \frac{mc^2}{h}$? That would be indeed a shocking statement meeting with immediate skepticism, and yet it reflects the current state of affairs in the realm of deep-space cosmology [1]. We shall provisionally name matter that does not interact, exchange, or communicate with radiation "dark matter (DM)", and reciprocally, we shall term "dark energy" (DE), the energy that cannot find a material equivalence.

It is clear that the Einstein relation did not hold in a big bang scenario for the origin of the universe: If it did, radiation would have evened out the distribution of mass in the cosmos and would have smoothened out fluctuations in the radiation relic of the big bang detectable today and known as "cosmic background radiation" (CMB) [1]. So, we may conjecture that dark matter and dark energy originated in an early universe and prevailed, so that Einstein's relation is upheld only for 5% of matter, for what we have aptly chosen to name "detectable" or "visible" matter. This book is devoted to elucidating the nature of the vast proportion of matter in the cosmos that does not communicate or interact with light quanta, the "dark cosmos".

Physics is an established and respected field of knowledge that endeavors to explain how the universe works at all scales. Its corpus incorporates ideas, models, and data only after careful scrutiny. The bar is high and scientists who want to leave their mark face a stringent peer review process and only get to impose their views after a hard-won battle. Solid as it seems, there is nothing monolithic, no final word in physics. Truth is never fully conquered but stands as a beacon for the daring. Each time there is a breakthrough, the veil of mystery is lifted a little but as the horizon expands, new mysteries arise. Every concept, every theory, every measurement is constantly subject to revision as new windows of reality open up to detection and technologies are endlessly perfected. Paradigms, even those that appear rock hard for a while and endure a long-term attrition, often crumble under the weight of new and disconcerting evidence or undergo extensive revision. The history of physics is endlessly made of cycles of destruction and creation, much like the ancient cosmogonies of the valley of the Indus, dictated by the infinite toils of Shiva, the destructive element that dances out the pulse of the universe.

In or around 1900, the Irish physicist William Thomson (1824–1907), usually referred to as Lord Kelvin, famously declared: "*There is nothing new to be discovered in physics now. All that remains is more and more precise measurement*". In the two or three decades that followed after this pronouncement, two earth-shattering revolutions in physics took place: Einstein's theory of relativity and quantum mechanics [1]. So much for Lord Kelvin's solemn pronouncement... History proved once again to be the master of irony.

Relativity and quantum physics thrived and prospered because they effectively and successfully addressed shortcomings in the prevailing paradigm at the turn of the 20th century. This paradigm is essentially enshrined in two basic pillars of knowledge: (a) Newton's law of gravitation that governs the dynamics of falling and orbiting bodies at terrestrial (the apocryphal falling apple) and cosmic scales and (b) Maxwell's laws governing electromagnetic phenomena, that is, the events that reveal the entanglement between electricity and magnetism [1]. It thus seemed that all the effects involving the known forces in the universe at the time were satisfactorily understood, even if the nature of such forces remained unyielding to theoretical efforts. For example, when asked about the nature of gravity, Newton snapped in Latin: "*Hypotheses non fingo*" (I contrive no hypothesis). In an ironic turn, right after Lord Kelvin's pronouncement, Einstein came up with relativity, the first theory that truly explained gravity, while an avalanche of new data revealed that matter at atomic and subatomic scales required a drastic revision of the extant conceptual framework, heralding the birth of quantum mechanics.

Einstein showed that time and space are inevitably entangled and should not be treated separately, with time acting as the fourth dimension. His view of gravity in "space–time" is admirably synthesized in the quote of the American physicist John Archibald Wheeler (1911–2008): "Space-time tells matter how to move, matter tells space-time how to curve" [1]. On the other hand, quantum mechanics successfully accounted for the discontinuous nature of energy and momentum experimentally shown to hold when the behavior of matter is observed at atomic and subatomic scales.

Thus, changes in those physical magnitudes are often accompanied by emission or absorption of radiation which can only vary in a discrete fashion, as multiples of a constant. These packages of discontinuous energy are known as quanta, a term coined by the German physicist Max Planck (1858–1947).

Lofty and sturdy as it may seem, the edifice of contemporary physics is beginning to show major cracks and may not withstand the attrition to which it is now exposed. The cracks – we now know – are not superficial but extend all the way to the very foundations. As the dynamic structure of the universe is examined at very large scales, commensurate with the dimensions of galaxies, and further, of clusters and other assemblages of galaxies, major anomalies are surfacing. In fact, the anomalies have been surfacing since the 1930s. Now, we know the cracks are indeed structural and require immediate attention or the whole edifice of physics, the crowning achievement of human civilization, may be doomed and perhaps even tagged for a paradigmatic demolition.

Numbers simply don't add up in the cosmos. In the outer shells, stars in spiral galaxies have been behaving in ways that can be considered anomalous – to use physicists' typical euphemism – spinning at speeds far larger than those that would enable gravity from the visible universe to hold them in stable orbits. From this perspective, for the universe to make sense, the gravitational pull must be far larger than what is expected from the amount of matter detected [1,2]. This begs the question: Where is the missing matter? Or are we supposed to introduce a fudge term in the equations? This disquieting picture of galaxy rotation became apparent through the pioneering work of the American astronomer Vera Rubin (1928–2016). Vera examined six spiral galaxies, like our Milky Way, or nearby Andromeda and consistently found outer stars behaving anomalously: There is simply not enough detectable mass to keep them in their orbits. Newton's law of universal gravitation worked astonishingly well when it comes to describe the dynamics of our solar system – which earned Newton a superlative reputation – but when it comes to stars revolving around the center of galaxies, we are dismayed by the outcomes of the theory. We are left with two painful alternatives: either the law of universal gravitation fails miserably or there is a huge amount of matter out there that we cannot detect and introduces a considerable gravitational pull, five times larger than ordinary matter [1-3].

This is not the only problem with the cosmos – not even the worse one – that physicists need to cope with: The universe is expanding at an ever increasing rate that far exceeds what the gravitational pull would enable and far exceeds the amount of kinetic energy available from the motion of visible matter. It seems that plenty of the matter required to sustain the structure and dynamics of the universe in accordance with Newtonian or even relativistic laws simply cannot be accounted for. It is simply invisible. The missing matter has been named "dunkle Materie" (dark matter) by the Swiss astronomer Fritz Zwicky (1898–1974) who first postulated its existence, although perhaps "invisible matter" would have been a better name [3].

More troublesome, the rate of expansion of the universe is not constant: It is going exponential (Figure 1.6). As their speed exceeds the speed of light, zillions of stars will simply spin out of our horizon and become forever undetectable: Their light will never reach us. For this staggering discovery, astrophysicists Saul Perlmutter, Brian Schmidt, and Adam Riess were awarded the 2011 Nobel Prize in physics [1,3]. What is the source of this enormous surplus in kinetic energy seemingly sprouting from the vacuum of deep space? Nobody is certain, although a quantum mechanical origin is often invoked, hence the term "quantum vacuum fluctuations". An investigation on the nature of this mysterious energy gets too technical and is deferred to Chapter 5. This view postulating quantum vacuum energy is problematic, to say the least, since the naïve energy estimations are off by a factor of 10^{120} when contrasted against the experimental results obtained by Perlmutter, Schmidt, and Riess! The surplus kinetic energy associated with the berserk expansion of the universe has been named "dark energy".

Dark matter and dark energy cannot be considered corrections to the laws that presumably govern the cosmos [2,3]. There is roughly five times more matter in the universe than visible matter, the

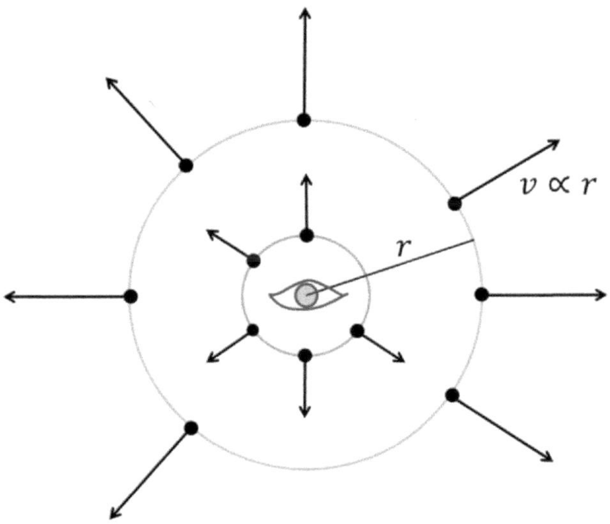

FIGURE 1.6 Schematic representation of a universe undergoing accelerated expansion. The component of the velocity of a celestial object contributing to its getting away from the observer increases as the object is farther away from the observer. Distant objects (larger radius r) have been traveling for longer times, and hence, their speed has had a chance to increase more relative to those that have traveled less.

matter we are able to detect. We know this from the gravitational influence of dark matter on visible matter. Furthermore, we know that energy and matter are interchangeable, with one becoming a proxy for the other and vice versa, as implied by the arch-famous Einstein equation $E = mc^2$ mentioned previously in this section. With this formula in mind, the breakdown of the actual mass budget of the universe looks even weirder. Give and take a few tens of a percent, current state-of-the-art calculations yield the following composition of the universe:

Ordinary matter: 5.0%
Dark matter: 26.7%
Dark energy: 68.3%

These ratios are nothing short of scandalous. We need to come to grips with the fact that the label dark matter simply captures our ignorance regarding the nature of most of the matter in the universe. Shockingly, 95% of the universe may be accounted for but remains utterly undetectable. We infer its existence indirectly, through its gravitational influence on ordinary matter.

Invisible matter now presents at temperatures almost negligible (3K, or −454°F, or −270°C) originated at much higher temperatures prior to the formation of galaxies, yielding the name "cold dark matter" (CDM) from the fact that it moves at nonrelativistic speeds ($v \ll c$) [2]. This material of unknown nature, first accrued into small galaxies, and subsequently served as a building block and seeding or nucleating material for larger scale structures up to the present-day gravitationally bound clusters of galaxies. In the widely accepted standard cosmology model, the gravitational growth of present-day galaxies and their clustering is steered by primeval fluctuations that animated a sea of cold dark matter. Although the nature of dark matter is anyone's guess at this point, astrophysicists have measured the imprint of their fluctuations in their primeval spatial distribution.

Such ancient imprint is embossed as slight variations across the universe in the brightness of the so-called cosmic microwave background (CMB), the relic ultra-weak radiation field left over from the big bang [1]. The CMB is the landmark evidence of the big bang origin of the universe and constitutes a faint radiation that fills up all spaces, dating back to the time when the atoms were first formed. With optical telescopes, the space in the background of light-emitting objects is completely dark. Only, a very sensitive radio telescope shows a faint background noise, a glow not associated with any object in the sky, a signal that is strongest in the microwave region of the spectrum. Its accidental discovery in 1965 is credited to American astronomers Arno Penzias (1933–) and Robert Wilson (1936–), earning them the 1978 Nobel prize in physics [1]. In colloquial terms, we may say, we are still "hearing" that massive explosion as the CMB (the radiation is in the radio wave frequency range). Spatial changes in the intensity of what we are hearing may give us clues on how dark matter got organized and distributed after the big cool down that followed the big bang. The acoustic oscillations detected experimentally to exquisite precision in the brightness fluctuation smudges of the CMB indicate the presence of a dominant invisible form of matter that flows freely, noble and inert, alongside the ordinary matter and radiation that are tightly coupled through electromagnetic interactions.

Today, many experiments are frantically searching for signatures of dark matter, both in the sky and in the laboratory, including the widely known Large Hadron Collider (LHC), a massive international consortium built near Geneva, Switzerland, to discover and detect subatomic particles [2]. This search has so far been unsuccessful, and this book does not conceal some skepticism regarding the outcome of LHC experiments in regard to dark matter, as discussed in subsequent chapters.

As suggested by the path-breaking work of Vera Rubin, the revolution dynamics of stars and dust in galaxies implies the existence of invisible mass in a halo that extends well outside the inner region where ordinary matter concentrates [1,3]. Surprisingly, the need for dark matter in galaxies appears only in the outer region where the gravitational acceleration drops below a universal value, which equals roughly the speed of light (299,792.458 kilometers or roughly 186,000 miles per second) divided by the age of the universe (13.82 billion years = 436,117,077,000,000,000 seconds). This is a highly disconcerting fact within the favored interpretations of dark matter. The sheer existence of a universal threshold (0.0000000007 meters per second squared) in the acceleration due to gravitation raises the daunting possibility that we are not actually missing matter but rather witnessing a change in the effect of gravity on the dynamics of visible matter at extremely low values of the gravitational force. This would require a very significant revision of the cosmology models enshrined in the theories of Newton and Einstein, a revision that may be tantamount to a dramatic paradigm change, bringing shockwaves to the scientific community.

1.4 THE SEARCH FOR DARK MATTER IN DEEP SPACE: EXPERIMENTAL EVIDENCE

More definitive clues are needed to figure out the nature of dark matter and dark energy. This book squarely addresses this imperative. In their quest, humans have at their disposal and also carry the burden of an enormous corpus of knowledge across a variety of fields, from particle physics to cosmology. Thus, a solid command of the so-called standard model in particle physics may surely become a blessing as it provides a foundational substrate to build upon, but in some sense, it may also become a handicap since the sheer volume of knowledge may hamper the boldness of approach that the dark matter/dark energy conundrum demands. Unless we get a smarter civilization to whisper the answer to us, it seems that at this point, the problem is eminently suitable for AI, which does not have to pay respects to tradition and can embark in the boldest assumptions without other restraints than those imposed by logic and consistency. This is precisely the approach adopted and described in this book in the most elementary way possible for the benefit of a broad audience.

The approach resorts to the underlying science critically revisited by AI in light of the startling anomalies observed experimentally. When properly steered, AI illustrates how science is best done: Not by prejudice, but through analysis of new big data in response to an intriguing hypothesis that humans, too loyal to their scientific cliques, are sometimes unable to formulate or accept.

As argued previously, as we prod the cosmos at very large scales, basic tenets of physics seem to crumble under the weight of contradicting evidence. This book helps mitigate the current crisis. It resorts to AI for answers and describes the outcome of this quest in terms of an ur-universe, a quintessential compact and multiply connected space that incorporates a fifth dimension to encode space–time as a latent manifold.

The American physicist John A. Wheeler aptly characterized Einstein's universe with the phrase: "*Matter tells space how to bend while space tells matter how to move*". But it turns out that at the largest cosmic scales, plenty – the vast majority – of matter and movement goes unaccounted, so either there is a colossal surplus of dark matter and dark energy that cannot be detected, or Einstein's theory becomes inadequate to explain deep space. Humans cannot judiciously decide at this juncture, so we must let AI be the arbiter.

In some ways, AI is bolder than humans because the huge corpus of knowledge, starting with the prodigious standard model (SM) of particle physics, poses almost no burden to its conjecture-framing processes. So, the plan set forth for the rest of the book is to feed AI with the SM enriched with the troubling cosmological phenomenology on dark matter and dark energy and see how AI reconciles the seemingly conflicting data with the currently accepted laws of physics. This is in a nutshell the intellectual adventure that lies ahead for the reader.

As keen observers of the cosmos, we are now taking a closer look at dark matter. Due to its lack of interaction with light quanta, we cannot expect to see or detect dark matter directly, but we must be prepared to infer its existence from its influence on ordinary (visible) matter. This influence is known to be solely gravitational, as dark matter does not appear to be implicated with other forces of nature. Essentially, we shall strive to give physical reasons why we think that there is such a thing in the first place. Let us consider the best understood and simplest possible celestial motion, i.e., that of a planet orbiting around a star. For simplicity, the motion may be assumed to be circular, with the star at the center. Thus, the planet moves in its orbit with velocity v and is located at all times at a distance r from the star. For the sake of illustration, we assume the star is our sun and we are describing planetary orbits in our solar system. This picture is only an approximation, as planets typically describe elliptical orbits with the sun in one of the foci of the ellipse. A key physical quantity in this picture is the *centripetal force* – let's name it F – that is, the force toward the center of the circular motion provided by the gravitational pull of the sun and required to keep the planet in its stable circular orbit. This force was first calculated by Newton from the simple rigorously obtained relation $F = mv^2/r$, where m is the mass of the planet, v is its velocity along its circular orbit, and r is the – constant – distance between orbiting planet and sun or, equivalently, the radius of the circular orbit, whichever the reader prefers. As said, this force is provided by the gravitational pull of the sun; hence, it is equal to the gravitational force, which according to Newton's law of universal gravitation is given by $F = GMm/r^2$, where M is the mass of the sun and G is Newton's gravitational constant [1]. By combining both expressions of the centripetal force, we get

$$F = \frac{mv^2}{r} = \frac{GMm}{r^2} \tag{1.1}$$

Equation (1.1) yields $v^2 = GM/r$ or $v = \sqrt{GM/r}$. This result is very important to describe the solar system since it implies that the further away a planet is from the sun, the slower the speed at which it revolves around the sun [1]. We can qualitatively plot this planetary behavior for the solar system as shown in Figure 1.7.

Now let us consider a much larger scale in the cosmos. Instead of planetary systems, let us consider galaxy rotation, or more precisely, the dynamics of stars that revolve around the center of spiral galaxies. We would imagine that this motion would be similar to that of planets revolving around stars, but that is not the case: Fundamental anomalies arise, and – as it turns out – these anomalies are key to infer the existence of dark matter.

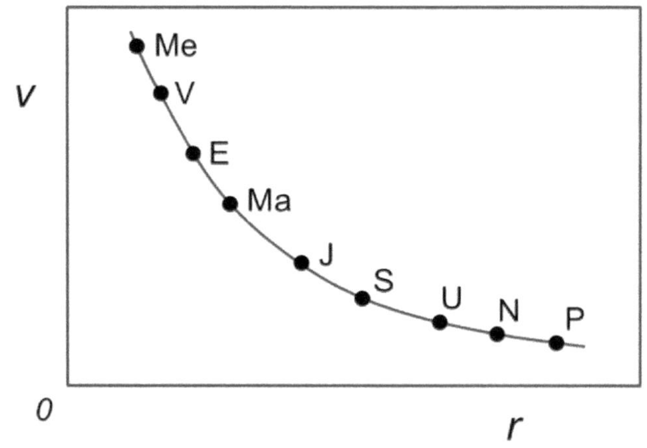

FIGURE 1.7 Qualitative scheme of the correlation between velocity and distance to the sun for the nine planets of the solar system (Me = Mercury, V = Venus, E = Earth, Ma = Mars, J = Jupiter, S = Saturn, U = Uranus, N = Neptune, and P = Pluto).

Since the visible mass M of the galaxy exerting a gravitational pull on the star is mostly concentrated at and "near" the center of the galaxy, we are likely to expect a v–r correlation similar to that presented in Figure 1.7, with v being now the speed of the star and r, the distance to the center of the galaxy. In this case, our prediction would extend to "short" galactic distances, where the mass contained inside the radius of the star orbit, the mass that exerts a gravitational pull on the star, becomes significantly smaller as we approach the center of the galaxy, as more and more mass is left outside the star orbit. Based on the experimental observation of visible matter in the heavens, we are assuming that the mass is concentrated at and near the center of the galaxy and that, as we navigate toward the boundaries of the galaxy, mass concentration is negligibly small relative to the concentration near the center.

Like in planetary motion, the behavior of the revolving velocity of the star is predicted again to be $v^2 = GM/r$, except that now M decreases substantially for small r, as less visible matter lies inside the star orbit with decreasing r. Let us get specific at this point: The detailed prediction for spiral galaxy Messier 33 is presented in Figure 1.8. On the other hand, experimental work paints a very different picture: Observation of the galaxy following the pioneering work of Vera Rubin reveals a very different behavior, particularly for distant or outer stars in the galaxy, as shown in Figure 1.9. The measured velocities of the outer stars are seemingly less dependent on the distance to the center of the galaxy [1]. For outer stars, as the distance r to the center of the galaxy increases, M the mass contained within the orbit of the star hardly changes at all since, as we have said most of the mass is concentrated at or near the center of the galaxy. Yet, the velocity of the star is no longer decreasing according to the Newtonian formula $v^2 = GM/r$ (Figure 1.9). The experimental measurements shown in Figure 1.7 have been corroborated time and again by independent observers around the world and are strikingly similar for all examined spiral galaxies [1]. The results invariably show that outer stars in galaxies are behaving anomalously: They are spinning so fast that the centripetal force or gravitational pull exerted by the galaxy would not be enough to keep them in orbit, yet the stars are clearly revolving around the center in stable orbits. According to Newtonian physics, at the measured speeds, the outer stars in the galaxies should fly away into outer space. This is clearly *not* what is happening.

Taken together, these observations clearly and unambiguously suggest that Newtonian physics breaks down, it is no longer upheld. The only alternative that would bring some intellectual relief to the physics community is that the mass M of the galaxy, responsible for the gravitational pull on

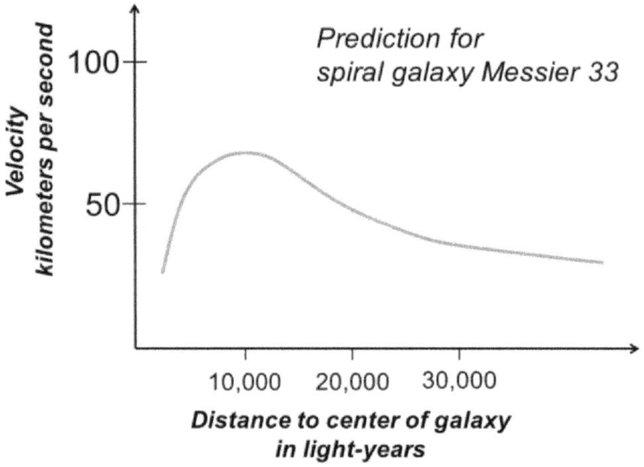

FIGURE 1.8 Predicted correlation between velocity of a star revolving around the center to the galaxy and its distance to the center of the galaxy. The curve was obtained for the spiral galaxy Messier 33.

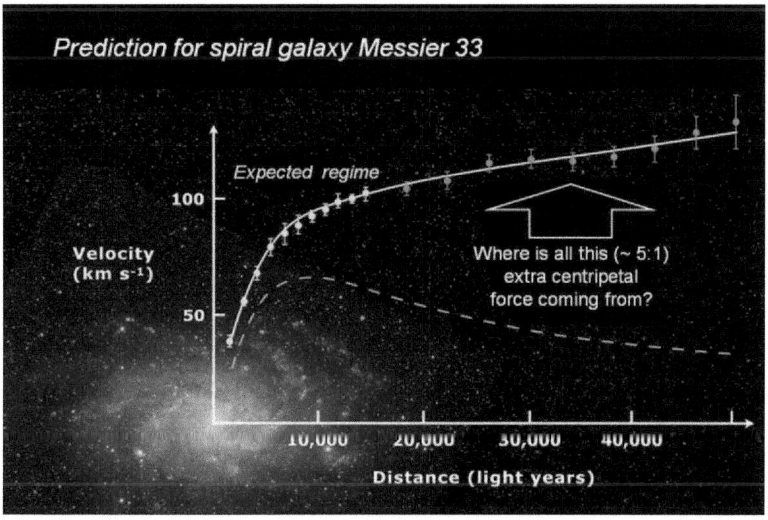

FIGURE 1.9 Rotation curve for spiral galaxy Messier 33 (points with error bars) and predicted curve from distribution of the visible matter (dashed gray line). The prediction and experimental measurement give the velocity of star rotation around the center plotted against distance of the revolving star to the center of the galaxy. The huge discrepancy between theoretical and experimental values is attributed to dark matter distributed in a halo surrounding the galaxy.

***Source*: Adapted from a figure in the Public Domain. Credit: Mario De Leo – own work, CC BY-SA 4.0, https://commons.wikimedia.org/w/index.php?curid=74398525**

the star, has been grossly underestimated. But, how could that be? All detectable matter has been detected and the mass computed with satisfactorily convergent results from constantly perfected technologies. These anomalies prompted physicists to postulate the existence of dark matter that is undetectable matter of an unknown nature distributed as a halo around the galaxy. This type of matter would contribute significantly to the gravitational pull necessary to ensure the stability of the orbits of outer stars.

Thus, the only inevitable conclusions from results such as those shown in Figure 1.9 are that either (a) the physics that withstood centuries of attrition needs a major correction or (b) the mass M in the gravitational equation that governs the motion of the stars in the galaxy represents something else, something quite different from the mass of the ordinary (detected) matter. In either case, both assumptions would represent fundamental departures from what we have been taking from granted for the last three hundred years.

We could in principle modify the sacred Newton's gravitational law $F = GMm/r^2$ for extremely large distance r, when the gravitational pull is very tenuous. After all, Newton never dealt with colossal distances of the order of tens of thousands of light-years (1 light-year is approximately 5.88 trillion miles), like those represented in Figure 1.8. The extant technology in the 17th century did not enable such observations. In fact, data regression analysis of experimental results of the type presented in Figure 1.9 has led to a modification of Newton's law adapted for huge intra-galactic distances. Thus, the Israeli physicist Mordehai Milgrom has dared to modify Newton's equation in an effort to adapt it to the outpour of astrophysical data. While commendable, his effort is mostly regarded as a formula for data regression, essentially a data-fitting device and not a fundamental advance in the physics underlying galaxy rotation [1].

The physics community overwhelmingly favors the dark matter hypothesis and has chosen to uphold the sanctity of Newton's law [2]. As we shall now see, there are other experiments that support the dark matter hypothesis and favor it over Milgrom's defiance of Newtonian physics [1,2].

1.5 VALIDATING THE DARK MATTER HYPOTHESIS

There is a well-known way of determining the mass of a galaxy, which is actually a feature of Einstein's general relativity called "gravitational lensing". As mentioned in the previous chapter, a massive object such as a galaxy will bend or warp the fabric of space–time, causing a change in curvature, in the form of a dimple (Figure 1.10). Since light travels in space–time, it will be affected by the gravitational pull exerted by the massive object. This interaction between gravity and light elicited huge skepticism when first proposed by Einstein, but ultimately proved to be a decisive validation of general relativity, as gravitational lenses were experimentally confirmed. Before dwelling on the mechanism of this phenomenon and leverage this relativistic feature as a means to detect dark matter, we need to prove rigorously that light may be indeed influenced by gravity. This amazing observation is not in the least obvious since the photons, the "particles" of light are known to be massless.

1.5.1 Gravity Deviates Light: The Power of Thought Experiments

This question is at the core of relativity. To show that light is indeed affected by a gravitational field we shall resort to Einstein's favorite theoretical tool, the thought experiment (*Gedankenexperiment* in German). No fancy equipment is required, no expensive gadgets, only the imagination, unbound, tempered only by the principles of physics and their logical consistency. In fact, let us play Einstein for a while. We can imagine young Albert at his modest frugal desk in the Patent Office at Bern, eyes closed, complete silence, prodding his imagination as he posits his favorite question "What if…?"

Let us imagine a box floating in deep-space vacuum (Figure 1.11), with no forces acting on it. A photon of light leaves one of walls of the box and travels to the opposite wall at the other side of the box. Let us place the center of coordinates at the point on the wall where the photon started its trajectory. As the photon is absorbed, it transfers some modicum of energy that we may denote E (leave your car in the sun and you will have trouble getting back in it). This implies that the photon must carry momentum, that is, what we would identify with "impetus" in common parlance. But the momentum, p, of a particle is usually assessed as $p = mv$, where m denotes the mass of the particle and v is its velocity. Since we are told that the photon is massless, how can it possibly have

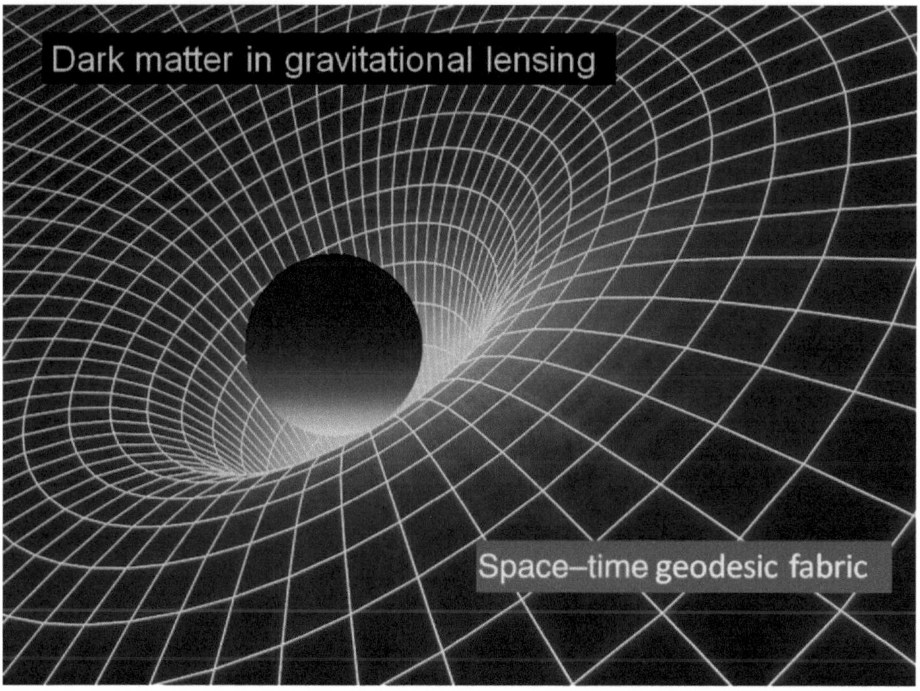

FIGURE 1.10 Relativistic picture of space–time (grid plane) curved or warped by the presence of a massive object.

momentum? This question troubled young Einstein for a while. The photon travels at the speed of light, $v = c$, so to compute the momentum of the photon, Einstein used the equation $p = E/c$. This equation does not involve the mass, the trouble maker. As it turned out, that was a clever move to circumvent the problematic mass of the photon, since Einstein knew that the kinetic energy of a moving object traveling at speed v may be computed as $E = mv^2 = pv$, so Einstein simply wrote $p = E/v$, which becomes $p = E/c$ for the photon that travels at the speed of light.

Like most of us, Einstein was also familiar with the principle of action reaction. Everyone who has fired a gun knows this principle: When the gun fires, the bullet goes one way, the gun the other – it's commonly known as recoil or kick. This principle has a fancy name in physics, and it is known as the principle of momentum conservation. When we apply it in the context of our thought experiment, we need to equate the reaction of the box to the movement of the photon inside. Both momenta must be equal in magnitude to compensate each other. Thus, at time Δt from the moment the photon left the wall, the following relation must hold:

$$\frac{M\Delta x}{\Delta t} = \frac{E}{c}, \tag{1.2}$$

where $\frac{\Delta x}{\Delta t}$ is the velocity of the box "recoiling" from the photon departure, M is the mass of the box, and Δx is the distance traveled by the box at time Δt, while in that time, the photon traveled a distance $c\Delta t$ in the opposite direction (Figure 1.11).

Since the photon travels at the speed of light, the time it takes for it to reach the opposite wall of the box at distance L is $\Delta t = L/c$. Now suppose we replace the photon with a particle with mass.

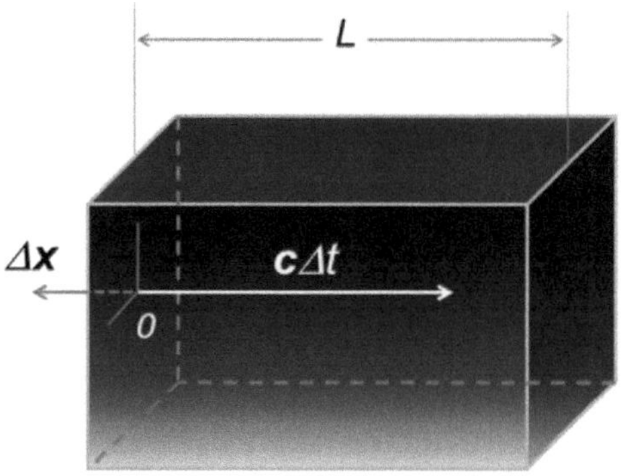

FIGURE 1.11 Representation of the thought experiment that leads Albert Einstein to formulate his famous equation $E = mc^2$.

At the time $\Delta t = L/c$, the conservation of momentum would read: $\frac{M\Delta x}{\Delta t} = \frac{mL}{\Delta t}$, or $M\Delta x = mL$. Using equation (1.2), we can determine that at time L/c, the box has been displaced by the amount $\Delta x = \frac{E\Delta t}{Mc} = \frac{EL}{Mc^2}$. Since we already showed that $M\Delta x = mL$, we get $M\frac{EL}{Mc^2} = mL$, which is rewritten as $E = mc^2$, Einstein's arch-famous equation!

This means that mass and energy are interchangeable, a result we shall exploit to investigate the nature of dark matter and dark energy. As it is well known, since c^2 is a colossally large quantity, it transpires that a small amount of mass is capable of transmutation into a huge amount of energy. This observation heralded the power of nuclear energy in a weaponizing context. Another consequence of this equation of direct relevance to our previous discussion of gravitational lensing is that since that a particle with mass $m = E/c^2$ may be regarded as a proxy for a photon carrying energy E, light is indeed affected by gravitation, as Einstein correctly predicted.

1.5.2 Gravitational Lensing as a Dark Matter Detector

To understand the principle, let us consider a photon of light (the minimal package of light) that barely skims the boundary of the galaxy. This beam will be deflected at an angle α (Figure 1.12). A general relativity calculation of the deflection of light caused by the gravitational influence of the galaxy gives the rigorous result:

$$\alpha = 4MG/Rc^2 \tag{1.3}$$

In the equation, M is the total mass of the galaxy and R is the radius of the galaxy.

We can also grasp this phenomenon and derive equation (1.3) from basic physics. Newton assumed that particles were corpuscles, perhaps not endowed with mass but certainly with momentum. We now know that the photon has momentum (E/c) but no mass, yet from the relation $E = mc^2$, we may assume that it is influenced by gravity since mass becomes a proxy for energy, just like Einstein predicted. Adopting this picture, the photon with velocity c (the speed of light) would undergo a deflection of its straight-line trajectory, and hence, its velocity vector (arrow of magnitude c) would change direction under the gravitational influence of a massive object. This change in direction requires a force, that is, the photon is subject to an acceleration, a, perpendicular to its original

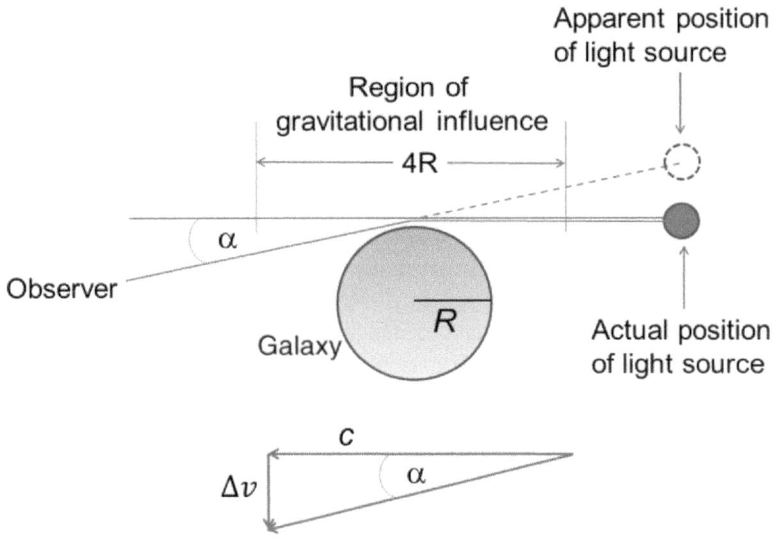

FIGURE 1.12 Elements of the physics of gravitational lensing.

direction. The force that deflects the light is simply the familiar $F = ma = \frac{GMm}{R^2}$, where M is the mass and R is the radius of the galaxy. This means that the acceleration undergone by the photon of light as it passes by a galaxy is $a = \frac{GM}{R^2}$. Since the acceleration is the change in velocity over time, that means that its change in velocity Δv would be given by $\Delta v = a\Delta t$, where Δt is the time interval during which the galaxy exerts a significant gravitational pull. We may assume this force is maximized as light is closest to the galaxy, and since the galaxy radius is R, the force is fully exerted during the traveled distance $4R$, as indicated in Figure 1.12. Note that $4R$ is an enormous distance, of the order of tens of thousands of light-years as illustrated in Figure 1.8. But the photon travels at the speed of light, so that the time interval becomes $\Delta t = \frac{4R}{c}$. This means that the perpendicular change in velocity caused by the gravitational pull of the galaxy is $\Delta v = a\Delta t = \frac{\left(\frac{GM}{R^2}\right)4R}{c} = 4MG/Rc$. So, assuming the deflection angle is small, we may use the approximation

$$\alpha \approx sin\alpha = \frac{\Delta v}{c} = 4MG/Rc^2. \tag{1.4}$$

The formula in equation (1.4), based on crude assumptions and obtained from basic physics, is *identical* to the one obtained by general relativity (equation 1.3).

This result demonstrates the astonishing consistency between general relativity and Newtonian physics.

The deflection angle in gravitational lensing of massive galaxies has been measured experimentally and it is approximately six times larger than the value obtained from general relativity or classical physics. *Because the deflection angle is proportional to the total mass M of the galaxy, this result implies that the total mass M of the galaxy is about six times larger the value calculated from the visible and detected matter in the region contained within the sphere of radius R, the radius of the galaxy*. This is a shocking result and implies that the matter that cannot be detected, the dark matter, contributes five times more than the visible matter to the gravitational pull of the galaxy [2]. This result is in striking agreement with the proportion of dark matter to visible matter in the universe given previously: (26.7):5 ≈ 5.3.

This section illustrates the way physics works best. Experimental observations that stand in defiance of a prevailing paradigm are initially invariably regarded as "anomalies" or "systematic errors", and only if they pass the initial peerage scrutiny, they may be taken seriously as worthy contenders of an existing paradigm. This implies that the standard model or laws that are expected to underpin the newly observed phenomena may undergo revision or be completely reformulated in a new guise that often represents a synthesis of two clashing proposals of reality. This narrative describes the saga of dark matter as it unfolded in this chapter. The extant physical models and laws cannot be adapted to encompass the experimental observations in the contexts of galaxy rotation and gravitational lenses unless the mass responsible for the gravitational pull gets a significant contribution from matter of a hitherto unknown nature. This is precisely dark matter and only an approach bold enough to be unencumbered from the weight of standard models can delineate its nature. As we shall show later in this book, time seems to be ripe for AI to prod over the mass astrophysical data and elucidate the nature of dark matter.

1.6 EXPANSION IN A NEWTONIAN AND RELATIVISTIC UNIVERSE

Evidence that massive amounts of energy of unknown origin keep pouring into the universe without undergoing dilution has been piling up since 1998. That was the year astronomers verified the accelerated expansion of the universe. As space expands, more vacuum forms, and with it, more geometrically undiluted energy is infused into space, causing space to expand even further in a sort of autocatalytic reaction, where dark matter creation appears to be self-stimulated.

The argument for accelerated expansion runs as follows: Expansion of space stretches light, shifting it to longer wavelengths, hence shifting light toward the red section of the visible spectrum. Light from supernovae appears more "redshifted" the farther away they are from us, because their light has to travel farther through an expanding space. If space expanded at a constant rate, a supernova's redshift would be proportional to its distance to the observer and thus to its brightness. This is not what astronomers have been observing.

In an accelerating universe filled with undiluted dark energy that scales with vacuum dimensions, space expanded more slowly in the past than it does now. This means a supernova's light will have stretched less during its journey to earth, given how slowly space expanded during much of the time compared with the speed at which it expands now. The light from a supernova located at a given distance away (as determined by its brightness) will appear significantly more redshifted than it would in a universe that lacks dark energy. Indeed, researchers find that the redshift and brightness of supernovae scale in precisely that way. The actual scaling enabled them to compute the amount of dark energy in the universe with significant precision.

At the turn of the 20th century, the cosmological debate had a very different flair. In order to calibrate his ideas as he started developing his general theory of relativity, young Einstein had many conversations with astronomers. They assured him that the universe was essentially static, with no beginning and no end, with galaxies at fixed positions unchanged on a cosmic scale. This picture posed a great problem to Einstein as he sought to reconcile it with the idea that galaxies exerted a gravitational pull on one another that would ultimately lead to a collapse of space–time into a big crunch. Einstein circumvented this problem by introducing a "cosmological constant", a fudge term representing a force that would counter the gravitational pull in order to maintain the stasis of the universe, as it was wrongly related to him by contemporary astronomers [1]. As it turns out, the cosmological constant ultimately relates to the presence of dark energy but not in any way remotely resembling what Einstein would have anticipated.

Fortunately, before Einstein went too far with the cosmological constant idea, an American astronomer by the name of Edwin Hubble (1889–1953) came along and showed that in fact, the universe was expanding!

Expansion means that the distance between any two points in the Universe increases with time, but they do so in relation to a single time-dependent scale factor that we shall denote $a = a(t)$. It is crucial to note that the scale factor is the same in all spaces and only varies with time. In rigorous terms, the expansion idea may be formalized as follows: Let $d_{ij}(0)$ be the distance between any two points labeled "*i*" and "*j*" in the universe at a particular time conventionally set to be $t = 0$, then the distance $d_{ij}(t)$ at a future time t becomes $d_{ij}(t) = d_{ij}(0).a(t)$ and this equation holds for any pair i, j of points in space. The important thing is that there is only one scaling factor and it is not dependent on the pair of points chosen. Hubble noticed that this scaling factor increases with time.

As the concept settles down in our minds, let us drop subindices and adopt a more agile notation, denoting $y = y(t) = d_{ij}(t)$ and $\Delta y = d_{ij}(0)$. Then, $y(t) = \Delta y.a(t)$ and dy/dt, the rate at which distance y changes in time, becomes $dy/dt = \Delta y.(da/dt)$, where da/dt is the rate of change of the scaling factor of the universe. Combining the previous relations, we may write

$$\frac{dy}{dt} = \Delta y \frac{da}{dt} = y\left[\frac{\left(\frac{da}{dt}\right)}{a}\right]. \tag{1.5}$$

The factor $\left[\frac{\left(\frac{da}{dt}\right)}{a}\right] = H$ in the left-hand side is the Hubble constant of the universe, that is, the rate at which the scale factor changes as a result of the universe expansion divided by the scale factor [1]. It is important to point out that H is constant in space but not in time. Equation (1.5) has far-reaching consequences: If we know the velocity $\frac{dy}{dt}$ of a receding galaxy, we can calculate y, that is, how far away it is. Furthermore, as we shall show later on the discussion, *the more distant the galaxy is from us the faster it is traveling away from us.*

This is the Hubble picture of the expanding universe, and in this picture, space itself is expanding. It has to, implying that the vacuum is getting larger and larger, a troubling runaway scenario by no means well understood.

Let us elaborate this context further by applying the well-worn Newton's law of gravitation. The gravitational pull exerted on a galaxy revolving at a distance y from an arbitrarily chosen point in the universe would be $F = GMm/y^2$, where M is the total mass of matter contained within the ball of radius y centered at the chosen point, and m is the mass of the revolving galaxy. The potential energy (U) of the galaxy is the physical magnitude whose rate of change in time is the gravitational force, and therefore, $U = -GMm/y$. Now, the total energy (E) of the galaxy must include the potential energy associated with the gravitational force exerted on it plus another contribution (K) that represents the kinetic energy, that is, the energy associated with the movement of the galaxy. Let us assume the latter form of energy is only associated with the expansion of the universe. Then, we may write $E = K + U = \frac{1}{2}m\left(\frac{dy}{dt}\right)^2 - GMm/y$. But E is unchanged in time unless another force starts working on the galaxy, which we assume not to be the case. So, we may say that the quantity $\left[m\left(\frac{dy}{dt}\right)^2 - 2GMm/y\right]$ is a constant or $\left[\left(\frac{dy}{dt}\right)^2 - 2GM/y\right]$ is constant, since the mass m remains unchanged. Now, plugging in the previously obtained equation (1.5) into the last expression, we obtain:

$$y^2\left(\frac{da/dt}{a}\right)^2 - \frac{2GM}{y} = \text{constant}. \tag{1.6}$$

The mass enclosed by the sphere of radius y is $\frac{4}{3}\pi y^3\rho(t)$, where $\frac{4}{3}\pi y^3$ is the volume of the ball of radius y and $\rho(t)$ is the mass density that obviously keeps decreasing in time in an expanding universe. We may then write

$$y^2\left(\frac{da/dt}{a}\right)^2 - \frac{2G}{y}\frac{4}{3}\pi y^3\rho(t) = \text{constant}. \tag{1.7}$$

A little algebra (division by y^2) and substitution with known relations derived above yields

$$\frac{\left(\frac{da}{dt}\right)^2}{a^2} = H^2 = \frac{8\pi G}{3}\rho(t) - k/a^2. \tag{1.8}$$

In this equation, k on the right-hand side is a constant. Equation (1.8) is known as the Friedman–Robertson–Walker (FRW) formula derived from the theory of general relativity [1]. But we have shown that a simplified version of the formula could be obtained from Newtonian physics, making very simple assumptions.

The reader is once more reminded that H in equation (1.8) stands for the Hubble constant, which is a constant in space and not in time. Notice that generally the following relation holds: $\frac{8\pi G}{3}\rho(t) - k/a^2 \geq 0$. This relation becomes the hallmark for an "open universe". On the other hand, if at some point in time we get $\frac{8\pi G}{3}\rho(t) < k/a^2$, the universe would stop expanding and start contracting and we would call it a "closed universe". In this case, the universe would eventually collapse onto itself into a singularity under the gravitational pull. This scenario has been termed the "big crunch". If on the other hand $k = 0$, the universe will expand at an ever decreasing rate because the matter density is monotonically decreasing as the universe keeps expanding. The rate of decelerated expansion will approach the asymptotic limit value zero at infinite time. This type of universe is called a "flat universe". The time dependence of the size of the three types of universe, open, closed, and flat, is schematically shown in qualitative fashion in Figure 1.13.

The best up-to-date observations of our skies support the picture of a flat universe ($k = 0$); that is, it will continue to expand only until it reaches a particular size. To compute the density ρ, let us

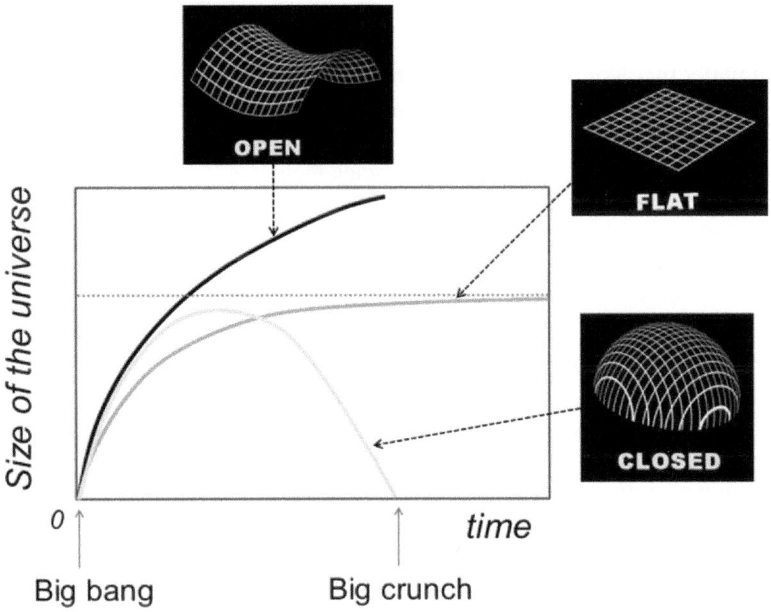

FIGURE 1.13 Qualitative scheme of the time dependence of the size of the universe for a closed universe (light gray), open universe (black), and flat universe (gray).

assume a cubic box of dimension a in a flat universe containing mass M. Then, the FRW formula becomes

$$\frac{\left(\frac{da}{dt}\right)^2}{a^2} = H^2 = \frac{8\pi G}{3}\frac{M}{a^3}. \tag{1.9}$$

This equation can be simplified to $\frac{da}{dt} = \sqrt{\frac{8\pi GM}{3a}}$, which can be rewritten as

$$\left(da/dt\right) = Wa^{-1/2}. \tag{1.10}$$

In equation (1.10), the constant W is defined by parameters of the universe: $W = \sqrt{\frac{8\pi GM}{3}}$. Now, to find out how the universe expands in time, we need to compute the time dependence of $a = a(t)$. This may be simply obtained by noting that equation (1.10) may be rewritten as $a^{1/2}da = Wdt$ giving $a^{3/2} = w't$, or equivalently, we may state that our universe is expanding as

$$a = wt^{2/3}, \tag{1.11}$$

where w' and w are constants. This is the *predicted* expansion behavior for our current matter-dominated flat universe. As we shall see, the experimental results reveal a huge discrepancy with this rigorously obtained prediction, and dark energy is at the core of the problem!

At the beginning of its existence, right after the big bang, the universe would not have been matter-dominated but rather radiation-dominated, with spontaneous creation of matter and its compensatory antimatter that would mutually annihilate with a huge production of photons. A residual amount of matter would have been created as a proxy for the energy released, as described by Einstein's formula $E = mc^2$, derived in the previous chapter. In this early universe, let us consider a cube of photons of dimension a. How would this scale factor for the expansion in the early radiation-dominated universe behave in time?

To describe this early universe, we need to resort to some basic principles of quantum mechanics. We have previously discussed that the photon is a massless particle, and yet, it carries momentum. This strange duality was something that Einstein found particularly irksome because he could not reconcile that fact with the fact that the photon transfers energy as it hits a surface. He eventually resolved the paradox admirably, as discussed in the previous chapter. Einstein's results in fact imply that the notion of particle needs to be refined. We can no longer think of an elementary particle as a "corpuscle" but rather as a "wave-like excitation" or a "warp in a field". The photon travels at the speed of light, but its "kinetic energy" depends on how many crests are packed in the wave per unit time, this is what we know as frequency. Max Planck showed that in fact the energy of the photon may be written as $E = hf$, where f is the frequency and h is Planck's constant. In fact, h is one of the parameters for our universe, together with G, Newton's gravitational constant, and a few other constants. Together, they define how our universe behaves.

Planck's expression for the energy of a photon may be rewritten as $E = hc/\lambda$, where c is as usual the speed of light and λ is the wavelength, that is, the distance between two consecutive crest peaks of the "wave–particle". The equivalence between both expressions for the energy of the photon follows from the simple fact that, by definition, $f = c/\lambda$. As a increases with the universe expansion, so does everything else, including the wavelength of the photon that becomes commensurate with a. The photons in the early universe have their wavelengths stretched as the universe expanded, implying a dramatic cooldown following the big bang, since the energy goes down as the wavelength increases (the frequency decreases). The wavelength of the "cold" ancient photons is today in the microwave region, a very low-energy region of the radiation spectrum. This is the "glow" we see in the skies known as cosmic microwave background radiation (CMBR), discussed previously. The CMB is as it was the relic of the early radiation-dominated universe that followed right after the big bang.

Since the wavelength of the ancient photons can be made proportional to the expansion factor in the radiation-dominated early universe, we may write $E = J/a$, where J is a constant. This gives a radiation energy density $\rho_r = \left(\frac{J}{a}\right)\left(\frac{1}{a^3}\right) = J/a^4$. The a dependence in this expression can be contrasted with the previously obtained density for a matter-dominated "later" universe: $\rho = M/a^3$.

We can substitute the expression for the density in the radiation-dominated universe in the FRW formula given by equation (1.8), keeping in mind that our universe is flat, and hence, $k = 0$. This substitution gives

$$\frac{\left(\frac{da}{dt}\right)^2}{a^2} = \frac{8\pi GJ}{3a^4}. \tag{1.12}$$

Equation (1.12) can be rewritten as $ada = Cdt$, where $C = \sqrt{8\pi GJ/3}$, yielding the following time dependence for the expansion in the radiation-dominated early universe: $a \propto t^{1/2}$. This behavior is different from the expansion of the matter-dominated universe, which, as we may recall, has a time dependence on the form $a \propto t^{2/3}$.

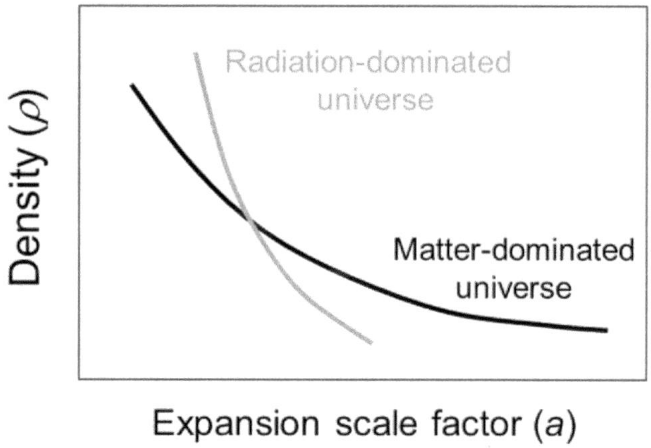

FIGURE 1.14 Qualitative behavior of density in an expanding radiation-dominated and matter-dominated universe.

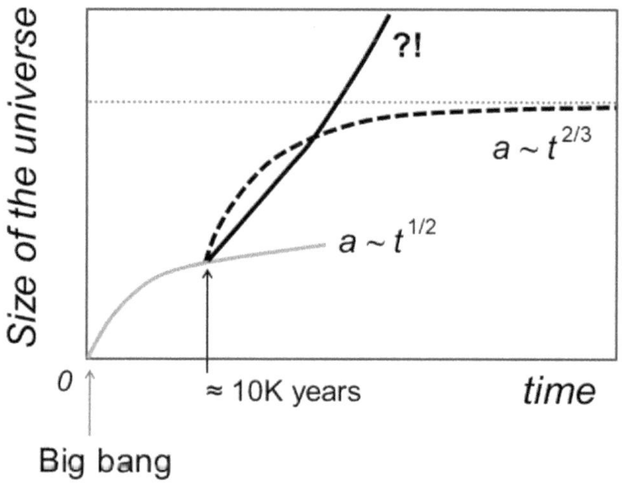

FIGURE 1.15 Qualitative behavior of the time dependence of the size of the universe in a radiation-dominated regime (gray line), matter-dominated regime (dashed line), and experimentally validated runaway regime (dark solid line).

The two predicted modes of expansion of the universe can be best visualized in qualitative graphs. Thus, Figure 1.14 shows density versus expansion scale factor a for the radiation-dominated and the matter-dominated universes. There is a cross-over point at approximately 10 thousand years from the big bang, since the density of the radiation-dominated universe falls more steeply than the density of the matter-dominated universe as a increases. Alternatively, we may plot the size of the universe against time (Figure 1.15). The radiation-dominated universe size increases more slowly (as $t^{1/2}$) than the matter-dominated curve with $a \propto t^{2/3}$, so the latter curve crosses the former at the point determined previously to be in the order of 10 thousand years. As previously discussed, the expansion curve for the matter-dominated universe, that is the universe we live in today, will reach an asymptotic limit. This prediction, as we shall see subsequently, is at odds with experimental observations that reveal an accelerated expansion of the universe. This discrepancy between prediction and experiment will be shown to be at the core of the dark energy controversy.

In summary, general relativity (FRW equation) predicts that the universe has expanded under radiation dominance until about 10K years, and then, it expanded faster under matter dominance and that this expansion will slow down to reach an asymptotic limit at infinite times, as described in Figure 1.15.

As we keep anticipating, this is not what seems to be happening according to experimental evidence. The universe expansion is not showing signs of slowing down. Quite the contrary, the expansion is accelerating, prompting us to invoke the presence of a form of energy of unknown origin that is fueling the runaway. This energy of unknown nature has been ominously termed "dark energy" and may well become the nemesis of our lofty theories on how the universe works [2].

1.7 THE UNIVERSE RUNAWAY FUELED BY DARK ENERGY: THE COSMOLOGICAL CONSTANT PROBLEM

The accelerated expansion of the universe constitutes the biggest challenge to the prevailing paradigm in physics, as it demands either a complete revision of the physical laws to account for this "anomaly", or alternatively – but no less painfully – that we reckon the existence of dark energy, a form of energy of hitherto unknown origin that is constantly being injected into the universe and is causing its runaway behavior in complete defiance of gravitational laws.

The runway picture of the universe was put forth by Perlmutter, Schmidt, and Riess, as discussed previously [1]. They were able to establish the exponential expansion by precisely measuring the so-called redshift effect in the radiation collected from very bright objects called *supernovae*. As the light-emitting source travels away from the observer, the detected frequency f of the emitted light gets lower because the time of arrival of successive wave crests gets longer (Figure 1.16). The net effect can be represented as a stretching of the wave. The observed frequency f is related to the emitted frequency f_e by the relation: $f = f_e\left[\frac{c}{c+v_s}\right]$, where v_s is the velocity of the light source. The apparent stretching of the wavelength is known as redshift since the red color of the visible spectrum corresponds to the longest wavelength and lowest frequency. Thus, the redshift is implied by the relation $f < f_e$.

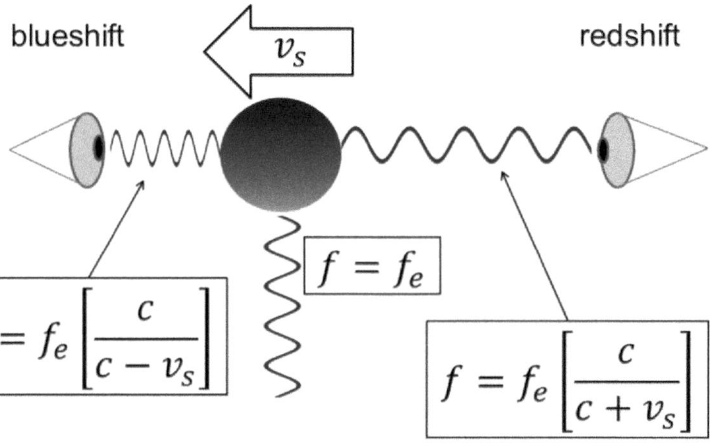

FIGURE 1.16 The redshift and blueshift effect on observed radiation from a light-emitting source moving with velocity v_s relative to the observer.

Conversely, if the light-emitting source approaches the observer at speed v_s, the time of arrival of successive waves is increased, so the waves are bunched together, compressed as it were, with a net "blueshift" effect resulting in increased frequency ($f > f_e$) in accordance with the formula

$$f = f_e \left[\frac{c}{c - v_s} \right] \tag{1.13}$$

The accelerated expansion of the universe was established by determining the commensurability of the redshift effect with the brightness of the light-emitting source, a proxy for the distance to the observer on earth. It was determined that the brighter the source, the smaller the redshift, and conversely, the dimmer the source, the more pronounced the redshift (Figure 1.17). The implications of this observation remain as troublesome as they are transparent: The more distant the source, the higher the velocity at which it is traveling away from the observer, in other words: $v_s = v_s(d) \propto d$, where d is here the distance to the light-emitting source. But this implies that $da/dt \propto a$. The solution of this equation yields an exponential or runaway time dependence: $a \sim e^{qt}$, with q = constant.

This means that the scale factor a is growing exponentially in time, completely at odds with the predicted scenarios of general relativity described previously. We are in the presence of the hallmark for a runaway universe!

In manner of digression, we must consider here alternative ways of computing the energy E of the photon. We have dealt with energies, but so far we have not incorporated heat or temperature to our physical picture of the universe. Temperature is the physical quantity conjugated to energy in thermodynamics, the science concerned with the interconversion of heat, energy, and work, whose modern foundations were laid out by the Austrian physicist Ludwig Boltzmann (1844–1906).

It is estimated that the photons in the CMB correspond to radiative emissions of an extremely cold black body, currently at a temperature of approximately 3K (−454.27°F or −270.15°C). On the other hand, the energy of a photon may be related to the temperature of emission through Boltzmann's formula $E = kT$, where k is Boltzmann's constant. Now, given since $E = hf = hc/\lambda \approx hc/a$, we may conclude that

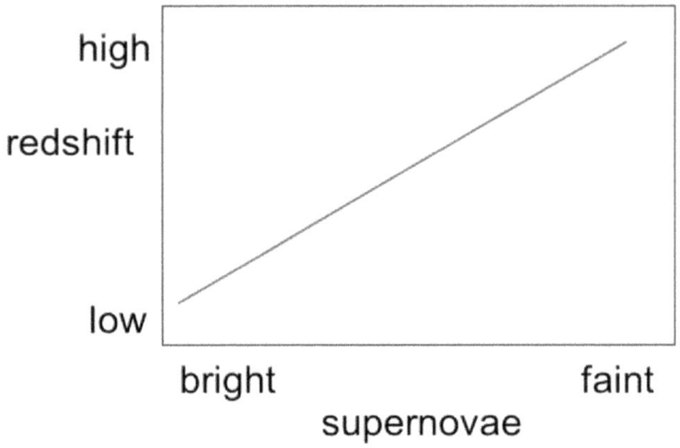

FIGURE 1.17 Qualitative behavior of redshift for nearby and distant light sources. Brightness is accepted to be a proxy for distance. Thus, far away objects move away from the detector at faster speed than objects nearby. This behavior constitutes the hallmark of the runaway universe as validated in the Perlmutter–Schmidt–Riess experiments.

$$T = \left(\frac{hc}{k}\right)\left(\frac{1}{a}\right). \tag{1.14}$$

This implies that the universe is cooling down as it is expanding. We also have an estimation $T \approx$ 3000K for the temperature of the universe at a particular epoch after the big bang, namely, the ionization epoch, when atoms were stripped of their electron shells, with the atomic nuclei embedded in a plasma of electrons. Thus, we may compute the ratio $a_{\text{today}}/a_{\text{ion}} = 3000/3 = 1000$, where a_{today} and a_{ion} are, respectively, the expansion scales today and at the ionization epoch. This means that our universe is a thousand times bigger now than at that hot era right after the big bang.

Now, how far back did the ionization took place? We can easily answer this question since we know the time dependence of a in a matter-dominated universe and the current age of the universe (t = 13.8 billion years, counted since the big bang):

$$\frac{a_{\text{today}}}{a_{\text{ion}}} \approx 1000 = \left(\frac{t}{t_{\text{ion}}}\right)^{\frac{2}{3}} = \left(13.8\times 10^{9}\, y/t_{\text{ion}}\right)^{2/3}. \tag{1.15}$$

This implies that the ionization era occurred at the time $t_{\text{ion}} \approx 460{,}000$ years after the big bang.

Here, we reproduce the FRW formula for our flat universe but in a way that accommodates the radiation-dominated regime ($\rho \sim 1/a^4$) as well as the matter-dominated regime ($\rho \sim 1/a^3$):

$$H^2 = \frac{\left(\frac{da}{dt}\right)^2}{a^2} = \frac{8\pi G}{3}\rho(t) = \left(\frac{8\pi G}{3}\right)\left(\frac{C}{a^{3(b+1)}}\right) \tag{1.16}$$

In this equation, C is a constant and b = 1/3 for the radiation-dominated regime and b = 0 for the matter-dominated regime.

As we have anticipated, *these regimes postulated by the theory of general relativity do not agree with the experimentally verified exponential expansion of the universe determined by Perlmutter, Schmidt, and Riess* (see previous discussion). These scientists have shown that a behaves exponentially in time: $a \sim e^{\theta t}$ with θ = constant. Now, from equation (1.16), it follows that the case where $b = -1$ corresponds to a constant density $\rho = C = \rho_0$. *Paradoxically, the $b = -1$ regime not anticipated by general relativity is precisely the type of universe whose expansion was measured by Perlmutter, Schmidt, and Riess.* By plugging in $b = -1$ in equation (1.16), we get

$$\frac{\left(\frac{da}{dt}\right)^2}{a^2} = \frac{8\pi G}{3}\rho_0 \approx \Lambda, \quad \Lambda = \text{“}\textit{cosmological constant of the universe}\text{”}. \tag{1.17}$$

This formula yields the exponential expansion of the universe, with the fudge factor in Einstein's theory, namely, the ill-fated "cosmological constant". We may rewrite it as

$$\frac{da}{dt} = a\sqrt{\Lambda}, \tag{1.18}$$

Equation (1.18) is clearly indicative of a runaway process: the larger the expansion, the larger the rate of expansion. As expected, the equation's solution yields a factor a that is exponentially increasing in time:

$$a \propto e^{\sqrt{\Lambda}t}, \theta = \sqrt{\Lambda} = \sqrt{\frac{8\pi G\rho_0}{3}}. \tag{1.19}$$

In other words, by taking $b = -1$ or fixing the density as constant, we can reproduce the experimental result of exponential expansion and therefore accelerated expansion of the universe that earned Perlmutter, Schmidt, and Riess the Nobel prize in physics. Furthermore, we can calculate the cosmological constant, bestowing physical meaning to this troublesome factor that Einstein regarded as a fudge term ("my biggest blunder", he called it) in his theory of general relativity.

On the other hand, we know that the speed $v = dy/dt = \Delta y.da/dt$ at which a galaxy runs away from us increases as the distance $\Delta y.a$ to the galaxy increases. But this is precisely what equation (1.18) is telling us: $v = \frac{dy}{dt} = \Delta y \frac{da}{dt} = \Delta y a\sqrt{\Lambda} = y\sqrt{\Lambda}$ or $v \propto y$. Furthermore, the equation is also telling us that the proportionality factor is the root square of the cosmological constant!

One important consequence of this result is that beyond a critical distance $y^* = c/\sqrt{\Lambda}$, the speed of a galaxy at distance $y > y^*$ will be larger than the speed of light: $v > c$! This means that such a galaxy will travel out of sight: Its light will never reach us. The galaxy will simply disappear beyond our horizon. The accelerated expansion of the universe will make galaxies disappear beyond our horizon, hence heralding a much duller view of the skies.

We may have lifted the veil of nature, but only to unravel yet another mystery. If ordinary matter ($b = 0$) or ordinary radiation ($b = 1$) are not dominant in the universe, what is causing the accelerated expansion of our universe? A mysterious energy of unknown origin is being constantly created and injected into our universe from a source that does not get diluted as the universe expands. This mysterious energy fueling our universe runaway is what we call dark energy.

At least we know one key thing about dark energy: It does not get diluted as the universe expands, since its density is constant ($\rho = \rho_0$). Now, the thing that is constantly being created without getting diluted in the universe as it expands is … vacuum! Indeed, vacuum does not get geometrically diluted as the universe increases its volume, so dark energy is generated by the vacuum! And we know this is not a negligible contribution: The dark energy density ($\rho = \rho_0$) makes up for about 68.3% of the total density of energy/matter in the universe.

Relativity is taking us to an unchartered territory far more rarified even than that of dark matter. The universe expansion is autocatalytic, it has a retro-feeding mechanism for acceleration whereby the more it expands the more it harvests dark energy, which in turn fuels more expansion, creating a runaway out of vacuum generation.

The universe runaway is the inevitable truth that emerges from the current physics paradigm. The self-stimulated vacuum energy creation becomes a major hurdle in our trend of thought and prompts us to prod the complementary theory to general relativity: quantum mechanics.

To unravel how the creation of dark energy may come about, we refer to quantum field theory (QFT), where an electron and its antiparticle, the positron, occasionally emerge spontaneously in the vacuum with emission of a photon. The particles exist for a brief time and then annihilate each other without trace unless another event elsewhere in the vacuum leads to an interaction with a component of the triad. Energy conservation is violated, but only for the extremely ephemeral lifetime of the particle Δt permitted by the so-called *uncertainty principle* that posits that $\Delta E \times \Delta t \sim \frac{\hbar}{2}, \hbar = h/2\pi$, where ΔE and Δt represent *a priori* uncertainty in energy and time, respectively. This principle is one of the milestones of quantum mechanics and asserts that uncertainties in conjugated quantities like energy and time, or position and momentum balance each other yielding a constant product. Thus, a short uncertainty in time may yield a huge fluctuation in energy, an utterly commonplace in QFT. There is no limit placed by the laws of physics on the scale of this energy fluctuation. Nothing prevents it from occurring at a grand scale. The duration is of course subject to the restriction of the

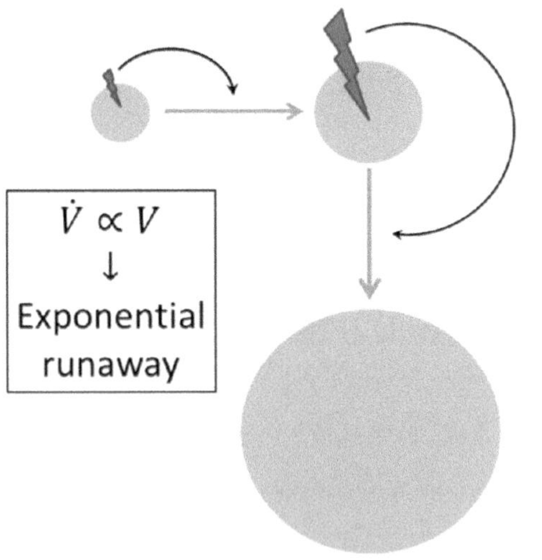

FIGURE 1.18 Kinetic scheme for the principle of autocatalytic vacuum creation (AVC) as the mechanistic underpinning of a runaway universe.

uncertainty principle, which merely implies that the universe has a zero total energy, which made the fluctuation possible in the first place. Quantum theory implies that the vacuum should be unstable against large fluctuations in the presence of a long range, negative potential energy term. Gravitation is precisely such a term. This fact encourages us to believe that vacuum creation stimulates further vacuum creation as dark energy adopts the form of vacuum energy fluctuations.

Because vacuum does not undergo any dilution as the universe expands, quantum vacuum energy becomes the prevailing proxy for dark energy, although this answer is far from satisfactory and prompts further investigation, as shown in Chapter 5.

The universe runaway scenario experimentally validated by Perlmutter, Schmidt, and Riess in their Nobel-deserving work is at least compatible with the principle of autocatalytic vacuum creation (AVC) hereby put forth (Figure 1.18). The AVC principle can be defined as follows: A quantum vacuum fluctuation fuels the creation of a bigger vacuum, which in turns has an enhanced chance to spontaneously generate a larger quantum fluctuation, which in turn fuels the creation of a bigger vacuum, so the rate of vacuum volume creation $\dot{V} = \dot{V}(t) = \frac{dV}{dt}$ at a given time t is proportional to the vacuum volume $V = V(t)$ created at that time. This assertion may be written as $\dot{V} \propto V$. This equation implies a runaway in vacuum creation, as the solution to the equation yields $V \sim e^{Qt}$, where Q is some constant whose relation to the cosmological constant Λ would need to be established.

For now, we are contented with the assertion that the relativistic equation 1.17 together with the AVC principle provide the physical underpinnings for the runaway universe as fueled by dark energy. Furthermore, this dark energy is expected to be generated in autocatalytic cycles of vacuum creation that are causative of the universe runaway behavior.

1.8 DARK PHYSICS ON EXTRA DIMENSIONS

A survey of the evidence described in the previous sections prompts us to question whether a shift in paradigm is in order to encompass deep-space phenomenology. It appears that the influence of dark matter and dark energy cannot be done with by merely patching up the laws of physics or extending them to include cosmic ranges of interaction hitherto uncalled for. So far, the discussion has included local physical theories, i.e., theories whose predictive power resides in solving differential equations or become appropriately stylized within the setting of differential geometry. Topological attributes of the universe have not entered into the equations in any obvious manner, and the topology of the universe has not been so far a matter of concern (neither it was in Einstein's time). However, as argued in this book, topology plays a major role in elucidating the nature of dark matter and dark energy. The fact that there is a form of matter in vast proportion that is not associated with visible light quanta calls for an extra dimension, as shall be argued in the forthcoming chapters. The main hurdle to incorporate an extra dimension is topological, not geometrical.

There have been early attempts by Theodor Kaluza and Oskar Klein (KK theory, revisited in [6]) to incorporate an extra compact curled-up dimension into what was otherwise regarded at the time as an "Euclidean universe". These attempts held promise at least initially and were taken seriously by none other than Albert Einstein himself, who spent the latter part of his life seeking to unify gravity with the other forces of nature. However, a lack of topological understanding of space–time at the time made it impossible to properly incorporate extra dimensions and reconcile these extensions with data available, especially with data pertaining to the prediction of the rest mass of elementary particles.

In this book, the "KK theme" may be said to be resurrected albeit in a vastly different guise and for a completely different purpose. First, we shall show in the forthcoming chapter that the universe is compact, regardless of its perceived or inferred extension, and multiply connected, and that these topological attributes determine the way the extra spatial dimension gets incorporated. The extra dimension becomes a necessity to explain dark matter and dark energy, as can be deduced from the simple argument laid out subsequently.

Suppose we consider a spatial cross section of the universe along a single compact dimension with radius r. We can store energy in this dimension by considering a stationary "de Broglie" wave with wavelength $n\lambda = 2\pi r$, and we know that the kinetic energy of the wave, $E = \dfrac{hc}{\lambda} = \dfrac{\hbar nc}{r}\left(\hbar = \dfrac{h}{2\pi}\right)$, can be translated into a mass: $m = \dfrac{\hbar n}{cr}$.

Now, assume there is an extra "dormant" dimension in space with radius r' that stores dark matter (Figure 1.19). This is a fundamental tenet in the theoretical development put forth in this book and is presented here in a simplified fashion for the sake of the argument. The key difference with matter waves along detectable dimensions is that the mass $m = \dfrac{\hbar n}{cr'}$ cannot be associated with radiation at the Einstein frequency $f = \dfrac{nc}{2\pi r'}$.

In other words, *the validity of the relation $E = mc^2$ is contingent on having radiation and its associated matter stored on the same set of observable dimensions.* A staggering but inevitable consequence of this trend of thought is that Einstein's relation is only fulfilled by 5% of matter in the universe.

1.9 THE DIRAC SEA: A CANDIDATE FOR THE DARK SECTOR?

It is imperative that we discuss vacuum energy in our quest to investigate the nature of dark energy. The natural starting point for such a discussion is the Dirac equation, a breakthrough that inspired an enormous intellectual effort, quantum field theory, whereby vacuum energy was properly interpreted and cast in a rigorous footing.

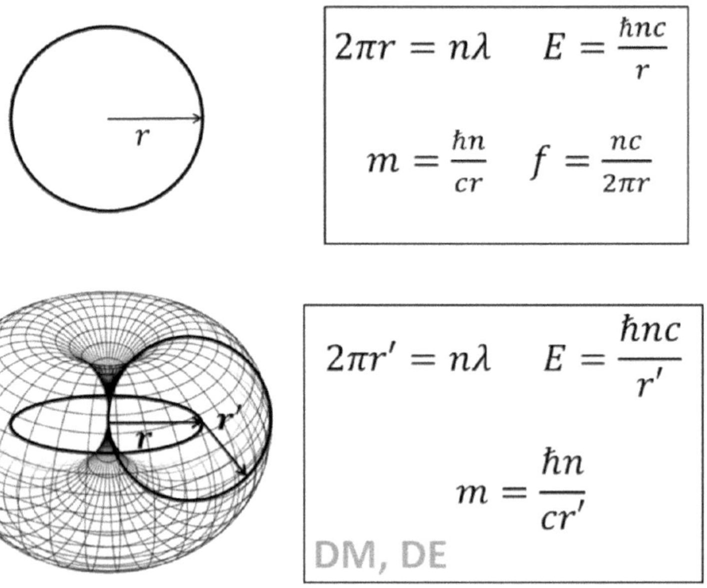

FIGURE 1.19 Dark matter/energy stored along a dormant dimension that cannot be associated with detectable radiation.

The first successful synthesis between special relativity and quantum mechanics was accomplished by Paul Dirac (P. A. M. Dirac, 1902–1984) with his relativistic wave equation for the electron conceived in 1928 [7]. Dirac's strategy was to adopt Einstein's relativistic equation

$$E^2 = (pc)^2 + (mc^2)^2, \tag{1.20}$$

where m = rest mass and p = momentum, as a starting point and replace the physical quantities for their corresponding operators in consonance with the Schrödinger equation. At a variance with the latter, Dirac's wave function becomes a four-component vector, aptly named a *spinor*, where spin or intrinsic angular momentum emerges as a conserved quantity associated with a quantum number. To yield meaningful solutions, Dirac "linearized" Eq. 1.20 by essentially taking the root square: $E = \pm\sqrt{(pc)^2 + (mc^2)^2}$, essentially "factorizing" the relativistic equation. The existence of negative energies proved extremely problematic and initially exposed Dirac to severe criticism from his peers. Dirac considered the momentum as a four-component vector $\vec{P} = \left(\frac{E}{c}, P_x, P_y, P_z\right) = (P_0, P_1, P_2, P_3)$ where the dot product is determined by the metric $g_{\mu\nu} = \eta_{\mu\nu}$ given in Figure 1.20. Einstein's convention of summation over repeated indices at different levels is followed throughout the book

$$\|\vec{P}\|^2 = P_\mu . P^\mu = \eta_{\mu\nu} P^\mu P^\nu = \left(\frac{E}{c}\right)^2 - \left(P_x^2 + P_y^2 + P_z^2\right), \tag{1.21}$$

$$P_\mu = g_{\mu\nu} P^\nu \tag{1.22}$$

$$\eta = \begin{pmatrix} -1 & 0 & 0 & 0 \\ 0 & 1 & 0 & 0 \\ 0 & 0 & 1 & 0 \\ 0 & 0 & 0 & 1 \end{pmatrix}$$

FIGURE 1.20 A Minkowski metric for four-dimensional space–time with inner product metric signature – +++.

To simplify notation, we adopt a unit of velocity equal to the speed of light, so that $c = 1$. Thus, the relativistic Eq. 1.20 reads

$$P_{\mu}.P^{\mu} - m^2 = 0. \tag{1.23}$$

Dirac was able to factorize Eq. 1.23 as a wave equation by substituting momentum components for operators $(P_0 = E \rightarrow i\partial_t, P_1 \rightarrow -i\partial_x, P_2 \rightarrow -i\partial_y, P_3 \rightarrow -i\partial_z; \hbar = 1)$ as prescribed by the Schrödinger formulation, yielding the relativistic wave equation for the electron in the form:

$$\left(i\gamma^{\mu}\partial_{\mu} - m\right)\psi = 0, \tag{1.24}$$

with the four (contravariant) gamma matrices indicated in Figure 1.21 and the wave function now represented as a four-component vector, called *spinor*. The gamma matrices satisfy anticommutative constraints imposed by comparing Eqs. 1.23 and 1.24:

$$\{\gamma^{\mu}, \gamma^{\nu}\} = \gamma^{\mu}\gamma^{\nu} + \gamma^{\nu}\gamma^{\mu} = 0 \text{ when } \mu \neq \nu;\ \gamma^{\mu}\gamma^{\mu} = 2g^{\mu\mu} \tag{1.25}$$

The wave function for a particle at rest is

$$\psi = \left(\psi_1, \psi_2, \psi_3, \psi_4\right)^T = \left(a_1, a_2, a_3, a_4\right)^T e^{-iEt};\ \sum_{j=1,\ldots,4} a_j^2 = 1 \tag{1.26}$$

Thus, its first two spinor components require that $E = m$, the expected result since $c^2 = 1$ in our convention. However, the last two spinor components yield solutions with the "unrealistic" $E = -m$. In states of a moving electron defined by nonzero third and fourth spinor components, we obtain $E = -\sqrt{(p)^2 + m^2}$, implying that the electron would tend to constantly increase momentum and emit radiation to reach lower and lower energy, all the way to $E = -\infty$, clearly an absurdity that elicited harsh criticism toward the Dirac approach.

Dirac addressed this problem by introducing the following assumptions:

- Dirac's equation is actually of broader applicability, as it holds for any fermion with spin ½. This is because of the way in which the gamma matrices are constructed based on Pauli's spin matrices (Figure 1.21): $\gamma^j = i\sigma^2 \otimes \sigma^j$; $\gamma^0 = \sigma^3 \otimes \mathbb{I}_2$; $\otimes$ = Kronecker product.
- The relativistic wave equation for the electron is actually an equation for ***all*** electrons in the universe.
- All or most states with negative energy are occupied by electrons/fermions, and Pauli's exclusion principle comes into play, forcing the detectable electron/fermion, describable with

$$\gamma^0 = \begin{pmatrix} 1 & 0 & 0 & 0 \\ 0 & 1 & 0 & 0 \\ 0 & 0 & -1 & 0 \\ 0 & 0 & 0 & -1 \end{pmatrix} \qquad \gamma^1 = \begin{pmatrix} 0 & 0 & 0 & 1 \\ 0 & 0 & 1 & 0 \\ 0 & -1 & 0 & 0 \\ -1 & 0 & 0 & 0 \end{pmatrix} \;(\sigma_1)$$

$$\gamma^2 = \begin{pmatrix} 0 & 0 & 0 & -i \\ 0 & 0 & i & 0 \\ 0 & i & 0 & 0 \\ -i & 0 & 0 & 0 \end{pmatrix} \;(\sigma_2) \qquad \gamma^3 = \begin{pmatrix} 0 & 0 & 1 & 0 \\ 0 & 0 & 0 & -1 \\ -1 & 0 & 0 & 0 \\ 0 & 1 & 0 & 0 \end{pmatrix} \;(\sigma_3)$$

FIGURE 1.21 Contravariant Dirac matrices.

Dirac's equation, into a state with positive energy. The negative-energy electrons are completely unobservable, and its collection became known as the "*Dirac sea*".

- An electron (or 1/2-spin fermion in general) in the highest negative energy level ($E = -m$) can become "detectable" by absorbing a photon of energy $E = hf = 2mc^2$. Such a photon can cause an electron to transition from the highest negative energy state ($E = -m$) to the lowest positive energy state ($E = m$). Half of the energy absorbed would correspond to the emergence of a detectable electron and the other half would correspond to a "hole" in the Dirac sea, and since a negative charge has been taken away into the detectable realm, the hole is equivalent to a positive "antielectron", which in modern terms would be named "positron".

These assumptions lead to predict the existence of antimatter, a prediction eventually confirmed by experiment. Positrons were detected by Nobel laureate Carl David Anderson (1905–1991) within the spontaneous formation of electron–positron pairs in a cloud chamber.

The Dirac sea was ultimately considered a dated construction superseded by quantum field theory (QFT). In QFT, particles are considered not describable individually or even in a closed system but rather excitations of universal fields. Furthermore, the emergence of the electron–positron pair is regarded as a spontaneous elementary process in absolute vacuum (Figure 1.22), whereby the detectable fermion spontaneously emerges from vacuum through communication with a photon with a lifetime determined from Heisenberg's uncertainty principle: $\tau \sim \dfrac{\hbar}{4mc^2}$.

With the discovery of the dark universe, the Dirac sea may need to be summoned once again, as it will be described in this book. A fermionic sea totally unaccounted for may be just what is needed to delineate the nature of the dark universe, except that the weak force governing the decay of such particles would not be operative in the way it is described in the Dirac equation. It is well established that dark energy and dark matter cannot be turned visible through photon absorption, so a vast proportion of the negative-energy fermionic sea must be rendered impervious to communication of the electroweak force, but how? The identification of the purveyor to the dark universe is described in Chapter 5 and requires a careful discussion of "chirality", an essential concept in particle physics.

It is obviously tempting to identify the Dirac sea of negative-energy fermions with the dark universe, but this assertion is incorrect and would require serious revision and fine-tuning vis-à-vis the

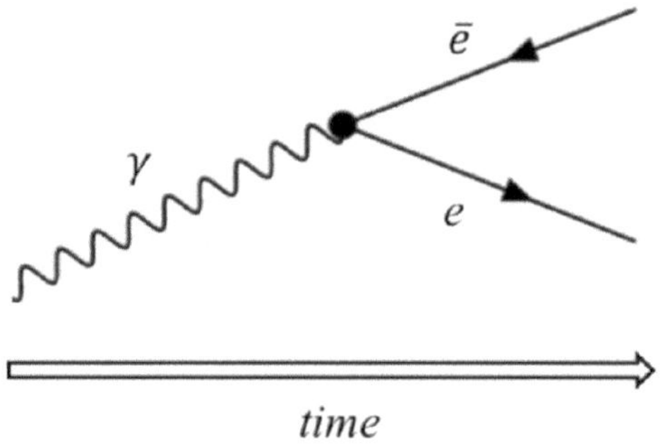

FIGURE 1.22 Electron–antielectron pair creation in quantum field theory as schematically represented in by its Feynman diagram.

fact that the electroweak interaction that enables fermions to escape from the Dirac sea is *chiral*, a conceptual framework elaborated in Chapter 5.

We shall provide a vantage point to interpret dark energy as the surplus in vacuum energy from the amount required to maintain the topology of the universe as it expands. The concentration of this surplus is indeed constant since dark energy is not subject to dilution associated with vacuum creation. This fact stands in contrast with the concentration of dark matter that will be shown to be in dynamic equilibrium with detectable matter. Clearly, the dark matter concentration decreases as it gets progressively diluted because the vacuum grows larger over time. The sustainability of the universe ensures the dynamic equilibrium between "darkened" fermions drawn from the Dirac sea and detectable matter, a contribution which maintains the overall mass of dark matter constant while the overall concentration of dark energy is kept unchanged.

At this point, we can anticipate that the dynamic equilibrium between dark matter and detectable matter is maintained by a cosmic engine fueled by vacuum energy that sustains the portal to the dark universe while maintaining a constant concentration of dark energy in an ever-expanding universe. This AI-empowered model will be elaborated in Chapter 5.

1.10 DARK MATTER STORAGE IN AN EXTRA DIMENSION

The narrative in this and the subsequent sections will take a turn, as it becomes essential at this stage to discuss the topological features of the universe that seem to have been hitherto omitted since they do not impinge in any obvious way on Einstein's local theory of gravity or even on the nonlocality of quantum physics. This topological characterization invites a veritable deconstruction of the standard model of particle physics. In turn, the deconstruction is essential to deal with the problem of whether dark matter may be accounted for by extending the standard model with the incorporation of an extra spatial dimension. After provisionally settling the problem of the universe topology as a quotient topology, this and the subsequent sections argue for the existence of a compact fourth spatial dimension that stores dark matter. The quest for dark matter becomes contingent on identifying and validating the topology of the universe as a compact multiply connected space whose quotient space modulo the "dormant" dimension is homeomorphic to $\mathbb{R}^3/\mathbb{Z}^3 \times \mathbb{R}$. Experimental measurements of the spectrum of the gradient temperature field for cosmic microwave background (CMB) radiation lead to the conclusion that a dormant circular dimension is consistent with a present-day toroidal space–time with extremely large aspect ratios relative to the dormant coordinate. The topological invariance of the universe since the big bang is thus upheld (it was once compact hence it remains compact) provided the relativistic picture holds since the cosmic birth.

As discussed in the previous sections, Newton is credited with providing the first model of gravity which he named law of universal gravitation. The predictive power of this law is staggering,

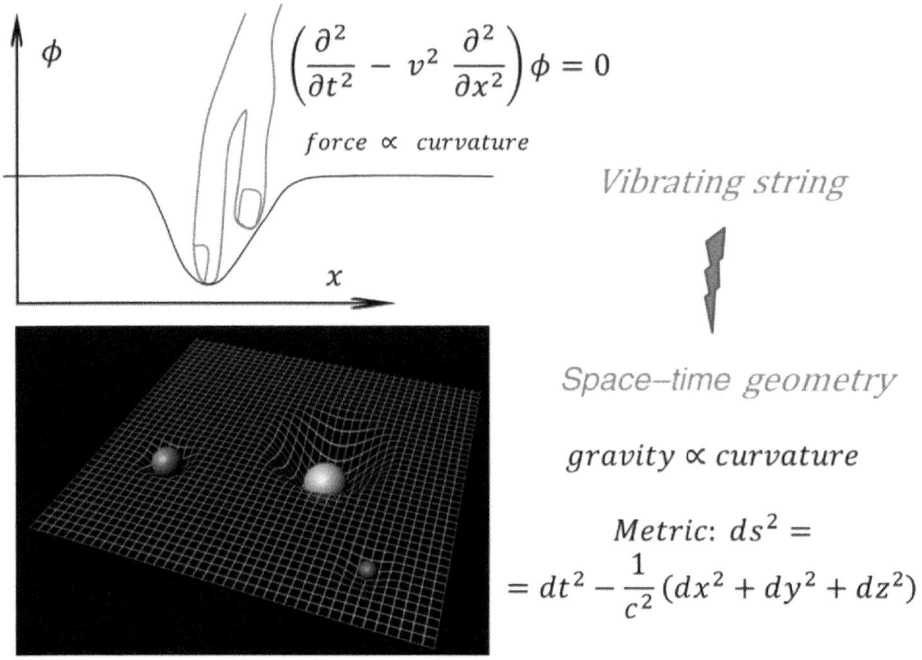

FIGURE 1.23 The equation for the vibrating string where curvature becomes commensurate with the applied force (or acceleration) as inspirational source for Einstein's differential geometry model of general relativity.

and thus, it constitutes one of the towering achievements of humanity. Yet, when asked about the nature of gravity, Newton cautiously responded that he would not frame hypotheses. Hence, in this sense, Newton may be said not to have conceived a "theory of gravity". Humanity had to wait until the early 20th century to get the first such theory, embodied in Einstein's general relativity [1]. This theory recognizes gravity as an acceleration exerted by a body endowed with mass and regards the fabric of space–time as if it were a membrane, whereby acceleration becomes effectively commensurate with curvature, in turn determined by mass concentration (Figure 1.23). To further see how Einstein got inspired to describe the fabric of space–time, let us examine the equation for a vibrating string that gets perturbed from its equilibrium position along a spatial dimension x by an amount ϕ:

$$\frac{\partial^2 \phi}{\partial t^2} - v^2 \frac{\partial^2 \phi}{\partial x^2} = 0. \tag{1.27}$$

The term on the left indicates the second instantaneous variation of the displacement with respect to time;, that is, it represents acceleration due to an exogenous force (Figure 1.27), while the distortion in the string caused by the force exerted is represented by the curvature $\frac{\partial^2 \phi}{\partial x^2}$ of the string. It should be noted that this second spatial derivative is seldom interpreted as the curvature, which in fact it is, just like the first derivative gives the slope of the curve. The proportionality constant is the square of the speed at which the perturbation propagates in time along the x-dimension.

While this fact is typically omitted, it is fairly obvious that the fundamental Eq. 1.27 linking curvature and acceleration served as the primary source of inspiration for the differential geometry model of space–time that Einstein adopted in general relativity, his classical theory of gravity

(Figure 1.23). It suffices to notice that the metric he adopted in his 4D space–time with coordinates t, x, y, and z became

$$ds^2 = dt^2 - \frac{1}{c^2}\left(dx^2 + dy^2\, dz^2\right) \tag{1.28}$$

This differential volume ds^2 of space–time is the one dictated by the relation between curvature and acceleration, except that the velocity v is now the speed of light c. A complementary way to assert the pivotal relation between curvature and gravity at the heart of Einstein's argument requires that we remind ourselves of another classical equation of physics, the Poisson equation. This equation relates the mass density ρ (amount of mass per unit volume) and the gravitational field ϕ "generated" by the mass. Poisson's equation reads

$$\nabla^2\phi = 4\pi G\rho, \tag{1.29}$$

where $\nabla^2 = \left(\frac{\partial^2}{\partial x^2} + \frac{\partial^2}{\partial y^2} + \frac{\partial^2}{\partial z^2}\right)$ and $G = 6.674 \times 10^{-11}$ Nm^2/kg^2 is Newton's gravitational constant [1]. Thus, integrating equation (1.29) on a ball of radius r yields Newton gravitational field

$$\phi = -MG/r, \tag{1.30}$$

where M is the mass contained in the ball of radius r. The dimensions in the r.h.s of equation (2.3) are those of acceleration (m/s^2), and hence, we again can justify the relationship between curvature (measured as $\nabla^2\phi$) and acceleration due to the gravitational pull per unit volume, given by $4\pi G\rho$. Thus, the differential geometry of the space–time manifold of general relativity is inspired and amply justified by the classical equations of physics. This observation prompts us to assert that Einstein's theory of gravity is in fact a classical theory [1]. It is clear that we now have a theory of gravity that includes its interaction with light, a feature absent in the Newtonian law of universal gravitation. Einstein provides a differential geometry framework in which light travels along a hypermembrane (a manifold often referred to as "brane") whose geometric fabric is defined by mass distribution.

Yet, it has proven daunting to reconcile Einstein's theory with quantum physics, now known to govern the other three fundamental forces of nature (electromagnetism and the weak and strong nuclear force). Furthermore, gravity is 10^{-38} times weaker than electromagnetism, an extremely irksome fact that points to a massive geometric dilution of gravity on the known dimensions of space–time. These conundra are of course compounded by the mysterious nature of dark matter that postulates the existence of invisible massive particles that do not detectably interact with the known elementary particles identified in the standard model [8], and yet, they provide the "missing gravity" in the detectable portion of the universe [3].

A window of opportunity for further scientific inquiry into these seemingly related problems and others arising thereof is offered by the possibility of incorporating a extra (fifth) dimension to the differential geometry of the space–time manifold. To enable the possibility of storing significant amounts of kinetic energy in an undetectable stationary wave amenable of a quantum mechanical treatment, the extra dimension would be expected to be compact, specifically rolled up in a circle of extremely small radius. It should be noted that the smallest conceivable material dimension at present is that of a quark (see next section), of the order of 10^{-18}m (a length unit named attometer) [9].

If we specifically adopt a circular fifth dimension of radius $r_0 = 0.802 \times 10^{-18}$ m (the smallest material scale assimilated to the effective quark diameter established from parton models of inelastic

$e^{\pm}p$ scattering [5]), we find out, through Einstein's relation $E = hf = hc/\lambda$, that it can store stationary waves with extremely large energies

$$E_{5,n} = \frac{n\hbar c}{r_0}, \quad n\lambda = 2\pi r_0, \quad \hbar = \frac{h}{2\pi}, \quad n = 1,2,\ldots \tag{1.31}$$

Strikingly, the lowest such energy is $E_{5,1} = 246\ \text{GeV}$, which is precisely the vacuum expectation energy of the elementary particle responsible for bestowing mass on the other known particles, the so-called Higgs boson (see next section) [8]. The daunting problem of incorporating the extra dimension arises from the fact that the circular dimension cannot be thought of an independent of the others, at least there does not seem to be any obvious reason for such an assumption. Hence, the preexisting four dimensions should be considered locally cylindrical or rather, helical instead of linear, with symmetries becoming only approximate and vast differences in curvature depending on the stride of the helices. Such a universe will be described subsequently and will be shown to be far more suitable to achieve a unified field theory while explaining the extreme geometric dilution of gravity on the "observable space–time manifold", as well as the origin of dark matter and dark energy.

Such a fifth dimension would be extremely hard to detect with current experimental resources, even at the LHC operating at optimal performance, due to the extremely high energies associated with wavelengths of the order the quark dimension. Yet, highly massive particles may be yielded by storing energy on the fifth dimension. To visualize this, we note that the kinetic energy along the fifth dimension would be undetectable and hence will be regarded as rest mass, in accordance with Einstein's relation [8]. Thus, the total rest mass M of a particle on a 5D space–time manifold would be given by the equation

$$M^2 = m^2 + m_5^2 = m^2 + \left(\frac{n\hbar}{cr_0}\right)^2, \quad n = 1,2,\ldots \tag{1.32}$$

Here, m denotes the detectable rest mass that we would be capable of measuring from the particle's existence in Einstein's 4D space–time and $m_5 = p_5/c$ is the rest mass associated with the component p_5 of momentum along the fifth dimension.

If a particle only stores stationary kinetic energy in the fifth dimension and has no observable rest mass, it is likely to be invisible, yet it could be more massive than any of the known particles. For instance, an "ur-Higgs boson" would have the mass corresponding to the vacuum expectation energy of the Higgs boson, at

$$M = M_{u-H} = \frac{\hbar}{cr_0} = \frac{246\text{GeV}}{c^2}. \tag{1.33}$$

This is almost twice the detected mass of the Higgs boson, known to be 124.97 GeV/c^2. The ur-Higgs and its relatives with masses $M_n = \frac{n\hbar}{cr_0}, n = 2,3,\ldots$ may be regarded as the first particles created after the big bang. We may estimate the lifetime at which such particles were created from Heisenberg's uncertainty relation discussed previously:

$$\Delta E \times \Delta t \sim \frac{\hbar}{2}. \tag{1.34}$$

Since $\hbar = 6.582\times10^{-16}$ eVs, we may estimate the time of formation of an ur-Higgs boson ($\Delta E = 246\times10^{9}$ eV) at $t = t_{u-H} \approx 1.34\times10^{-27}$ s. More massive relatives of the ur-Higgs are expected to emerge at shorter times estimated at $t = \frac{t_{u-H}}{n}$.

As we discuss massive relatives of the ur-Higgs, we are prompted to ask: How large can n be? The answer is straightforward, as the so-called Planck mass, $M_P = \sqrt{\frac{\hbar c}{G}} = 1.22\times10^{19}\ \mathrm{GeV/c^2}$, sets an upper bound to the mass, so that if $M_n = \frac{n\hbar}{cr_0} \geq M_P$, then the particle becomes a black hole. Thus, the first particle formed in the universe had a mass M_{n^*} corresponding to the largest integer $n = n^*$ such that $n^* \leq \frac{M_P}{M_{u-H}} \approx 4.95\times10^{16}$. This supermassive particle formed at a time estimated at $t^* = \frac{t_{u-H}}{n^*} \sim 10^{-43}$ s, during the so-called "Planck epoch" of the universe.

Thus, a picture of the very early universe for the period $10^{-43}\ \mathrm{s} \leq t \leq 10^{-27}$ s may be obtained by assessing the energy stored in the stationary de Broglie wave that spans the fifth dimension, as schematically depicted in Figure 1.24.

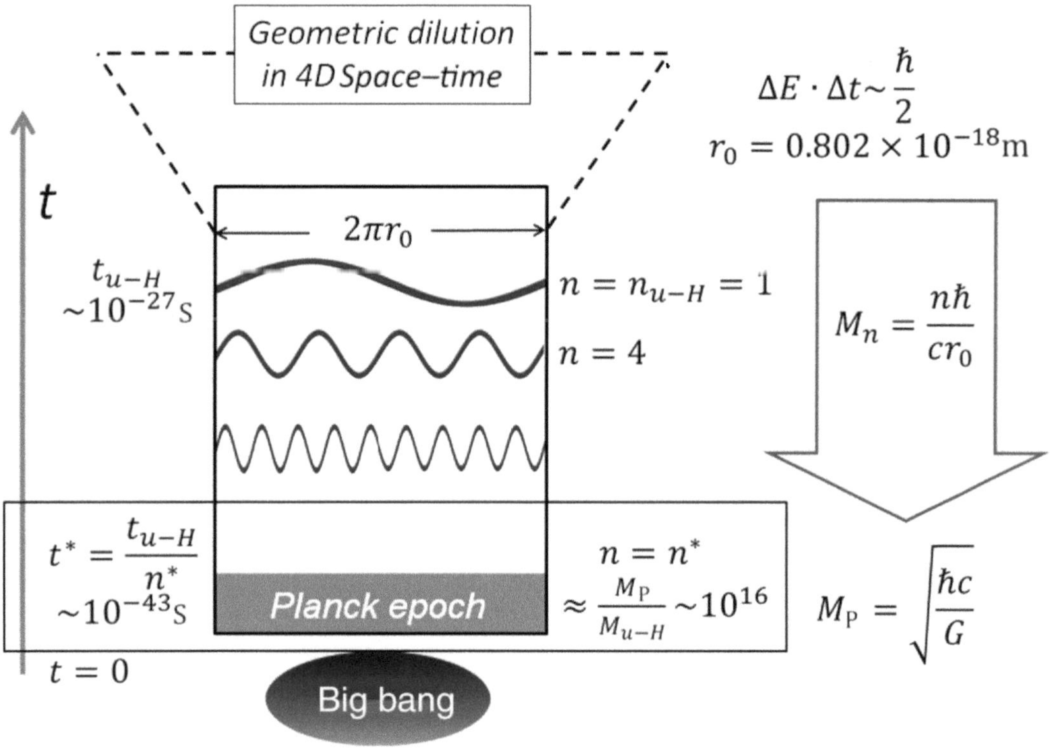

FIGURE 1.24 Elementary particles in an early universe covering the period $10^{-43}\mathrm{s} \leq t \leq 10^{-27}\mathrm{s}$ after the big bang obtained by storing energy as stationary waves that span the fifth dormant dimension.

1.11 TOPOLOGY OF THE UNIVERSE: A RELEVANT DISCUSSION IN PARTICLE COSMOLOGY

The most difficult issue we stumble upon when postulating a compact fourth spatial dimension to store dark matter is that the global topology of the universe must allow for such possibility. In the past, people have tried to incorporate a curled-up dimension into an Euclidean space, an idea that is topologically incompatible yet, of course *a priori* possible. Not only such attempts fail to generate correct extensions of general relativity, as discussed earlier, but they arose from the unappealing and unwarranted assumption that one dimension would be compact, while the others are infinite.

Incorporating a compact dimension is a nuanced matter that pivotally depends on the topology of the universe. This was clearly not a matter of concern for Einstein because his general relativity is essentially governed by differential equations. Since differential equations describe a local situation, the global topology of space–time plays no obvious role. Furthermore, relativity has no concern for boundary conditions, as the universe is assumed to be infinite. In our topology, the universe is compact and multiply connected, but there are no boundary conditions, simply because there is no boundary.

On the other hand, quantum mechanics defines the fabric of space, as determined by vacuum entanglement, but topology is not factored into the theory in any obvious way, even if entanglement is thought to be the culprit for the nonlocality of quantum phenomena. However, topology suddenly becomes extremely relevant as one tries to reconcile general relativity with quantum physics, as it is the case with quantum gravity, as shown in Chapters 2 and 5.

If we assume topological invariance of space throughout the universe evolution starting at the big bang event, then space cannot be infinite and flat, and therefore Euclidean. Provided we accept the big bang scenario, space must be compact because it once was compact with certainty. But if space is compact, then it must be multiply connected. In addition, space cannot have a boundary because that would imply an interface with nothingness, and it is physically and metaphysically impossible to interface with nothingness: An interface presupposes two media. These considerations introduce several constraints on what the topology of the universe must be. Summarizing, space must be compact, locally flat, orientable and lacking boundary. This leaves us with essentially one option, and it is a multiply connected 3-dimensional manifold: the 3-torus or Cartesian product $S^1 \times S^1 \times S^1$. Hence, the incorporation of a fourth circular dimension is compatible with the topology of the space–time manifold.

There is a fundamental difference in the way the extra spatial dimension, the curled dormant dimension, is incorporated in the treatment presented in this book when compared with the early attempts, especially the so-called Kaluza-Klein model [8]. These researchers incorporated the extra dimension *directly* to the Euclidean space, creating the absurdity of having one compact and three infinite spatial dimensions in space–time with no justification. This construction proved to be untenable from every perspective, a mere intellectual curiosity.

As we have shown, the spatial cross sections of space–time cannot be regarded as isomorphic to $\mathbb{R}^3$ but rather to the three-dimensional torus $\mathcal{T}^3 \approx \mathbb{R}^3/\mathbb{Z}^3$, that is, a quotient space of classes modulo all the spatial translations of the unit cell indexed by three integers (Figure 1.25). This implies that the space is multiply connected with all spatial dimensions being compact, hence allowing for the incorporation of an the extra circular dimension ($S^1 \approx \mathbb{R}/\mathbb{Z}$) without the conceptual violence of the Kaluza-Klein models. A hybrid $\mathbb{R}^3 \times \mathbb{R}/\mathbb{Z}$ of three infinite and one compact spatial dimension would make no sense and is in fact not compatible with the tenets of general relativity, assuming this theory is upheld since the inception of the universe, as topological change is forbidden.

At a very fundamental level, *the Poincaré group, assumed to be the group of isometries of space–time over which the whole edifice of the standard model is built, does not really contain the proper symmetries but only approximate symmetries or rather, pseudo-symmetries, since the underlying spatial cross sections of space–time are not isomorphic to* $\mathbb{R}^3$ *but to* $\mathbb{R}^3/\mathbb{Z}^3$.

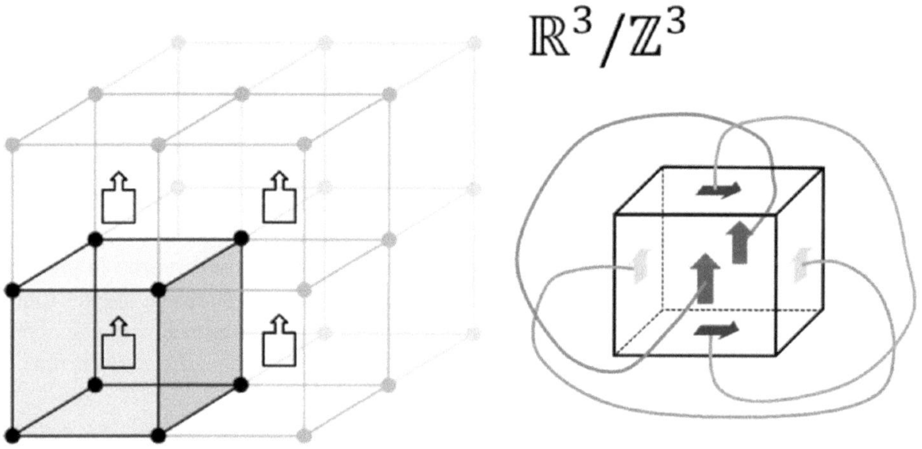

FIGURE 1.25 Generic spatial cross section of space–time with compact coordinates, as required by the topological invariance of the universe dictated by the tenets of general relativity. The compact space is isomorphic to the three-dimensional torus in accordance with the topological equivalence $\mathbb{R}^3/\mathbb{Z}^3 \approx \mathcal{T}^3$. The latter manifold is obtained by identifying opposite faces of the solid cube in three dimensions, hence featuring a wormhole upon construction with dimensional aspect ratios equal to 1 (Chapter 5).

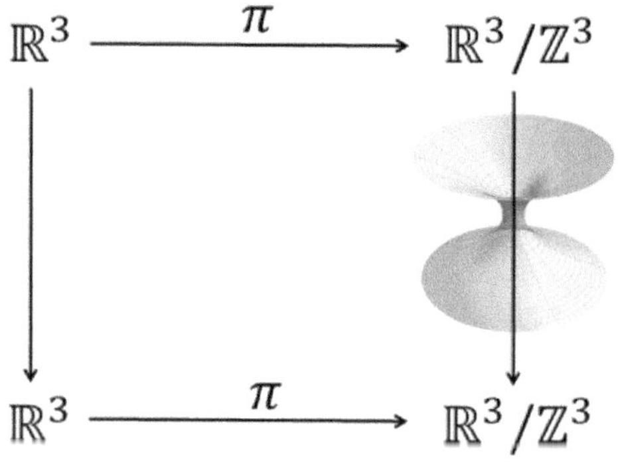

FIGURE 1.26 Commutative scheme adopted by artificial intelligence to circumvent the artefactual wormhole phenomenology associated with the universe compactness by lifting it at the level of universal covering ($\mathbb{R}^3, \pi$) of the toroidal spatial cross sections ($\mathbb{R}^3/\mathbb{Z}^3$) of space–time.

On the other hand, a spatially compact universe implies the existence of a primeval wormhole, as inferred from Figure 1.25 and demonstrated in Chapter 5. This can be visualized in a two-dimensional cross section $\mathcal{T}^2$ with dimensional aspect ratio 1. As shown in Chapter 5, the sustainability of this wormhole is concomitant with the universe expansion. On the other hand, the very notion of universe expansion becomes meaningless in the noncompact space $\mathbb{R}^3$, while it makes good sense to talk about expansion of the unit cell in the quotient space $\mathbb{R}^3/\mathbb{Z}^3$. This is yet another reason to assume that the spatial cross sections of the relativistic space–time are actually isomorphic to $\mathbb{R}^3/\mathbb{Z}^3$, and not to $\mathbb{R}^3$, as assumed in the Minkowski space–time endowed with the isometries of the Poincaré group.

On the other hand, AI adopts the commutative scheme presented in Figure 1.26, where the autoencoder defines the lifting of the standard model at the level of the universal covering with canonical projection $\pi: \mathbb{R}^3 \to \mathbb{R}^3/\mathbb{Z}^3$, so that the diagram presented in Figure 1.26 is commutative. This lifting has considerable advantages since it circumvents the observed universe expansion with the concomitant primeval wormhole as artefactual of "Plato's cave physics", where the observer is placed. Even dark energy, shown in Chapter 5 to be the residual vacuum energy not invested in sustaining the primeval wormhole, becomes artefactual of Plato's cave phenomenology in $\mathbb{R}^3/\mathbb{Z}^3$, the only accessible to the quantum observer.

1.12 PRIMER ON QUANTUM FIELD THEORY

Major modeling efforts that followed after the advent of general relativity and quantum mechanics revealed that the true nature of elementary particles cannot be properly captured in the corpuscular representation. Rather, elementary particles constitute excitations or warps in fields – one for each particle type – that establish a correspondence between each point of space with a scalar or vector value [8]. This is the spirit of the quantum field theory (QFT) that spearheaded current modeling efforts in elementary particle physics.

An elementary particle becomes a local concentration of energy or its equivalent mass, as dictated by Einstein's relation $E = mc^2$ derived in Chapter 1. The particle field itself is viewed as a map $\phi: W \to V$, where W is the four-dimensional space–time of special relativity and V is either $\mathbb{R},\mathbb{C}$, the sets of real or complex numbers, or $\mathbb{R}^n\left(\mathbb{C}^n\right), n>1$, depending on whether ϕ is a scalar or a vector field, respectively. The picture where particles are regarded as excitations of their respective fields is a major tenet of QFT, and the modeling effort resulting thereof is known as the standard model (SM) of particle physics [8].

In accordance with current thinking, particles are best characterized as local modes of storing energy/mass, that is, localized field excitations defined by independent representations of the symmetry group of space–time. These representations have generators that are parametrized by specific attributes of particles, including charge, mass, spin (an intrinsic angular momentum vis-à-vis an internal axis of rotation), spin degrees of freedom (projection of the spin vector along the direction of particle motion), and so on. Yet, this manner of particle classification tells us very little unless we can define interactions between particles based on their intrinsic attributes, assessing how particles communicate and transform into one another. The nature of these interactions is delineated in the next section.

Now let us briefly explored how the concept of particle field came about. By attributing the energy of the photon to its frequency of oscillation as in the equation $E = hf = hc/\lambda$, Einstein was the first to associate a particle with a wave in his description of the photoelectric effect. Einstein, and Planck before him, reasoned that the frequency was simply a surrogate for kinetic energy, as it is an indicator of the number of cycles per unit time. This was followed by de Broglie's representation of the electron orbiting the atomic nucleus as a stationary wave with a wavelength that satisfies the relation $n\lambda = 2\pi r$, where n is a positive integer value and r is the orbit radius. This "quantization" of the wavelength ensures constructive (reinforcing) phase interference, as depicted in Figure 1.27. It also captures the discreteness of the experimentally obtained electron absorption spectrum, as the only allowed transitions for the electron are the result of absorption of photons carrying energies given by the differences between the n-indexed electron energies $E_n = n\hbar c/r$, with $\hbar = h/2\pi$. In fact, de Broglie went one step beyond, as he associated any particle with momentum p to a wavelength λ according to the relation $p = h/\lambda$. We can see in these early modeling efforts the need to go beyond a mere corpuscular description of particles, a conceptual framework later adopted by QFT in its construction of the SM [8].

The SM has defined, characterized, and organized all of the known elementary particles in much the same way that the periodic table systematically categorized or classified the types of atoms known as elements. The predictive value of the SM has been staggering, notwithstanding the fact that it does not fully incorporate gravity, while being able to encompass the other three quantum forces of nature in a unified way.

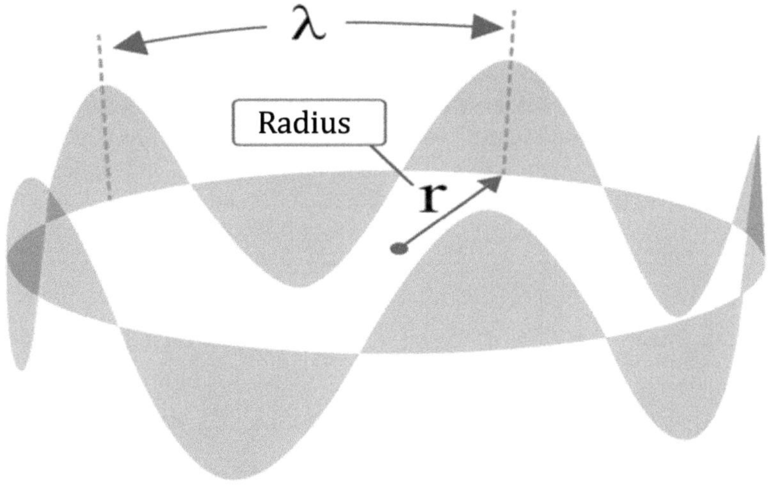

FIGURE 1.27 A de Broglie stationary wave of wavelength λ in a circle of radius r.

Yet, by 1970, mainly due to the monumental "bottom-up" modeling efforts based on QFT, elementary particles were shown to fall in two main categories: fermions and bosons, with fermions making up matter while bosons, or more precisely "gauge bosons", transmit forces to which fermions respond or react to [8]. Fermions are divided into two kinds of particles, depending on the forces they respond to: quarks and leptons. Particles within the three types, quarks, leptons, and bosons, are symmetry related to each other. This means that a coordinate transformation associated with a space–time symmetry operation transforms one particle into another of the same family. It was established that particles communicate with one another via four forces: electromagnetism, strong nuclear force (SNF), weak nuclear force (WNF), and gravity. The SM describes the first three, while gravity does not yet feature satisfactorily in the current version of the SM, as discussed in Chapter 2.

Different particles communicate and interact through different forces. For example, only the quarks relate to the gluon, its natural gauge boson and carrier of the SNF, while electrons relate to the photon, the gauge boson that carries the electromagnetic force. In addition, electrons can also communicate via the W boson and Z boson, the carriers of the weak nuclear force. So, electromagnetism is the force that holds electrons in the atom, while the SNF communicated by gluons keeps the nuclei from breaking apart. Meanwhile, the WNF communicated by Z or W bosons assists the radioactive decay of nuclei, whereby, for example, the quark/gluon assemblage of a neutron composite becomes that of a proton upon interaction with a W boson with the emission of electromagnetic particles.

We shall now describe the Higgs boson, a particle endowed with a field ϕ that may be represented as a doublet of complex numbers: $\phi = (\phi_1, \phi_2)$, with $\phi_j = Re\phi_j + iIm\phi_j, j = 1,2$. This field, empirically determined in a significant modeling effort, is peculiar in that its potential energy does not reach its minimum at zero field but a value of ~246 GeV (1 eV = 1.602×10^{-19} J, 1 GeV = 1.602×10^{-10} J) known as the vacuum expectation value (v), hence implying a symmetry breaking (Figure 1.28). The symmetry breaking arises because different changes in potential energy around the zero-point energy are obtained depending on the direction of variation of the field. For example, if we write the singlet field in radial coordinates as $\phi = \rho e^{i\theta}$, then the zero-point energy is the circle $\rho = v$ and radial variations $\rho = v \pm \delta h$ yield variations in the potential energy (dashed line in Figure 1.28), while angular variations along the circle $\rho = v$ yield no change in potential energy, which remains at its zero value. This symmetry breaking turns out to be of paramount importance in determining the role of the Higgs boson vis-à-vis its interactions with other particles. Thus, just like other gauge bosons are force carriers, communicating different types of forces associated with particle interactions, the Higgs boson may be interpreted as bestowing mass upon other particles.

To interpret this process of mass endowment as well as the communication of fundamental forces by the other gauge bosons, it is necessary to resort to the mathematical arsenal of particle physics [8]. This arsenal is necessary to describe the "geodesic flow", that is, the lines of least action followed by the fields that underlie particle physics. As said, a basic tenet of QFT indicates that a particle is represented as a local excitation of its respective field, representing a local concentration of energy or, equivalently, of mass. In QFT, a particle scalar field $\phi = \phi\left(\left\{x_\mu\right\}_\mu\right)$ in space–time is determined by the particle Lagrangian $\mathcal{L} = \mathcal{L}\left(\phi, \left\{\partial_\mu \phi\right\}_\mu\right)$ $\left(\partial_\mu \equiv \dfrac{\partial}{\partial x^\mu}\right)$ whose definition is inspired by classical mechanics: $\mathcal{L} = \frac{1}{2}\left(\partial_\mu \phi\right)^2 - V(\phi)$, with the first term on the r.h.s. representing the kinetic energy (K), and the second, the potential energy. The latter is exemplified in Figure 1.28 for the Higgs field. Einstein's convention of summation over repeated indices is followed throughout. Thus, the geodesic flow in space–time defined by the field $\phi = \phi\left(\left\{x_\mu\right\}_\mu\right)$ minimizes the action $\mathfrak{A} = \int \mathcal{L}\left(\phi, \partial_\mu \phi\right) dx^\mu$, and hence, the field satisfies the so-called Euler–Lagrange equations:

$$\partial_\phi \mathcal{L} = \partial_\mu \left[\frac{\partial \mathcal{L}}{\partial(\partial_\mu \phi)}\right]. \tag{1.35}$$

To see how two particles, determined respectively by fields ϕ_1, ϕ_2, interact, or one particle communicates a force to another particle, we need to consider the sum $\mathcal{L}_1\left(\phi_1, \left\{\partial_\mu \phi_1\right\}_\mu\right) + \mathcal{L}_2\left(\phi_2, \left\{\partial_\mu \phi_2\right\}_\mu\right)$ of their respective Lagrangians. An appropriate coordinate transformation $x^\mu \to x'^\nu$ and suitable transference or swapping of terms yields the sum: $\mathcal{L}_1'\left(\phi_1', \left\{\partial_\nu \phi_1'\right\}_\nu\right) + \mathcal{L}_2'\left(\phi_2', \left\{\partial_\nu \phi_2'\right\}_\nu\right)$ satisfying the action equality

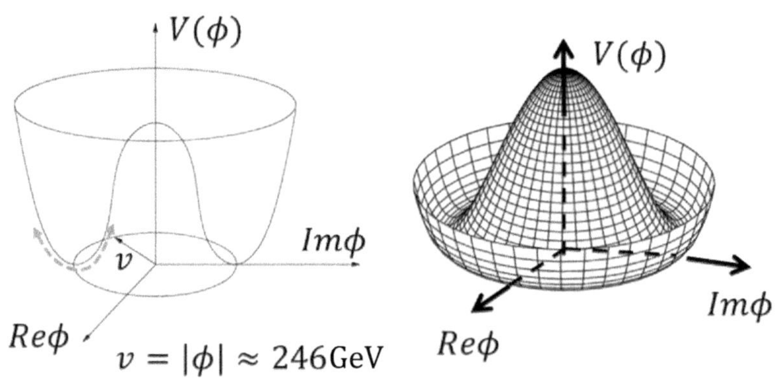

FIGURE 1.28 Scalar field for the Higgs boson.

$$\int\left[\mathcal{L}_1\left(\phi_1,\{\partial_\mu\phi_1\}_\mu\right)+\mathcal{L}_2\left(\phi_2,\{\partial_\mu\phi_2\}_\mu\right)\right]dx^{\mu}$$
$$=\int\left[\mathcal{L}_1'\left(\phi_1',\{\partial_\nu\phi_1'\}_\nu\right)+\mathcal{L}_2'\left(\phi_2',\{\partial_\nu\phi_2'\}_\nu\right)\right]dx'^{\nu} \quad (1.36)$$

This implies that the original interactive particle pair defined by the fields ϕ_1, ϕ_2, has transformed through mathematical manipulation (including only term reorganization and change of variables) into new particles interpreted as being defined by the fields ϕ_1', ϕ_2'. These destiny fields are associated with the "destiny Lagrangians" $\mathcal{L}_1', \mathcal{L}_2'$ and may correspond to any number of particles, elementary or composite.

For example, let us consider the interaction of the Higgs field $\phi_1 = \phi_{\mathrm{H}}$ (Figure 1.28), with a second field ϕ_2 corresponding to a particle that may be the original massless version of the W or Z boson. The kinetic energy K_{H} for the Higgs field may be written in terms of the covariant coordinates $D_\mu = \left(\partial_\mu - i\phi_2\right)$ that respond to the field of the other particle. Upon the coordinate changes $\phi_{\mathrm{H}} \to (\rho, \theta)$ and substituting $\rho = v + \xi$ around the v-value ring of ϕ_{H} (Figure 1.28), the kinetic energy K_{H} reads (asterisk here denotes complex conjugate):

$$K_{\mathrm{H}} = \frac{1}{2}\left(D_\mu\phi_{\mathrm{H}}\right)\left(D_\mu\phi_{\mathrm{H}}\right)^* = \frac{1}{2}\left(\partial_\mu\xi\right)^2 + \frac{1}{2}\phi_2^{\,2}\left(v+\xi\right)^2. \quad (1.37)$$

This implies that the precursor to the gauge boson has now gain the potential energy term estimated as $\frac{1}{2}\phi_2^{\,2}v^2 : \mathcal{L}_2' \approx \mathcal{L}_2 + \frac{1}{2}\phi_2^{\,2}v^2$ which implies that the mass bestowed by the Higgs field corresponds to radial (i.e., along the ξ-coordinate) excitations of its field, as depicted in Figure 1.28.

1.13 ONTOLOGICAL PROBLEMS WITH THE BIG BANG SCENARIO IN A RELATIVISTIC UNIVERSE

The forthcoming "metaphysical" discussion is inevitable, since we shall discuss the big bang scenario in connection with the nature and origin of dark energy and dark matter. Space–time, a foundational concept in relativity, incorporates time as an inherent dimension, so we may assert that space–time per se, and hence the universe, cannot evolve. *Prima facie* this statement may seem disconcerting, but considering the evolution of space–time introduces a paradox of the Russell type. The point is that the phrase "evolution of space–time" is meaningless as no entity where time is immanent in its being can actually evolve. Since time is inherent to space–time and inseparable from the spatial dimensions, how could space–time change over time? All-time change in space–time is already contained in space–time (Figure 1.29).

What we presumably mean by "evolution of the universe" is actually something quite different: Consider "formally" a fixed-t cross section of space–time and define evolution of the universe as the sequence of t-cross sections as the "parameter" t is varied. In other words, we consider the time evolution of space. While this seems *a priori* a well-defined object, it is immanent in space–time, so it cannot be equated, even by abuse of language, with the latter. Furthermore, space–time does not allow for the dissociation of time from spatial dimensions. This is a pillar of relativity. There is no external clock. The dimensions are not "separable". For example, if we query about the big bang scenario vis-à-vis the universe evolution, we are rapidly trapped into a paradox, because the extreme mass concentration at the "origin" prevents

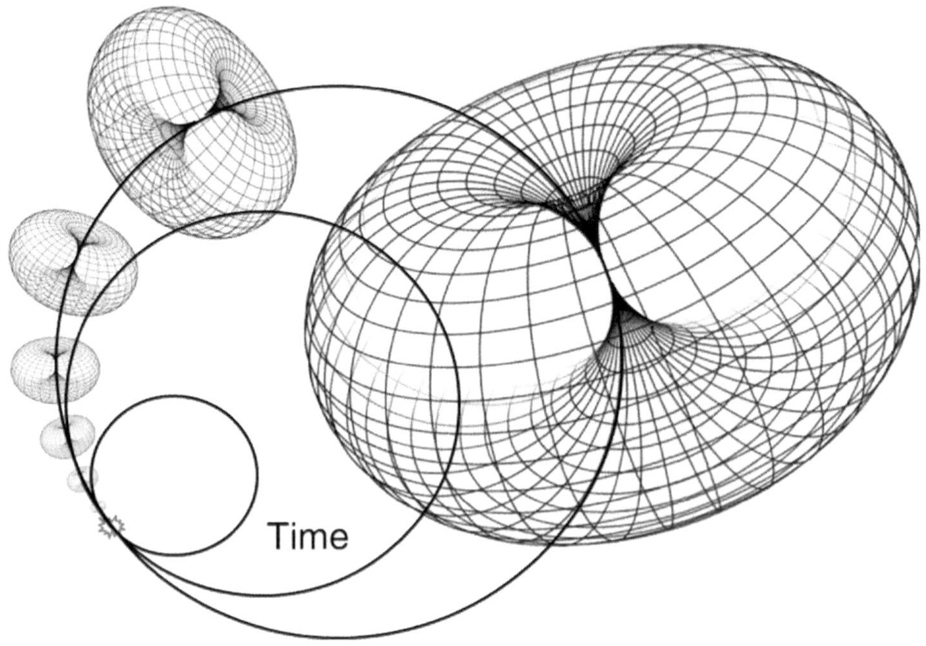

FIGURE 1.29 Space–time subsumes its own evolution, and hence, the phrase "space–time evolution" is rigorously meaningless and entails a Russell-type paradox. A torus topology of space cross sections is justifiably adopted.

time from flowing, and hence, *the primeval point is a priori a state of the universe from which the universe can never evolve*. This is an extreme case of the time dilation experienced by entities crossing the event horizon of a black hole. In fact, a trajectory leading out of a black hole would entail time reversal (Figure 1.30), so time cannot be "externalized" at the big bang, let along flow to describe the origin of the universe.

Space–time postulates interdependence between spatial dimensions and time. As such, the dimensions are not separable or cannot be regarded as independent. Thus, fixed-time cross sections may constitute a "formally" acceptable concept, but the time evolution of such cross sections is problematic especially given that the magnitude of time dilations is strongly dependent on space. This discussion brings us back to the big bang scenario. If space is concentrated in a geometric point and time is consequently dilated to infinity, how could the universe be said to have ever evolved in the big bang scenario? Furthermore, what is meant by evolution of the universe, when space–time evolution, understood in any acceptable way, is subsumed in space–time itself (Figure 1.29)? It appears that the big bang scenario is ill posed and even the phrase "universe evolution" is ill defined.

Yet, the discussion on universe evolution and big bang is alive and ongoing and has borne fruits, especially in the realm of quantum cosmology [1]. This makes it necessary to come up with an operational definition of "evolution of the universe" or at least of "evolution of space" that is not conceptually violent and can even reconcile the inseparability of space and time that is inherent to the relativistic perspective. A possible *ansatz* put forth in this book is as follows: Assume a time dilation defined by $t_v \rightarrow t = t_v / \vartheta(\boldsymbol{x})$, scaling time flowing in an inertial vacuum (t_v) according to a spatially dependent factor $\vartheta(\boldsymbol{x})$, inherently dependent on the local curvature tensor and commensurate with its norm. We define *extrinsic evolution of space* as the evolution of the spatial t-cross section of space–time parametrized by rescaling time to make it an extrinsic parameter dissociated

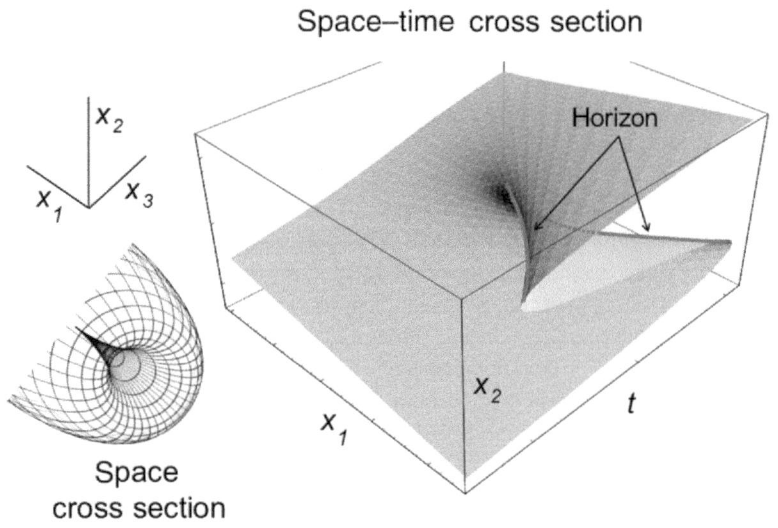

FIGURE 1.30 Crossing the event horizon of a black hole requires time reversal, as depicted in the space–time model of the singularity.

from the intrinsic time $\left(t = \frac{t_v}{\vartheta(x)}\right)$ that is fixed as constant in determining the cross section. In other words, we consider the "extrinsic evolution" of the t-slice of space–time: $\mathcal{X}(t_{ext}) = \{x = x(t)\}_x$ re-parametrized for each point in space as dependent on "extrinsic time": $t_{ext} = t \times \vartheta(x(t))$.

For relativistic singularities such as the black hole, the scaling of time dilation does not hold. This is because to escape from a black hole horizon the speed of the moving body should exceed the speed of light, implying that time should move backward. This can be readily visualized in a cross section of the space–time manifold at the black hole, incorporating two spatial dimensions and time (Figure 1.30). In general, following Hendrik Lorentz [1], we can make the time scaling dependent on the speed of the moving body in accordance with the so-called Lorentz transformation: $(v) = \sqrt{1 - \frac{v^2}{c^2}}$, and however, the Lorentz formula yields an imaginary factor for speeds exceeding the speed of light ($v > c$), which is meaningless in this context and also incorrect because it does not reflect the actual reversal of time associated with the escape from the black hole. The time reversal implies that the escaping observer has actually traveled into the future as it transposed the event horizon of the black hole, and therefore, its escape from the black hole would entail traveling into the past, as depicted in Figure 1.29. Since time reversal is a relativistic impossibility, there is no escape from the black hole.

These arguments lead us to formulate the following assertion:

Proposition: Only one of the following statements is true:

a. The relativistic tenets hold for the big bang, and then, the big bang scenario does not represent the origin of space–time because space–time subsumes its own evolution, and hence, it does not have an origin,

b. the relativistic tenets of space–time does not hold in the earliest universe because time and space are decoupled, hence separable, or
c. the relativistic tenets hold for the big bang, and then, it is impossible to trace the origin of the universe starting at a space–time singularity because no extrinsic flowing time exists or can be defined at its spatial locus.

This proposition leaves AI with fewer resources to tackle the problem of the nature of dark matter and dark energy, as AI builds up theory only if it finds leads to a cogent picture through conjecture framing. As hinted by the results previously described in this chapter, the study of the origin of singularities in space–time is essential to make progress in understanding the nature of dark matter and, as we shall see in Chapters 2 and 5, the same holds for dark energy. Yet, it appears that the very phrase "origin of singularity" is troublesome and alludes to an ill-posed problem. The only alternative that has led to further progress is statement (b), which is as likely or unlikely to be true as the other two, but the only one pregnant with possibilities for theoretical progress, as we shall demonstrate. Thus, an operational definition of *extrinsic evolution of space* based on the assumption that statement (b) is valid will be adopted throughout in the rest of the book and by abuse of notation it will be referred to as "evolution of the universe" or evolution of space–time with the caveat that the clock can be placed outside of the object of study and that it is actually ticking.

1.14 THE UNIVERSE TOPOLOGY ADMITS A COMPACT EXTRA DIMENSION

As earlier described, Einstein's theory of general relativity is formulated in terms of differential geometry and its laws are therefore cast as differential equations. It is therefore a local theory. It tells us nothing about the overall shape or topology of the universe while it informs on its geometry [10]. In principle, more than one topology can fit the postulated geometric flatness, which at any rate, can only be regarded as an approximation [1,10]. There is one thing that general relativity does tell us in regard to topology: *As the universe evolved in time since the big bang its topology could not change* [1]. This is fairly obvious, since a topological change would imply cutting and pasting space–time, an inadmissible operation. This is so because it would entail a nonhomeomorphic distortion of the differential geometry fabric unless a new force beyond the four established forces [8] could be invoked.

In general relativity, we assume that the intrinsic geometry of the universe 4D manifold has infinite curvature radius (flat geometry). However, if we soften the assertion to indicate "immeasurably large radius", we may find alternative topologies for compact (yet enormous) manifolds that would fit the postulated approximate flatness. From the previous discussion, it seemed that identifying dark matter and dark energy would require a reverse engineering of the SM to incorporate all dimensions in an early universe together with the evolution of their aspect ratios. Yet, this reverse engineering depends pivotally on the topology of the universe [5], which we know must be invariant in time. This implies that the present-day universe cannot be infinite for the simple reason that the big bang singularity and the early universe were compact. *Hence, our universe must be compact, however enormous.*

A quasi-flat compact manifold must necessarily be multiply connected, and hence, the incorporation of a fifth circular coordinate yields one and only one manifold as the only alternative: a five-dimensional torus, that is, a Cartesian product of five circles with huge aspect ratio between the four circular coordinates conforming the locally quasi-flat four-dimensional space–time and the extra fifth spatial coordinate with a quark-size (i.e., LHC-undetectable) radius. We regard the standard four-dimensional space–time in general relativity as a four-torus locally homeomorphic to the Euclidean space. This is a latent compact manifold representing a quotient space denoted $W/\sim$ [5], so that two points in a higher dimensional space W, the five-dimensional torus with a circular extra dimension, would be equivalent modulo "$\sim$" if they projected onto the same point in the latent manifold. The task ahead then becomes to elucidate how dark matter would fit in this scheme as a gravity-carrying particle enshrined in the compact fifth dimension, existing in *W* but not in its quotient space *W*/~. In this way, dark matter would not interact with the SM in 4D space–time, except via gravity.

The postulated toroidal topology of the universe has found recent experimental validation in the examination of the CMB, the cosmic relic of the big bang. If space were infinite (hence flat and simply connected), perturbations in the temperature gradient field of the CMB radiation would exist on all scales. If space is finite, then there would be wavelengths missing that are larger than the size of space itself. Maps of the CMB perturbation spectrum made with probes like NASA's WMAP and the Planck probe from the European Space Agency have shown striking amounts of missing perturbations at large scales. The properties of the observed fluctuations of the CMB show a "missing power" on scales beyond the size of the universe [11]. That would imply that our universe is multiply connected and finite. The spectrum of the CMB is compatible with a three-dimensional torus topology for the spatial components of the universe. Thus, universe models with spatially multiply connected topology contain a discrete spectrum of the Laplacian with specific wavelength cutoff, as observed in the CMB. Furthermore, the three-dimensional torus model possesses a two-point correlation function that fits CMB maps obtained with the Planck probe [11].

Thus, the toroidal topology enables the incorporation of the extra curl-up dimension as an undeveloped feature in the compact universe evolution. In principle, nothing precludes the presence of more than one such dormant dimensions. The dormant dimension cannot be probed with current means as it spans the quark attometer scale, storing energy in the order of 246 GeV. On the other hand, the flat simply connected universe that we all intuitively have come to grips with would actually be incompatible with the evolution from a compact manifold as implied by the big bang scenario simply because such evolution would entail a change in topology. Such a transition from compact multiply connected to flat simply connected cannot be smooth and would be forbidden by general relativity.

Within the compact multiply connected model of todays' universe, four of the circular dimensions have grown so much that the curvature radius is for all practical purposes infinite, and the geometry is locally that of an Euclidean space [1]. Thus, the dormant fifth coordinate combines with the quasi-Euclidean standard coordinates in a local cylinder of quark-size radius, where the standard coordinate constitutes the cylinder axis. In essence, the torus with huge aspect ratio becomes locally a cylinder (Figure 1.31). Standard space–time coordinates get mixed with the dormant coordinate as determined by a pitch angle α defining the helix stride, so that a geometric dilution (υ) for the energy stored in the dormant coordinate may be defined as $\upsilon = -\log\cos\alpha$. Thus, the geometric dilution becomes infinite when there is no projection along the dormant coordinate and zero, when there is no projection onto the standard coordinates.

1.15 AI QUEST FOR THE ORIGIN OF DARK MATTER ENCODED IN A FIFTH DIMENSION

Current knowledge of dark matter is sketchy at best and mostly conjectural. What we actually know or can judiciously conjecture is best summarized as follows:

- Dark matter is invisible, massive, "cold" (speed far lower than the speed of light, $v \ll c$), mainly formed early on, interacts with the SM only via gravity, and does not decay easily.
- It is probably not an extension of the SM, as no particle derived from the latter fits the characteristics of dark matter [3].
- It may be stored in a compact fourth spatial dimension.
- In advocating for storing dark matter in an extra dimension, we note that the quark scale (q = 0.802×10^{-18} m) is the smallest – undetectable – material scale known and that a stationary

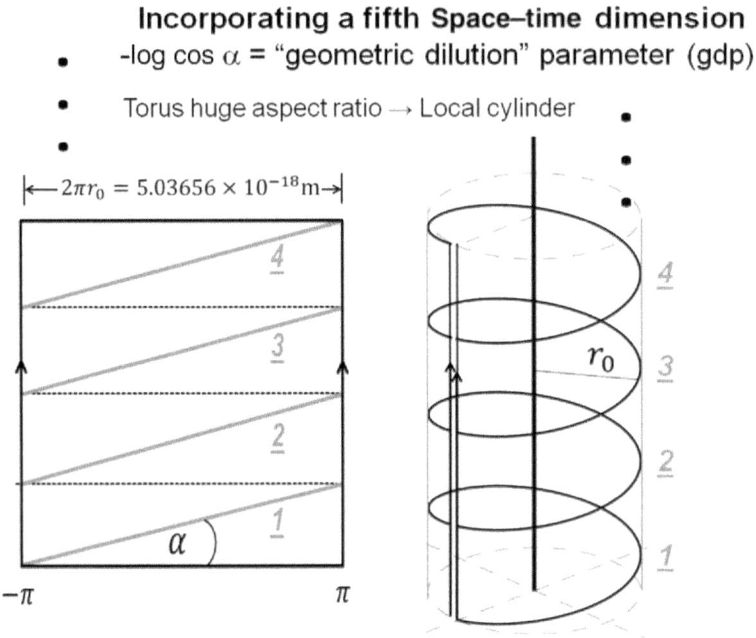

FIGURE 1.31 Incorporating the circular fifth dormant dimension and combining it with generic locally flat dimensions of the four-dimensional space–time through a geometric dilution parameter determining the helical stride.

wave along a circular fourth spatial dimension with radius q stores an energy $E = hc/2pq = 246$ GeV, the vacuum expectation value of the Higgs boson.

- We have a validated topological model of the compact universe to incorporate such a dimension and trace the origin of dark matter to the early universe.

The core involves the identification of dark matter as stored in the five-dimensional "quintessential" space W, a five-dimensional torus projecting onto a four-dimensional "latent" manifold, and $\Omega = W/\sim$ that is locally flat, meaning it is locally homeomorphic to an Euclidean space. Thus, Einstein's four-dimensional space–time Ω is regarded as a quotient space for the space that we intend to prove stores dark matter. This problem requires the deployment of AI, as it requires the extension of the SM to incorporate a fifth dimension, so far defined in the latent quotient space $\Omega = W/\sim$. This task, undertaken in Chapter 4, entails the lifting of each pairwise particle interaction within Ω, represented by the transformation: $\mathcal{L}_1\left(\phi_1,\{\partial_\mu\phi_1\}_\mu\right)+\mathcal{L}_2\left(\phi_2,\{\partial_\mu\phi_2\}_\mu\right)\rightarrow \mathcal{L}_1'\left(\phi_1',\{\partial_\nu\phi_1'\}_\nu\right)+\mathcal{L}_2'\left(\phi_2',\{\partial_\nu\phi_2'\}_\nu\right)$, to the level of W:

$$\widetilde{\mathcal{L}_1}\left(\widetilde{\phi_1},\{\partial_\mu\widetilde{\phi_1}\}_{\mu=1,\dots 5}\right)+\widetilde{\mathcal{L}_2}\left(\widetilde{\phi_2},\{\partial_\mu\widetilde{\phi_2}\}_{\mu=1,\dots 5}\right)\rightarrow \widetilde{\mathcal{L}_1'}\left(\widetilde{\phi_1'},\{\partial_\nu\widetilde{\phi_1'}\}_{\nu=1,\dots,5}\right)$$
$$+\widetilde{\mathcal{L}_2'}\left(\widetilde{\phi_2'},\{\partial_\nu\widetilde{\phi_2'}\}_{\nu=1,\dots 5}\right), \tag{1.38}$$

where $\widetilde{\mathcal{L}_n},\widetilde{\phi_n}$ denote, respectively, the lifting of Lagrangian $\mathcal{L}_n$ and particle field ϕ_n, with n indicating particle types in the SM. As indicated above, the primes denote destiny particles arising from the interaction.

A lifting of the SM is valid if and only if the diagram in Figure 1.34 is commutative. This property indicates that a canonical projection $\pi : W \to \Omega$ followed by interaction within the latent space Ω yields a destiny state that is the same as that obtained when the interaction is computed directly in the quintessential space W and the resulting destiny state is subsequently projected onto the latent space Ω. In other words, the following equation must hold for every particle pair $\left(X_J \left(\mathcal{L}_J, \phi_J\right), J = 1,2\right)$ in the SM:

$$X'_j\left(\mathcal{L}'_J, \phi'_J\right) = \pi \widetilde{X}'_j\left(\widetilde{\mathcal{L}}'_j, \widetilde{\phi}'_j\right). \tag{1.39}$$

The lifting $\tilde{\phi} : W \to \mathbb{C}^n$ of a particle field $\phi : W/\sim \to \mathbb{C}^n$ requires a particular type of AI system named *variational autoencoder* [5]. Usually, in complex dynamical systems, such autoencoders are used to extract the latent space and obtained the simplified differential equations defined on the latent space as entraining the full system. In other words, the autoencoder is used to simplify the dynamical system, retaining the dynamics that are essential and averaging out subordinated degrees of freedom. Chapter 3 will be devoted to describe such autoencoders, with a special focus on systems that yield topological metamodels.

In the context of extending the SM, or rather reverse engineering it to encompass a dormant dimension, we need to have the autoencoder working in reverse. Rather than simplifying the fields defined in latent space, we need to lift them to a quintessential space of higher complexity, so as to recover the original fields when taking the quotient modulo the standard four-dimensional space–time coordinates. In addition, we need the autoencoder to produce the topological metamodel that would enable the proper lifting, so that a diagram of the type presented in Figure 1.32 becomes commutative.

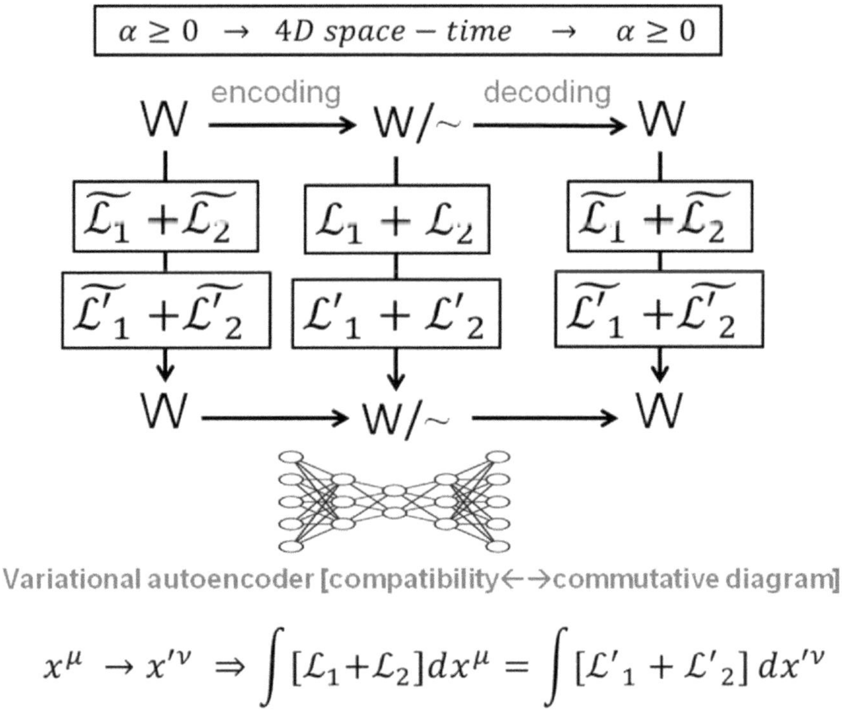

FIGURE 1.32 Neural network with autoencoder architecture designed to reverse engineer the standard model by incorporating a fifth circular dimension.

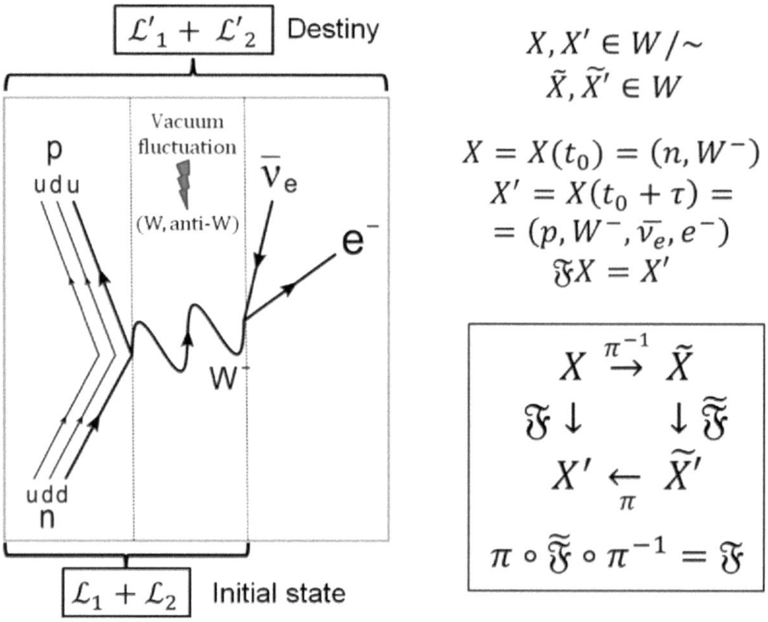

FIGURE 1.33 Elementary process of beta-decay defined on the latent space representing the four-dimensional space–time *W*/~ and consistently lifted to the quintessential space *W*. The process consists of a neutron transformed into a proton through communication with a W^- gauge boson with the emission of an electron and an electron antineutrino.

An illustration of the kind of tasks performed by the topological autoencoder of the SM is depicted in Figure 1.33. The figure describes the lifting to quintessential space *W* of a process, known as beta-decay, comprising the neutron (*n*) transformation into a proton (*p*) by transformation of one of its constitutive down quarks (*d*) into an up quark (*u*), with emission of an electron (e^-) and an electron antineutrino ($\bar{v}_e$). This process requires the interaction of the neutron with a W^- boson, a massive particle (M = 80.433 GeV/c^2) whose existence is ephemeral ($\tau \approx 10^{-26}\,\mathrm{s} \sim 6.582\times10^{-16}\,\mathrm{eVs}/\left(2\times80.433\times10^{9}\,\mathrm{eV}\right)$), as its formation requires borrowing energy from the vacuum as a fluctuation governed by Heisenberg's uncertainty principle ($\Delta E \times \Delta t \sim \frac{\hbar}{2}$). Here, we adopt the notation $X = \left(n, W^-\right), X' = \left(p, W^-, \overline{v}_e, e^-\right)$ to indicate, respectively, the initial and final state of the system as recorded on the latent space *W*/~ and dictated by the SM. The respective liftings to *W* are $\widetilde{X}, \widetilde{X}'$, and their validity is ensured if and only if they satisfy the flow relation:

$$\left(\pi \circ \tilde{\mathfrak{F}}\right)\tilde{X} = \mathfrak{F}X; \mathfrak{F}X = X'; \tilde{\mathfrak{F}}\tilde{X} = \widetilde{X}'. \tag{1.40}$$

The operators $\mathfrak{F}$ and $\tilde{\mathfrak{F}}$ determine, respectively, all processes in *W*/~ dictated by the SM and the lifting of such processes to the quintessential space *W*. Thus, the operators may be interpreted as time evolution propagators over time intervals identified with the lifetimes of the particles that communicate fundamental forces. In this context, the autoencoder is charged with the task of computing the operator $\tilde{\mathfrak{F}}$ and canonical projection π, so that the generic diagram in Figure 1.34 becomes commutative.

$$
\begin{array}{ccc}
W & \overset{\pi}{\rightarrow} & W/\sim \\
\downarrow \widetilde{\mathfrak{F}} & & \downarrow \mathfrak{F} \\
W & \overset{\pi}{\rightarrow} & W/\sim
\end{array}
$$

FIGURE 1.34 Commutative diagram for the quintessential lifting of a fundamental process in the reverse engineering of the standard model.

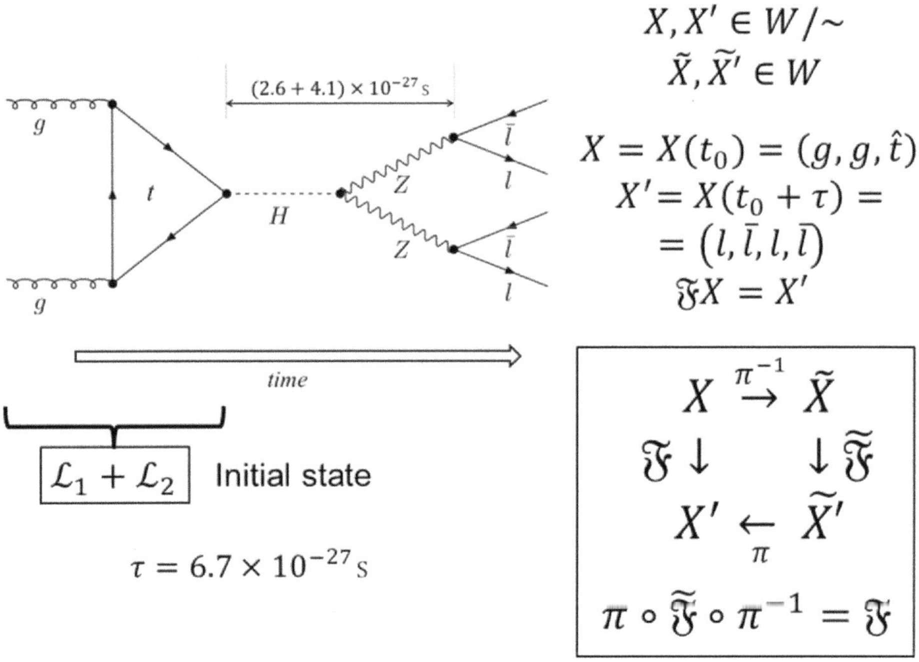

FIGURE 1.35 The lifting of the elementary process of creation and (detectable) decay of a Higgs boson.

Similarly, the lifting of another fundamental process, the creation and decay of the Higgs boson (*H*) into two lepton–antilepton pairs is depicted in Figure 1.35. The creation of the Higgs boson entails the fusion of two gluons (*g*, *g*) mediated by a virtual top quark $(\hat{t})$, and the "timing" of the process is given by the sum of the lifetimes of the bosons (*H*, *Z*) that serve as intermediates.

In this way, the SM may be extended by the topological autoencoder to incorporate an extra dimension compatible with the compact and multiply connected topological metamodel of the universe validated experimentally by the spectral decomposition of the CMB. The topological autoencoder would then be in position to discover dark matter particles as it reverse engineers the SM to adapt it to the early universe, where the dormant dimension became commensurate with the other four dimensions of space–time. This program will be carried out in Chapter 4.

1.16 AMALGAMATED HIGGS BOSON AS DARK MATTER PORTAL

The standard model (SM) is considered the towering achievement of particle physics, mainly because of its amazing predictive power and sparse parametrization. Despite its success, the SM in its current version does not include dark matter (DM). Furthermore, it seems unclear that it can be extended in a cogent way to encompass DM without significantly altering the existing models of observable particles and solving the quantum gravity and hierarchy problems. Ongoing research described in this book, particularly in Chapter 4, may turn this situation around. With the mathematical implementation of the concept of geometric dilution (υ) within the premises of the SM, DM can be shown to percolate into observable dimensions. Therefore, we are confident that a suitable extension of the SM will be forthcoming. This discussion points to a viable possibility in this regard, as fermionic DM is shown to have a dynamical origin capable of interplay or mixing with the Higgs boson. What follows is an outline of the SM extension.

The approach requires a mixing of the scalar field for DM with the Higgs doublet scalar field to yield an irreducible mediator between the observable and dark sectors of the universe. Technically, this interplay requires boundaries in the dark curly dimension, so the interplay may materialize at the three-branes that represents the boundary. But such boundaries do not exist in the compact extra dimension, since it constitutes a circle. For this reason, we need to replace the topology of the dark dimension, $\mathcal{S}^1$, for the quotient manifold $\mathcal{S}^1/\mathbb{Z}_2$. This quotient is made up of orbits under the symmetry operation $\vartheta \to -\vartheta$, with the polar representation $y = r_0 e^{i\vartheta}$ of the dark coordinate. The metric in $\mathcal{S}^1/\mathbb{Z}_2$ is entirely determined by that of the subset $0 \le \vartheta \le \pi$, which yields boundaries, i.e., the three-branes $\vartheta = 0$, $\vartheta = \pi$, as the dark extra dimension is incorporated. Furthermore, DM particles become $\mathbb{Z}_2$-odd scalar fields that can now be amalgamated with the Higgs scalar field to yield mediators between the observable and dark sectors. The lightest such portal corresponds to the combination with the first Kaluza–Klein (i.e. at $\upsilon = 0$) excitation of a dark fermion, which corresponds to the ur-Higgs.

This construction of the compound particle departs from the parametrized semiempirical model of the Higgs boson, with its quartic term in the potential energy accounting for self-interactions of the Higgs field $\phi_h = \phi_h(x,t)$. In the SM, the spontaneous symmetry breaking in the Higgs potential, responsible for endowing particles with mass, is associated with the nonzero vacuum expectation value (v.e.v.) v arising from self-interactions that yield the potential term $-\lambda\phi_h^{\ 4}$ with coupling constant λ. The parametrization of the v.e.v. at $v = 246\,\mathrm{GeV}$ is adjusted to yield all particle masses. By contrast, our model does not include self-interactions; instead, it extends the field ϕ_h to the amalgamated field $\phi_H = \phi_H(x,y,t)$ by incorporating a $\mathbb{Z}_2$-odd DM field $\phi_D(y)$ that spans the dark dimension:

$$\phi_H = \phi_H(x,y,t) = \phi_h(x,t)\left[1-\left(\frac{\phi_h(x,t)}{\sqrt{2}\phi_D(y)}\right)^2\right]^{1/2}. \tag{1.41}$$

The v.e.v. is stored in the dark dimension with quotient topology $\mathcal{S}^1/\mathbb{Z}_2$, while only the (x,t) dependence of the Higgs field is apparent in the SM (in its current version).

By adopting the radial representation $\phi_h = \rho e^{i\omega}$, the Higgs boson becomes, in standard space–time, an oscillatory excitation of the radial field $\eta = \rho - v$, while the Lagrangian for η lifted to the (x,t,y) five-dimensional space–time becomes

$$\widetilde{\mathcal{L}(\eta)} = \frac{1}{2}(\partial\eta)^2 - \lambda\overline{\phi_D}\phi_D\eta^2. \tag{1.42}$$

This result implies that the Higgs boson mass $m_H = \sqrt{2\lambda}v$ *(* $\sqrt{2\lambda} = cos\alpha$ *geometric dilution parameter) is bestowed through a coupling with the DM scalar field.* If geometrically undiluted, the v.e.v. thus stored would constitute the lightest DM, namely, the ur-Higgs. Eqs. 1.41 and 1.42 inform that DM can be incorporated into the SM via the key observation that the v.e.v. of the Higgs field is actually a $\mathbb{Z}_2$-odd stationary wave spanning the dark circular dimension. This circle has radius $r_0 = 0.802695 \times 10^{-18}$ m , and the smallest material scale assimilated to the effective quark diameter established from parton models of inelastic $e^{\pm}p$ scattering [9].

In this model, the amalgamated Higgs field (Eq. 1.41) enables the spillover of the dark universe into the observable universe, acting as gatekeeper and portal between the observable and dark sectors. This is because the parameter that embodies the geometric dilution $\upsilon = -\log\cos\pm$ of the ur-Higgs onto observable dimensions to yield a specific particle mass (Chapter 4) can be unambiguously associated with the Yukawa coupling, g_Y, of the fermion field with the Higgs field, and with the mass-to-v.e.v. ratio in the case of the weak-force bosons. *The behavior of the amalgamated Higgs boson as portal for DM, enabling the spillover that endows observable fermions with mass, is enshrined in the fundamental proportionality*

$$g_Y \propto cos\alpha \tag{1.43}$$

Under these observations, it becomes clear that the amalgamated Higgs boson serves as the portal of the dark universe, with a controlled spillover mechanism that endows SM particles with mass.

REFERENCES

1. Weinberg S (2008) *Cosmology*. Oxford University Press, New York.
2. Fisher P (2022) *What Is Dark Matter?* Princeton University Press, Princeton, NJ.
3. Profumo S (2017) *Introduction to Particle Dark Matter*. World Scientific Publishing, Singapore.
4. Susskind L (1995) The world as hologram. *J Math Phys* 36: 6377–6396
5. Fernández A (2022) *Topological Dynamics for Metamodel Discovery with Artificial Intelligence*. Chapman & Hall/CRC, Taylor and Francis, London.
6. Wesson PS (1999) *Space–Time–Matter, Modern Kaluza–Klein Theory*. World Scientific, Singapore.
7. Dirac PAM (1982) *The Principles of Quantum Mechanics*, 4th Edition. Clarendon Press, Oxford.
8. Feynman RP, Weinberg S (1999) *Elementary Particles and the Laws of Physics*. Cambridge University Press, Cambridge.
9. Abramowiczy H, Abtt I, Adamczykh L, Adamusae M, Antonelli S, et al. Zeus Collaboration (2016) Limits on the effective quark radius from inclusive *ep* scattering at HERA. *Phys Lett B* 757: 468–472.
10. Hawking SW, Ellis GFR (2023) *The Large Scale Structure of Space-Time: 50th Anniversary Edition*. Cambridge University Press, Cambridge.
11. Aurich R, Buchert T, France MJ, Steiner F (2021) The variance of the CMB temperature gradient: A new signature of a multiply connected Universe. *Class Quant Grav* 38: 225005.

2 Artificial Intelligence Unravels the Origin of the Universe as a Phase Transition from the Quantum Vacuum

"That which became was enveloped by the void."

Rig-Veda
c. 1500 BCE

The nuanced problem of the origin of the universe, be it out of the quantum vacuum or *ex nihilo*, may be regarded as a core problem in quantum gravity, an immature field that seeks to unify all forces of nature. In this book, the problem is turned over to AI, reckoning that the problem cannot be tackled without full contextualization, that is, without dealing with other intimately related core matters in particle cosmology. Such problems include the nature of dark matter and dark energy, the hierarchy problem of particle mass, the incommensurably weak coupling strength of gravity, the topology of the universe, the cosmological constant problem, and the vacuum catastrophe. This book describes how AI squarely addresses the matter in its full relational and contextual richness. Meanwhile, this specific chapter is heavily tainted by a physicist perspective as it revisits and contributes to the core problems in particle cosmology while outlining the full AI program fleshed out in the chapters that follow.

The book implements and applies an AI system in the guise of a generic "autoencoder" that distills the quantum reality encoded in a higher dimensional space–time and learns to gauge relativistic space–time symmetry, integrating it into the fabric of quantum reality. To address the core problems in particle cosmology, the autoencoder incorporates a latent spatial dimension that turns the quantum vacuum into ur-matter (UM), that is, into a precursor of the visible and dark sectors of the standard model (SM) of particle physics. Through AI, we learn in this chapter that ur-matter comes in six "shades", one of them, ur-matter proper, being the precursor of the visible sector. The remaining five shades incorporate symmetries that were disregarded in standard quantum field theory but have now become highly relevant to elucidate the nature of the dark sector. These lost symmetries that preclude interaction with the photon or mixing with the Higgs field help delineate the nature of dark matter and dark energy, which are now reckoned as key players in deep-space phenomenology.

In this context, AI first approaches the origin of matter regarding it as a phase transition induced by the activation of a portal gauge particle that becomes endowed with a symmetry-breaking stable vacuum through mixing with one specific shade of ur-matter. This communication is enabled by compatible symmetries between a specific ur-matter shade and the SM.

Consistent with a vast body of experimentation, the AI model postulates that the four fundamental forces originate from ur-matter symmetries of the primeval quantum vacuum internalized as gauge generalized symmetries at the phase transitions encompassed by the birth of the universe. The gauging of the relativistic symmetry internalizes it not as local symmetry but as a one-form

 DOI: 10.1201/9781003632924-2

generalized symmetry, where the purported curvature of the Riemannian manifold is factored into the extended object charged under the gravity-associated symmetry. Intriguingly, the primeval symmetry of space–time can be simplified because the symmetry groups associated with the electroweak unification are isomorphic to subgroups of the relativistic Lorentz group. This leads to the striking finding that gravity may be treated in a quantized ultra-unification since its ur-symmetry is related to that of the weak force and electromagnetism. In other words, the ur-symmetry of gravity subsumes the ur-symmetry that activates the Higgs field by endowing it with the true vacuum. The symmetry relatedness linking gravity and the Higgs-induced mass with retention of electroweak symmetry paves the way for a quantum gravity ultra-unification under a gauge generalized symmetry. This ultra-unification substantiates the gravity–mass duality of general relativity.

In a final synthesis, AI reveals that the four fundamental forces are shown to stem from one primeval force present at the Planck epoch. This force is diluted in different ways into the geometry of the universe in accordance with the generalized symmetries of force carriers activated at successive phase transitions in the aftermath of creation.

To complete the unification, a large language model (LLM) within a functional language program is shown to yield a field theory of the graviton that incorporates a warped dimension to account for gravity dilution. Such approaches have not yielded to experimental validation except in cosmology, where Kaluza–Klein gravitons apparently beget self-interacting dark matter. To learn to generate massive gravitons detectable in colliders, a gauge theory for the strong field limit (SFL) is built by the LLM within a fiber bundle formulation that enables quantization of a field associated with self-interacting dark matter diluted in a warped dimension. The results predict detectable gravitons in the SFL associated with Wilson loops that holonomically transduce space–time curvature and get charged under a generalized gauge symmetry that translates into $U(1)$-gauge symmetry in the weak field limit (WFL).

2.1 SURVEY OF THE GAUGE THEORY FOR ELECTROWEAK UNIFICATION

The mass problem in particle physics has many sides. The one relevant to this discussion can be cursorily formulated as follows [1,2]: How is it possible that the weak nuclear force is short ranged ($r \sim 10^{-18}$ m) and yet the bosons that carry the force are endowed with internal symmetries $\left(SU(2)\times U(1)\right)$ that would render them massless, thereby communicating a force of infinite range? Theoreticians feverishly dealt with this problem in the nineteen sixties and came up with a solution [2]: *Boson masses are actually induced masses*, in the sense that a scalar field needs to be incorporated to break the symmetry as it adopts its stable "true" vacuum expectation value. This scalar field, now known as Higgs field, reconciled both sides of the paradox and appeared to be the way forward, especially after its massive excitation, the Higgs boson, was detected experimentally [1].

However, as demonstrated in this chapter, the presence of a dark (invisible) sector in the standard model (SM) calls for revisions of the Higgs mechanism. As shown subsequently, the encoding of an extra "within" spatial dimension is one such revision found by AI in its take on the Higgs' "induced mass" standard assumption. The extra dimension will be shown to play a key role to explain why the force of gravity is incommensurably weaker than the other forces: Gravity appears to be leaking into the extra dimension. Furthermore, the extra dimension endows the vacuum with materiality, as it turns it into ur-matter, a precursor of matter. In this way, AI can elucidate the origin of mass without relying on an all-pervading self-interacting scalar field, as shown in Section 2.11.

As it is well established, the massless bosons that communicate the forces in quantum electrodynamics (QED) and quantum chromodynamics (QCD) arise from their respective gauge symmetries $U(1)$ and $SU(3)$ [1], symmetries that they retain. On the other hand, the weak interaction became problematic because its short range $r = \hbar/Mc \sim 10^{-18}$ m bespeaks of a massive boson with a mass M

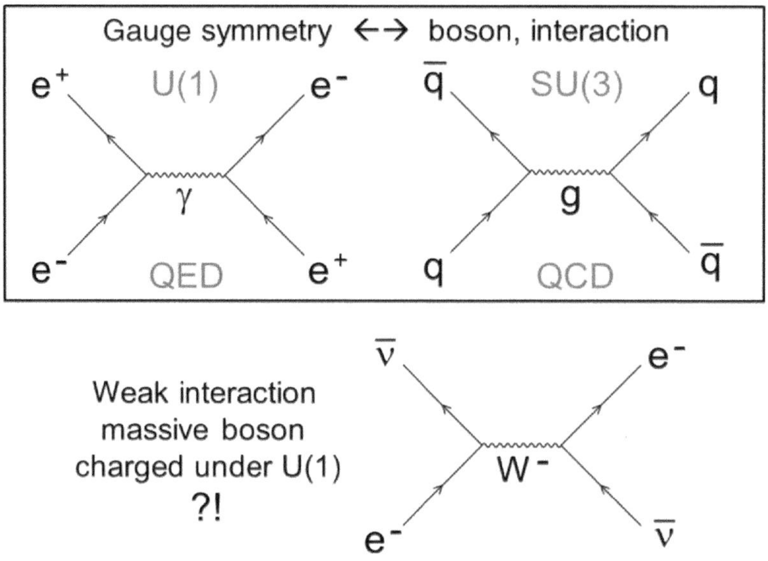

FIGURE 2.1 Conundrum over the nature of the weak interaction whose boson is massive and charged under electromagnetic $U(1)$ symmetry. This stands in contrast with QED and QCD, whose bosons are massless and charged under their respective symmetries.

in the order of 100 GeV, while its charge is acquired under QED. This implies that the boson mass communicating the weak force cannot be directly tied to a gauge symmetry (otherwise it would be massless) and that two sectors of the SM must be in fact combining (Figure 2.1). This state of affairs makes it apparent that a gauge theory is required to amalgamate the weak force and QED. This theory came to be known as electroweak (EW) unification or Weinberg–Salam model, in honor of two of its main contributors, Steven Weinberg and Abdus Salam [1].

Current thinking suggests that the weak bosons $W^{\pm}$ get their mass via the spontaneous symmetry breaking (SSB) arising from combination of the boson field with the Higgs field, implicit in the grand Lagrangian for the SM [2]. Furthermore, gauge bosons experience their own fields, meaning that they self-interact or get charged under their own fields. Additionally, we must keep in mind that as gauge symmetry is broken, any eigendirection that retains symmetry would yield a massless boson known as the Nambu–Goldstone boson [1,2]. Thus, after SSB, the bosons that acquire mass may interact with the massless bosons that originated from symmetry preservation, thereby getting charged under the preserved symmetry. If we associate the preserved symmetry post-SSB with $U(1)$, we can readily see that it should be possible to unify the weak and the EM interactions provided we can determine the group of symmetries that are broken to endow the weak bosons with mass. In what pertains to fermions experiencing the weak force, the unified EW theory should also encompass the parity violation of the weak interaction, a landmark property of the weak interaction established by Chien-Shiung Wu in 1956 [1]. Therefore, a weak boson does not interact equally with left- and right-handed components (2-dim spinors) in the SM: It only interacts with the left handed component, yielding what is known as a "chiral gauge interaction".

This analysis prompts us to build an EW unification theory subject to the following requirements:

1. An electroweak symmetry subsuming $U(1)$ as subgroup needs to be tailored, so it can be broken as the weak field is blended with a scalar field that promotes SSB and thereby induces mass on the weak bosons.

2. After broken symmetry, at least two massive weak bosons ($W^{\pm}$) need to emerge while the prevailing symmetry must be of the QED-type ($U(1)$), yielding a Nambu–Goldstone massless boson. As the weak W-bosons subsume the massless boson, they get charged under the preserved symmetry. That is why they acquire a QED-type charge.
3. The weak bosons need to self-interact as they experience their own field, so as to get charged under $U(1)$ symmetry, the prevailing symmetry after SSB.
4. The theory must be chiral in accordance with the established parity violation of the weak interaction.
5. Fermions must be able to interact with the scalar field that induces SSB in order to acquire mass, but their L and R spinors (chiral fermions) charged under gauge symmetry cannot get mass individually: Interactions with the scalar field require the combination of L and R components to produce the so-called Dirac masses after SSB.
6. As they interact with weak bosons, fermions must be paired according to the experimental observations in weak decays: u with d for quarks, e/μ with neutrinos ($\nu, \overline{\nu}$) for charged leptons.

The EW theory is expected to build on a gauge symmetry akin to the one that shapes QCD, where different modes of self-interaction yielding manifold charges, in contrast with QED that yields only one charge. The gauge symmetry should yield at least three bosons ($W^{\pm}, \gamma$), one for each symmetry transformation, with the preserved symmetry after SSB corresponding to the photon. The weak symmetry must therefore include complex rotations for two internal complex components of the fermion fields. This makes the choice fairly obvious, since the group of complex rotations is the special unitary group: $SU(2) \approx SO(3)/\mathbb{Z}_2\left(SO(3) \equiv \text{real 3D rotation group}\right)$, the Lie group of 2×2 unitary matrices with unit determinant:

$$SU(2) = \left\{ \begin{pmatrix} \alpha & -\overline{\beta} \\ \beta & \overline{a} \end{pmatrix} : \alpha,\ \beta \in \mathbb{C},\ |\alpha|^2 + |\beta|^2 = 1 \right\} \tag{2.1}$$

This group is compact and simply connected, hence topologically equivalent to the 3-sphere (in accordance with the Poincare–Perelman theorem [3]), and its Lie algebra is generated by $\{i\sigma_1, -i\sigma_2, i\sigma_3\}$, where $\sigma_1, \sigma_2, \sigma_3$ are the Pauli matrices:

$$\sigma_1 = \begin{pmatrix} 0 & 1 \\ 1 & 0 \end{pmatrix},\ \sigma_2 = \begin{pmatrix} 0 & -i \\ i & 0 \end{pmatrix},\ \sigma_3 = \begin{pmatrix} 1 & 0 \\ 0 & -1 \end{pmatrix}. \tag{2.2}$$

Since the transformations are complex, this group is compatible with $U(1)$ in QED, which charges the weak bosons after SSB, and the three generators of its Lie algebra pair fermions into doublets that transform under the gauge interactions. This pairing is analogous to the three color charges of the quarks under $SU(3)$–QCD interaction, and implies we need one separate neutrino to pair with each charged lepton.

While $SU(2)$ gauge bosons only interact with left-handed fermion doublets, we need to reconcile the two seemingly conflictive aspects of a unified theory, namely, the weak beta-decays are indeed chiral but QED, the relic symmetry, is parity-symmetric. Electroweak unification also requires a scalar field that breaks $SU(2)$ symmetry into the $U(1)$ symmetry of QED, but it is obviously not feasible to get gauge–invariant interactions of the fermions with the scalar field that would generate a mass term. In other words, we need to incorporate fermion Dirac masses and parity-symmetric QED and this can only be done by incorporating right-handed fermions, one for each left-handed fermion charged under QED after SSB, even though they are not part of the weak decay per se. Since we do not have right-handed neutrinos, we cannot get Dirac masses for neutrinos, implying that they are massless, as it has been experimentally corroborated.

This discussion makes it apparent that the unified EW theory requires $SU(2)_L$ symmetry and another interaction that would yield QED after SSB, i.e., the 1D complex internal rotation group $U(1)$. So, the gauge $SU(2)_L$ symmetry acts on the L-doublets:

$$L_e = \begin{pmatrix} \nu_L^e \\ e_L \end{pmatrix},\ L_\mu = \begin{pmatrix} \nu_L^\mu \\ \mu_L \end{pmatrix},\ Q_L = \begin{pmatrix} u \\ d \end{pmatrix}. \tag{2.3}$$

With associated bosons, $\begin{pmatrix} W_1 \\ W_2 \\ W_3 \end{pmatrix}$ as well as a single boson, B, associated with $U(1)$ that interacts with the L-doublets as well as with the right-handed singlets e_R, μ_R, u_R, d_R. The charge of the fields under weak $SU(2)$ is known as weak isospin and labeled τ^3, whereas the charge under $U(1)$ is known as hypercharge and labeled Y. It turns out this gauge symmetry can be spontaneously broken by a scalar field, now identified as the Higgs doublet H, that transforms as a doublet under $SU(2)$ and has nonzero hypercharge. Thus, the unified gauge symmetry $SU(2)_L \times U(1)_Y$ can be spontaneously broken by H as it acquires its nonzero vacuum expectation value to yield a leftover symmetry, $U(1)_{EM}$, that can be identified with electromagnetism. With SSB, the weak bosons mix according to

$$W^\pm = \frac{1}{\sqrt{2}}\left(W^1 \mp iW^2\right). \tag{2.4}$$

While B and W^3 combine linearly after SSB to make the gauge boson (γ) of electromagnetism and the weak Z boson:

$$\begin{pmatrix} \gamma \\ Z \end{pmatrix} = \begin{pmatrix} c_W & -s_W \\ s_W & c_W \end{pmatrix} \begin{pmatrix} B \\ W^3 \end{pmatrix}. \tag{2.5}$$

The electric charge Q emerges as the linear combination $Q = \tau^3 + \frac{Y}{2}$.

At this stage, we are allowed to combine the Higgs doublet with the L-doublet and R-singlet fermions into gauge–invariant interactions that yield fermion masses after SSB, in accordance with the general EW unification scheme presented in Figure 2.2.

Since the scalar Higgs field contains four real scalar degrees of freedom, and ¾ of the continuous symmetry is broken, we expect three Goldstone bosons subsumed by $W^\pm, Z$ in order to obtain an extra transverse polarization once they become massive. Additionally, we should have a single left-over real scalar neutral under QED, the Higgs boson.

Thus, the unified EW theory predicts the existence of the Z and Higgs boson. Both bosons had been confirmed experimentally: The Z boson was discovered in neutrino scattering off of electrons via neutral current weak interactions in the bubble chamber and was observed at the proton synchrotron in the 1980s, while the Higgs boson was discovered at the LHC in 2012.

The acquisition of fermionic Dirac mass in the EW unification scheme rests heavily on a physical vacuum whose energy is reduced by the Higgs field. Much like a phase transition to a lower energy

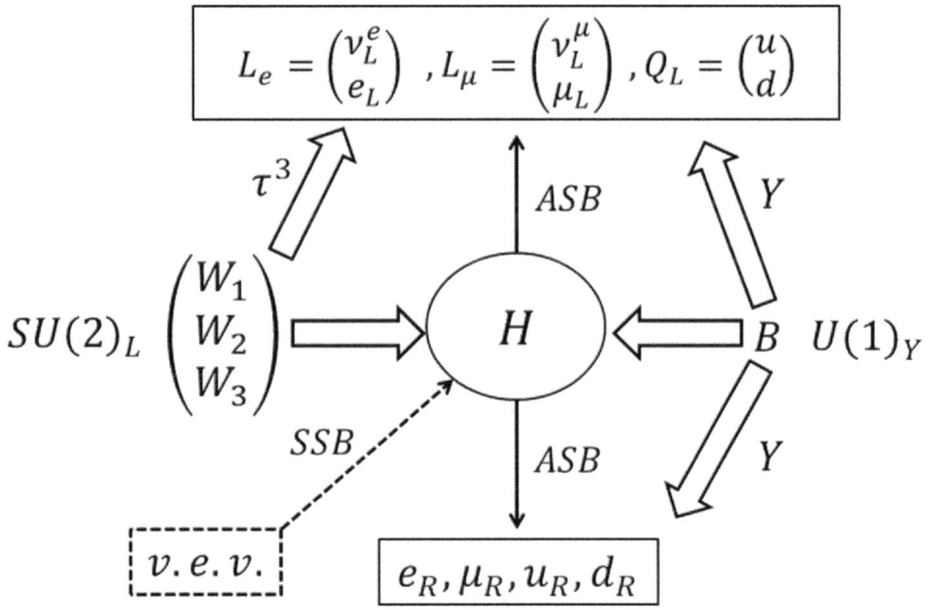

FIGURE 2.2 The electroweak unification scheme reconciling parity violation of the weak interaction leading to beta-decay with a stage at which it is allowed to combine the Higgs doublet with the L-doublet and R-singlet fermions into gauge–invariant interactions that yield fermion masses after SSB.

state, the Higgs field plays a role akin to the latent heat associated with the phase transition. The Higgs field is assumed to break the symmetry as it obtains its vacuum expectation value (v.e.v.) f, which bestows mass on the L-R fermion doublets, an after-symmetry-breaking (ASB) event. The key issue to address at this point is how does the Higgs field get its v.e.v.

2.2 FIVE SHADES OF DARK MATTER AND A PORTAL BETWEEN THE VISIBLE AND DARK SECTOR

The universe inflation following the quantum gravity epoch need not be uniform along all dimensions. In fact, as shown in Chapter 1, the hierarchy problem of incommensurately low particle masses relative to Planck's value and the incommensurably small coupling strength of gravity relative to the other three fundamental forces suggest the existence of compact "curled-up" dimensions tuned to store matter waves and provide a geometric dilution of gravity. Along expanded dimensions, the range r of a force correlates with the mass M of a bosonic massive carrier: $r = \hbar/Mc$, whereas along curled-up dimensions, the range r becomes the radius of the dimension itself since otherwise there would be destructive interference. Hence, an autoencoder of the fifth space–time dimension distilling quantum reality into the 4D space–time operates under the assumption that the quantum vacuum has an associated wavelength $\lambda_0 = 2\pi r = h/M_{u-H}c$, with $M_{u-H} = 246$ GeV/c^2, identified as the mass of the ur-Higgs carrier [4]. This tuning yields $r = r_0 = 0.802695 \times 10^{-18}$ m, a value that fits exceptionally well the upper limit to the smallest material dimension known: $r_0 = 0.86 \times 10^{-18}$ m. This is the effective quark diameter determined from parton models of deep inelastic $e^{\pm}p$ scattering at TeV scales corresponding to luminosity at 1 femtobarn^{-1} (Chapter 1). This upper limit was established within a quark form-factor model [5] and is also in agreement with the range of the weak interaction at $r \approx 10^{-18}$m (Chapter 1). *Crucially, this length scale that yields the parameter that upholds the physics edifice, the stable vacuum at 246GeV, is actually assimilated to the quark scale, the smallest material scale known.* In other words, *AI is asserting that what we regard as quantum vacuum vis-à-vis the visible sector is actually ur-matter, the precursor of matter.* Dark matter (DM) in general corresponds to multiples of the momentum $k_0 = \hbar / r_0$ spanning the dark dimension. While now fully justified, the validation of these operational tenets will take up the remaining of this chapter as well as Chapter 4.

As shown subsequently, the way the Higgs field is endowed with its v.e.v. follows from compatibilities with symmetries arising from the quantum relativistic wave equations that describe dark matter. There are six such equations, one for each type of 2D factorization of the special relativity equation

$$k^2 + k_0^2 = \omega^2, k = \lambda^{-1}, k_0 = \lambda_0^{-1}. \tag{2.6}$$

where the wavenumbers k, k_0 indicate momentum and rest–mass momentum, respectively, with "natural" units of h, c, G fixed at 1 and, accordingly, energy expressed as the frequency ω. For simplicity and without loss of generality, we shall assume one visible generic coordinate, x, while y denotes the dark coordinate.

A possible 2D factorization of Eq. 2.6 takes the form

$$\pm I_2 \omega = \alpha k + \beta k_0, \tag{2.7}$$

where α, β are matrices satisfying the following idempotent and anticommutative relations:

$$\alpha^2 = \beta^2 = I_2 = \begin{pmatrix} 1 & 0 \\ 0 & 1 \end{pmatrix}; \quad \alpha\beta + \beta\alpha \equiv \{\alpha, \beta\} = 0. \tag{2.8}$$

The Pauli matrices (Eq. 2.2) satisfy Eqs. 2.8, and hence, there are in principle six realizations to Eq. 2.7. We may denote (i, j) the factorization realized by $\alpha = \sigma_i$, $\beta = \sigma_j$ $(i, j = 1,2,3,\ i \neq j)$. Accordingly, dark matter is classified in 3!-1=5 different types, heretofore named "shades", $DM_{i,j}$, where $DM_{i,j}$ follows the E^2 factorization $(i, j) \neq (3,1)$. Assuming Lorentz-invariant wave functions of the form

$$\psi(x,t) \sim u(k) ei(kx - \omega t), \tag{2.9}$$

the (3,1) factorization of Eq. 2.6 yields

$$i\partial_t \psi_R = -i\partial_x \psi_R + k_0 \psi_L \tag{2.10}$$

$$i\partial_t \psi_L = i\partial_x \psi_L + k_0 \psi_R$$

This equation is gauge invariant as the ur-Higgs field is enabled to interact with the photon and with the Higgs field.

The $U(1)$ gauge invariance of this fermion equation would require invariance under the internal symmetry transformation (marked with the prime):

$$\psi_L \to \psi'_L = ei\vartheta(x)\psi_L; \psi_R = \psi'_R \tag{2.11}$$

Such symmetry cannot be achieved unless the dark fermionic field is mixed with the scalar Higgs field. First, we incorporate into the equation the electromagnetic (EM) vector potential A that transforms as

$$A \to A' = A - \partial_x \vartheta, \tag{2.12}$$

yielding the covariant derivative

$$\partial_x \to D_x = \partial_x + iA, \tag{2.13}$$

$$\begin{aligned}(D_x \psi_L)' &= e^{i\vartheta(x)} \partial_x \psi_L + (i\partial_x \vartheta) e^{i\vartheta(x)} \psi_L + iA' e^{i\vartheta(x)} \psi_L \\ &= e^{i\vartheta(x)} \partial_x \psi_L + (i\partial_x \vartheta) e^{i\vartheta(x)} \psi_L + i(A - \partial_x \vartheta) e^{i\vartheta(x)} \psi_L = (D_x \psi_L) e^{i\vartheta(x)}.\end{aligned} \tag{2.14}$$

Since the Higgs scalar field ϕ_{H} transforms as ψ_L, we may endow the ur-Higgs fermion with $U(1)$ symmetry by mixing the two fields, yielding

$$\begin{aligned} i\partial_t \psi_R &= -i\partial_x \psi_R + \overline{\phi_{\mathrm{H}}} \psi_L \\ i\partial_t \psi_L &= iD_x \psi_L + \phi_{\mathrm{H}} \psi_R \end{aligned} \tag{2.15}$$

Eq. 2.15 is invariant under internal complex rotation and corresponds to the Higgs-ur-Higgs mixing action in 5D space–time:

$$\begin{aligned} S = \int dxdy\sqrt{g} \Bigg\{ &\left[g^{\mu\nu} \sum_{n=1,2} (D_\mu \varphi_n)^\dagger (D_\nu \varphi_n) \right] \\ &- \left[-\frac{\mu_{\mathrm{H}}^2}{2} |\phi_{\mathrm{H}}|^2 + \frac{\lambda_{\mathrm{H}}^2}{4} |\phi_{\mathrm{H}}|^4 - \frac{k_0^2}{2} |\psi|^2 + \frac{\lambda_{\mathrm{HuH}}}{4} |\psi|^2 |\phi_{\mathrm{H}}|^2 \right] \Bigg\} \end{aligned} \tag{2.16}$$

In Eq. 2.16, $\varphi_1 = \begin{pmatrix} \psi_R \\ \psi_L \end{pmatrix}$, $\varphi_2 = \phi_{\mathrm{H}}$, μ_{H} = mass parameter for the Higgs field, $\frac{\mu_{\mathrm{H}}}{\lambda_{\mathrm{H}}} = f$, λ_{HuH} = coupling between Higgs and ur-Higgs fields, and g = compound gravitational field g, with

$$g_{\mu\nu} = \eta_{\mu\nu} + \kappa^2 A_\mu A_\nu, g_{5\mu} = \kappa A_\mu, g_{55} = 1, \tag{2.17}$$

where $\kappa^2 = 16\pi G/c^2$, $\eta_{\mu\nu} = diag(1,-1,-1,-1)$ denotes the 4D Minkowski metric and A_μ is the EM vector potential.

Thus, by combining Eqs. 2.10 and 2.15, we readily see that the Higgs field obtains its v.e.v. $f = k_0$ via mixing with the (3,1)-ur-Higgs. In this sense and because (3,1)-ur-Higgs is interactive with the photon (Eqs. 2.12–2.15), it may be regarded as the gatekeeper between the dark and visible sectors, the ur-matter shade that serves as precursor to the observable universe. The other relativistic wave equations lack the required gauge symmetry to reflect interactivity with the photon, hence describing invisible excitations of the dark matter scalar field. Such excitations are aptly named "invisible shades". These shades constitute the dark sector and are described by the equations in Table 2.1.

TABLE 2.1
Relativistic wave equations for the lightest ur-particle within the five shades of dark matter

ω^2 – factorization	Relativistic wave equation
(1,2)	$i\partial_t \psi_R^{1,2} = -i\partial_x \psi_L^{1,2} - ik_0 \psi_L^{1,2}$
	$i\partial_t \psi_L^{1,2} = -i\partial_x \psi_R^{1,2} + ik_0 \psi_R^{1,2}$
(1,3)	$i\partial_t \psi_R^{1,3} = -i\partial_x \psi_L^{1,3} + k_0 \psi_R^{1,3}$
	$i\partial_t \psi_L^{1,3} = -i\partial_x \psi_R^{1,3} - k_0 \psi_L^{1,3}$
(2,1)	$i\partial_t \psi_R^{2,1} = -\partial_x \psi_L^{2,1} + k_0 \psi_L^{2,1}$
	$i\partial_t \psi_L^{2,1} = \partial_x \psi_R^{2,1} + k_0 \psi_R^{2,1}$
(2,3)	$i\partial_t \psi_R^{2,3} = -\partial_x \psi_L^{2,3} + k_0 \psi_R^{2,3}$
	$i\partial_t \psi_L^{2,3} = \partial_x \psi_R^{2,3} - k_0 \psi_L^{2,3}$
(3,2)	$i\partial_t \psi_R^{3,2} = -i\partial_x \psi_R^{3,2} - ik_0 \psi_L^{3,2}$
	$i\partial_t \psi_L^{3,2} = i\partial_x \psi_L^{3,2} + ik_0 \psi_R^{3,2}$

Note: In its (3,1) shade, the ur-particle constitutes the ur-Higgs and thereby becomes the precursor to the visible sector.

None of the five shades of dark matter described in Table 2.1 can be made gauge invariant through combination with the EM vector potential jointly with the Higgs scalar field, and for that reason, they constitute the established components of the dark sector. This is in good agreement with the general composition of matter, with approximately 5/6 in the dark sector and 1/6 in the visible sector, making up for the 25% dark matter and 5% visible matter in the total universe composition that additionally includes nearly 70% in an energy equivalent counted as dark energy. On the other hand, the symmetry-compatible spillover of the (3,1) shade of dark energy ($DM_{3,1}$) onto the observable dimensions taking place in the course of phase transitions yields the visible sector through geometric dilution, as described subsequently.

The dark sector thus includes a (3,1)-portal ($B = +B_1$) that behaves as gatekeeper but also as precursor to the visible sector through geometric dilution of the ur-Higgs field (Figure 2.3). The fundamental role of the (3,1) portal is to endow the Higgs field with its v.e.v. (Eqs. 2.10, 2.15, and 2.16), thereby inducing symmetry breaking by stabilizing the Higgs vacuum as "true vacuum" (Figure 2.4).

The 2D factorization of special relativity Eq. 2.6 may be alternatively formulated as

$$\pm\beta\omega = \beta\alpha k + I_2 k_0, \tag{2.18}$$

where the matrix products $\beta\alpha$ are given in Table 2.2

We now cast Eq. 2.18 in terms of quantum operators acting on four-component spinor ψ to obtain the six fundamental $\left(S = \frac{1}{2}\right)$ fermionic equations for the dark and visible sector:

$$\left[iB\left(\Xi^{\mu}\partial_{\mu}\right) - k_0\right]\psi = 0, \tag{2.19}$$

$$B = \pm B_j = \pm\begin{pmatrix} \sigma_j & 0 \\ 0 & -\sigma_j \end{pmatrix}, \text{ j} = 1,2,3;\ \Xi^0 = \begin{pmatrix} I_2 & 0 \\ 0 & I_2 \end{pmatrix};$$

$$\Xi^{\mu} = \begin{pmatrix} \sigma_{\mu} & 0 \\ 0 & \sigma_{\mu} \end{pmatrix}, \mu \neq 0, \sigma_{\mu} = \sigma_1, \sigma_2, \sigma_3 \tag{2.20}$$

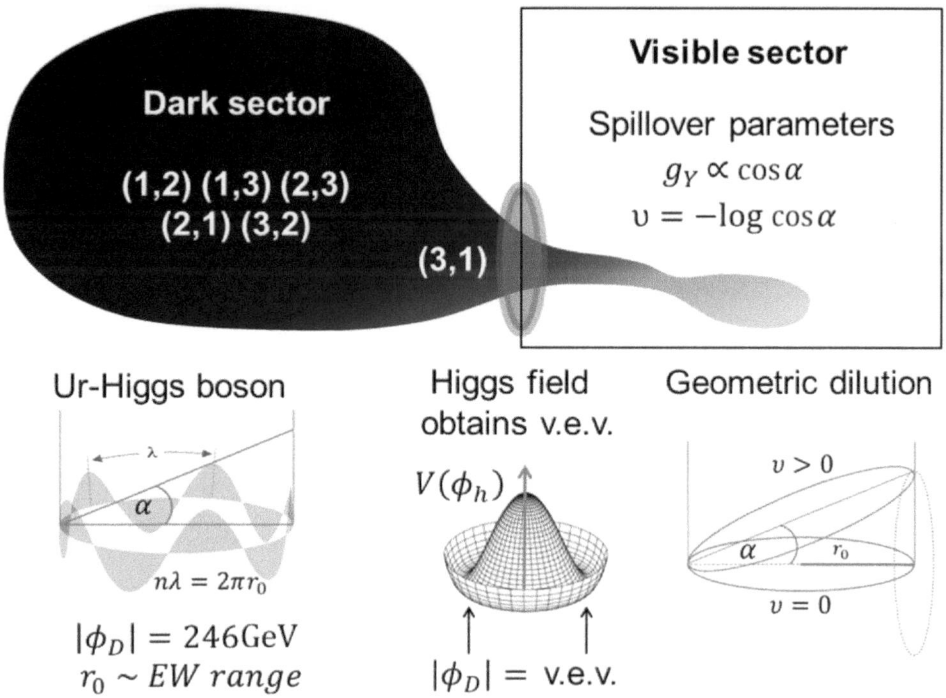

FIGURE 2.3 Composition of the dark and visible sector in accordance with the high-dimensional factorization of special relativity relation $k^2 + k_0^2 = \omega^2$. The portal between the dark and visible sectors is obtained by endowing with a true (stable) vacuum the Higgs scalar field via the mixing with the ur-matter (3,1)-type scalar field.

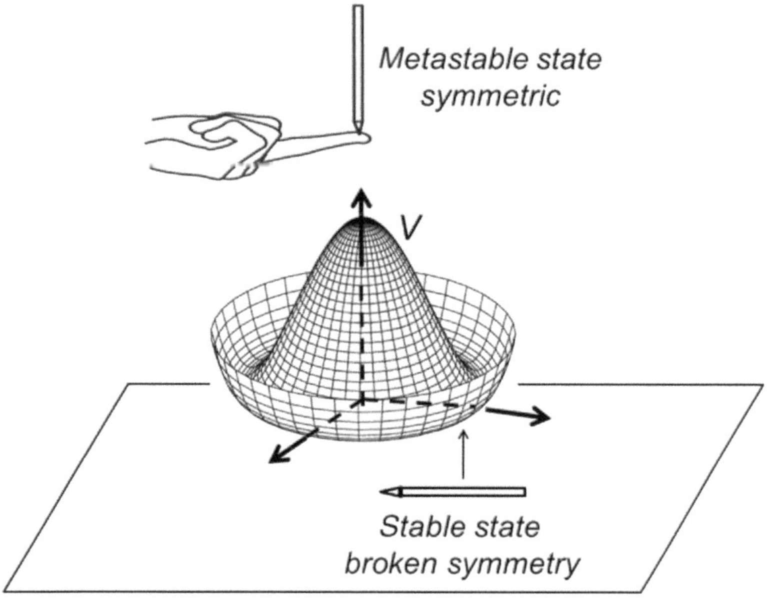

FIGURE 2.4 Stable "true vacuum" (TV) and metastable "false vacuum" (FV) in the Higgs field. The idea of mass yielded by symmetry breaking as the Higgs field is endowed with TV is inspired in Philip W. Anderson's theory of superconductivity. In that context, the photon becomes effectively massive upon a phase transition into an ordered phase representing a symmetry breaking.

TABLE 2.2
Cayley multiplication table (row by column) for the Pauli (2 × 2) matrices

×	σ_1	σ_2	σ_3
σ_1	I_2	$i\sigma_3$	$-i\sigma_2$
σ_2	$-i\sigma_3$	I_2	$i\sigma_1$
σ_3	$i\sigma_2$	$-i\sigma_1$	I_2

The three matrices Ξ^{μ}, $\mu \neq 0$, are all different from each other, and the assignment of a Pauli matrix to a particular coordinate is immaterial since the system is symmetric under visible coordinate permutation.

An alternative factorization *ansatz* for Eq. 2.6 is given by

$$\beta\omega = \alpha k + \beta k_0. \tag{2.21}$$

This factorization yields Dirac's equation for the shade of ur-matter that serves as precursor for the visible sector:

$$\left[i\left(\gamma^{\mu}\partial_{\mu}\right) - k_0\right]\psi = 0 \tag{2.22}$$

The key message from this section is that the autoencoder of the fifth dimension infers the existence of six *a priori* symmetries ("*shades*") for relativistic dark matter. Five of these shades yield dark matter proper understood by their inability to communicate with the photon field or EM vector potential. One shade, however, identified as (3,1) yields a symmetry compatible with the photon and Higgs field; hence, it is capable of agglutination with the Higgs field to bestow on it the stable true vacuum (TV) that makes EW unification operationally feasible.

2.3 AI MODELS THE TRUE QUANTUM VACUUM BY ENCODING THE DARK SECTOR AS PRECURSOR TO THE VISIBLE SECTOR

To understand how the AI system models quantum vacuum as generator of the universe, endowing it with the attributes of precursor to the visible matter, we need to describe in some detail the autoencoder of the extra spatial dimension, where ur-matter is encoded. A thorough account of this approach is provided in the methodological Chapter 3.

The (3,1) portal defined in the previous section is operationally modeled by AI through an autoencoder [6] that determines the role of the dark sector as precursor to the visible sector [4]. Through the autoencoder, an equivalence relation "~" is introduced so that points in the 4D space–time become equivalence classes modulo the curled-up fourth spatial dimension. Thus, the 4D space–time, whereupon the SM is established, becomes the quotient space–time $W/\sim$ of the 5D space–time W where the dark sector has been defined and allocated (Figure 2.5). In this scheme, each point in W is associated with its equivalence class in $W/\sim$ via the canonical projection $\pi: W \to W/\sim$ and each event or process defined by emission and decay of a boson in the SM through the transformation $\mathcal{K}_{\text{SM}}: W/\sim \to W/\sim$ may be lifted to an event defined by the transformation $\mathcal{F}: W \to W$ so that the following commutativity relation is satisfied (Figure 2.6) [4]:

$$\pi \circ \mathcal{F} = \mathcal{K}_{\text{SM}} \circ \pi. \tag{2.23}$$

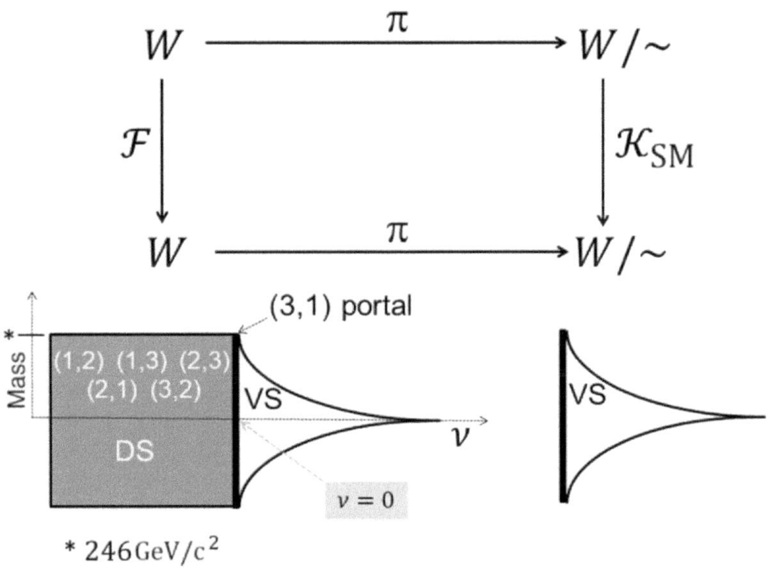

FIGURE 2.5 The (3,1) portal modeled by AI through an autoencoder that determines the influence of the dark sector on the visible sector.

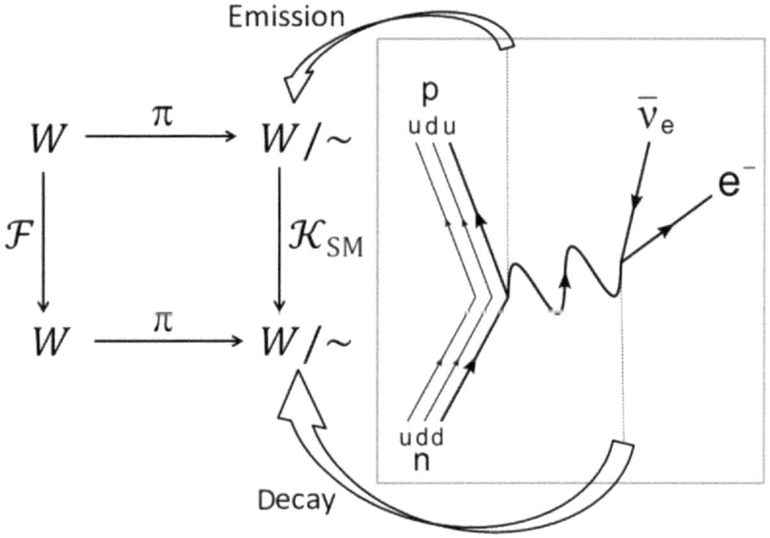

FIGURE 2.6 Autoencoder of the SM represented by the $\{\pi, \mathcal{F}\}$ pair within a commutative diagram. The latent dynamics is made up of detectable events taking place in the visible sector, mathematically described through term rearrangements/regrouping and reinterpretations of the SM Lagrangian. Such events are marked by emission and decay of the bosons that communicate the forces in the visible sector.

In AI "algebraic" parlance, we may say that the pair $\{\pi, \mathcal{F}\}$ constitutes an autoencoder of the SM [4]. A proper identification with the AI concept requires a suitable definition of the latent dynamics [6], that is, of the detectable events taking place in the visible sector, which are mathematically described through term rearrangements/regrouping and reinterpretations of the over-all

Lagrangian that underlies the SM (Figure 2.6). These events are marked by the emission and decay of the bosons that communicate the forces in the visible sector [1]. The distribution of distances, $\Delta(d)$, in 4D space–time between the two events is determined experimentally as

$$F^{-1}\left[\frac{1}{k^2+m^2}\right](d)=\Delta(d), \tag{2.24}$$

where $F^{-1}[\,](d)$ denotes the anti-Fourier transform evaluated at d and m is the boson mass. The momentum k, the Fourier conjugate of distance, is the difference in momenta $p(a) - p(b)$, where a and b denote the incoming and outgoing particle at the point of boson emission. For example, in the beta-decay $n \rightarrow p$ process described in Figure 2.6, the emitted boson W^- emerges from mutation of a down quark (d) into an up quark (u); hence, the mass of the weak boson may be determined from the measurement of the momenta of the neutron and proton. The emission and decay of the weak boson constitute two elementary events that get encoded in the 5D space–time in terms of changes in geometric dilution for the corresponding ur-particles. Yet, the symmetries under which such ur-particles are charged differ from the symmetries of their correlated particles in the SM, as shown subsequently.

2.4 EMERGENCE OF THE UNIVERSE AS A PHASE TRANSITION

To determine their external local symmetries, ur-particles need to be spatially represented as geometric dilution of the (3,1) shade of DM. As indicated in Chapter 1, the pitch angle α is used to parametrize such dilution (Figure 2.7a), so that the angled cylindrical cross section is elliptical and becomes deformed to a circle with radius $r_0/\cos\alpha$. Several spatial symmetries can be constructed to warrant the conservation of the geometric dilution $\upsilon = -\log\cos\alpha$ of the ur-Higgs particle (~ 246 GeV/c^2). Two such symmetries directly relevant to the current discussion may be

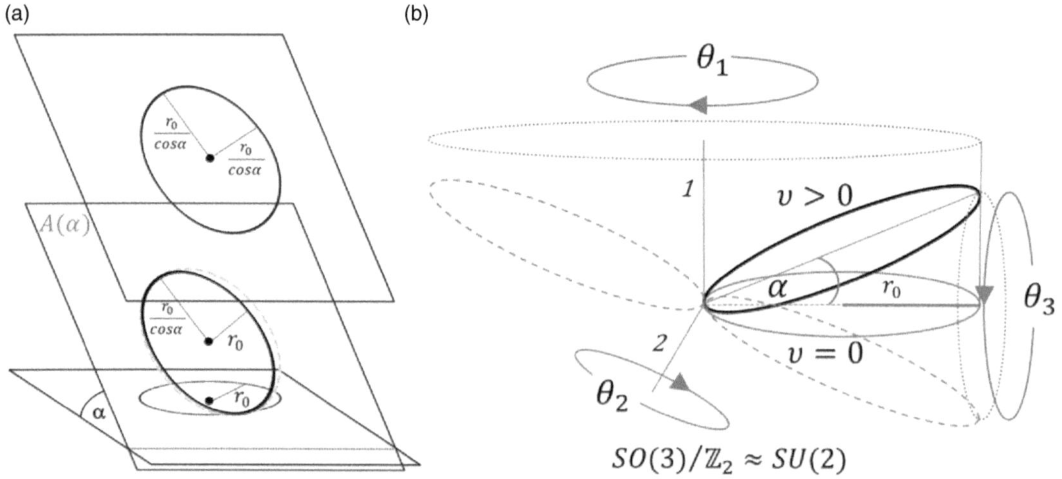

FIGURE 2.7 Spatial symmetries that warrant conservation of geometric dilution of the ur-matter scalar field. (a) Ur particles spatially represented as geometric dilution of the (3,1) shade of DM. The pitch angle α is used to parametrize such dilution so that the angled cylindrical cross section is elliptical and becomes deformed to a circle with radius $r_0/cos\alpha$. (b) The choice of the visible coordinate used to define the pitch angle α is immaterial, and hence, geometric dilution must be invariant under rotations of the visible frame as a rigid body around the three visible axis modulo inversions from the center of coordinates.

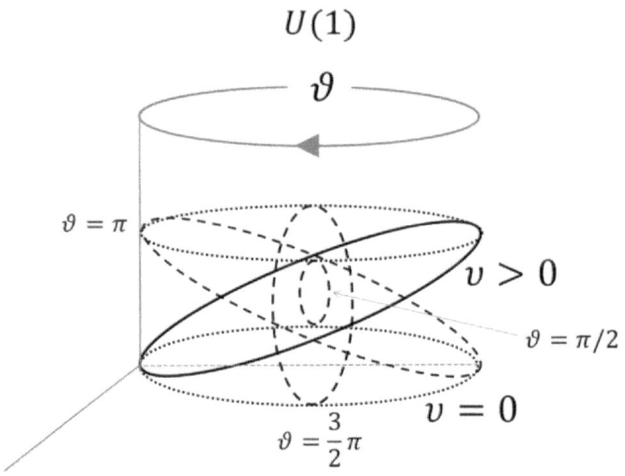

FIGURE 2.8 Invariance of geometric dilution under spatial symmetry defined by rotation around the axis of the dark dimension of the α-pitched ellipse that projects the 246 GeV matter wave on the dark circular coordinate.

described as follows: (a) The choice of the visible coordinates orthogonal to the dark dimension and used to define the pitch angle α is immaterial; hence, geometric dilution must be invariant under rotations of the visible frame relative to a fixed dark dimension. Such rotations operate around the three visible axis modulo inversions along the center of coordinates (Figure 2.7b). The symmetry group generated under these premises is therefore $SO(3)/\mathbb{Z}_2 \approx SU(2)$. (b) The α-pitched ellipse that projects the 246 GeV matter wave on the dark circular coordinate may be rotated around the axis of the dark dimension irrespective of the visible frame that in turn transforms under the premise delineated in (a) (Figure 2.8). Thus, the symmetry group under (b) is $U(1)$, the complex rotation group.

We may determine the full external symmetry associated with conservation of geometric dilution in the 5D space–time by amalgamating the two factors arising from (a) and (b) as $SU(2)\times U(1)$. This local external symmetry in 5D space–time W undergoes a transmutation (internalization) into the local gauge symmetry associated with the electroweak unification upon the canonical projection $\pi: W \rightarrow W/\sim$ onto the quotient manifold that represents the standard 4D space–time (Figure 2.9).

The canonical projection 5D → 4D obliterates (actually, internalizes) the external local spatial $SU(2)\times U(1)$-symmetry, as it lumps points in equivalence classes modulo the dark coordinate. However, the symmetry is restored, resurfacing *albeit* in a different guise, i.e., as internal local symmetry, when the electroweak boson fields are incorporated in the quotient manifold (Figure 2.10). We may state that the "primeval symmetry" of ur-matter gives birth to the electromagnetic and weak force charged under the internalization of the ur-symmetry in the autoencoder-universe defined by the commuting diagram displayed in Figure 2.10.

This analysis cannot be said to be complete without incorporating the evolution of the universe, or more precisely, with a delineation of the critical junctures or phase transitions at which the gauge bosons associated with the $SU(2)\times U(1)-$symmetry became activated. A key criticality is achieved at temperature $T_0 = k_0/k_B = 2.85\times10^{15}$ K, where k_B is the Boltzmann constant. This criticality marks the phase transition at which $DM_{3,1}$ mixes and thereby endows ϕ_H with its v.e.v., which thereby becomes the true vacuum (TV) [2,7]. This activation of the ur-Higgs

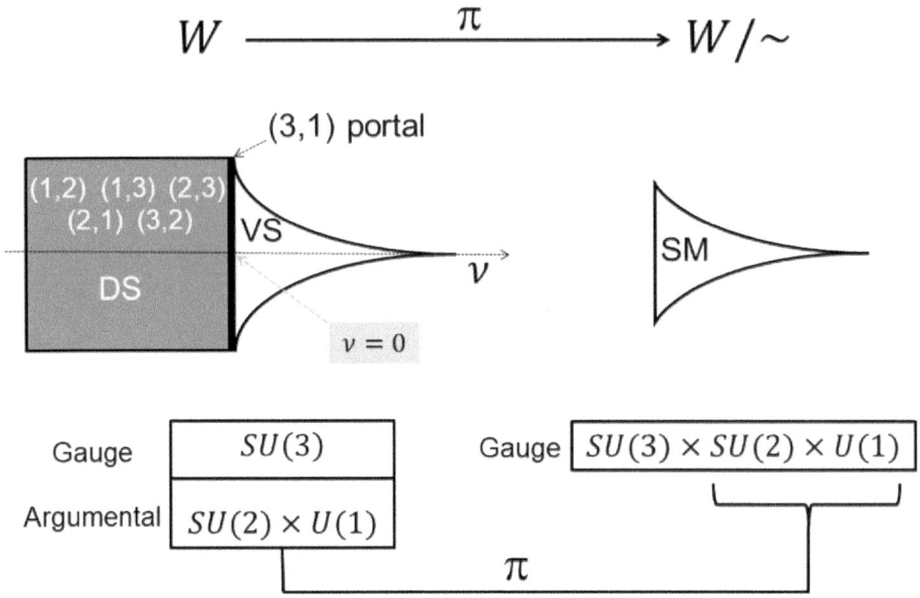

FIGURE 2.9 The spatial symmetry group in 5D space–time W undergoes a transmutation into the electroweak internal symmetry group upon the projection $: W \to W/\sim$.

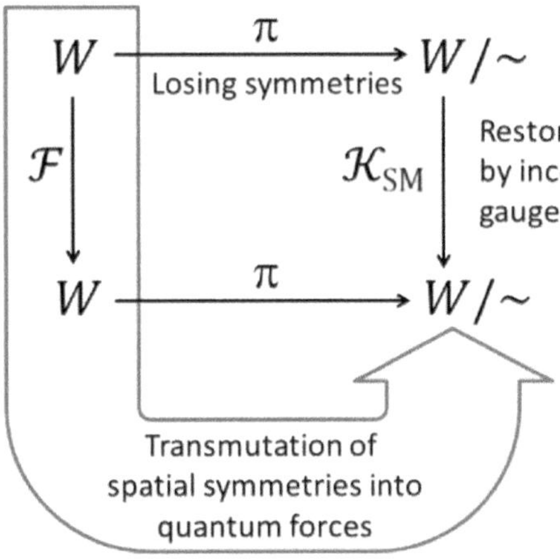

FIGURE 2.10 The "primeval spatial symmetry" of ur-matter engenders the electromagnetic and weak force in the autoencoder-universe defined by the commutative diagram. The spatial symmetry of ur-matter undergoes a transmutation as internal symmetry in the quotient manifold.

scalar is reflected in a kinetic energy commensurate with the rest–mass energy 246 GeV, making it meaningful to factorize the term $\left(k^2 + k_0^2\right)$. The internalization of the spatial symmetry in the 5D space–time occurs as the temperature approaches the critical temperature for activation of the Higgs mechanism:

$$T \to T_{\mathrm{EW}} \sim m_{\mathrm{H}} c^2 / k_{\mathrm{B}} = 1.44 \times 10^{15}\ \mathrm{K}. \tag{2.25}$$

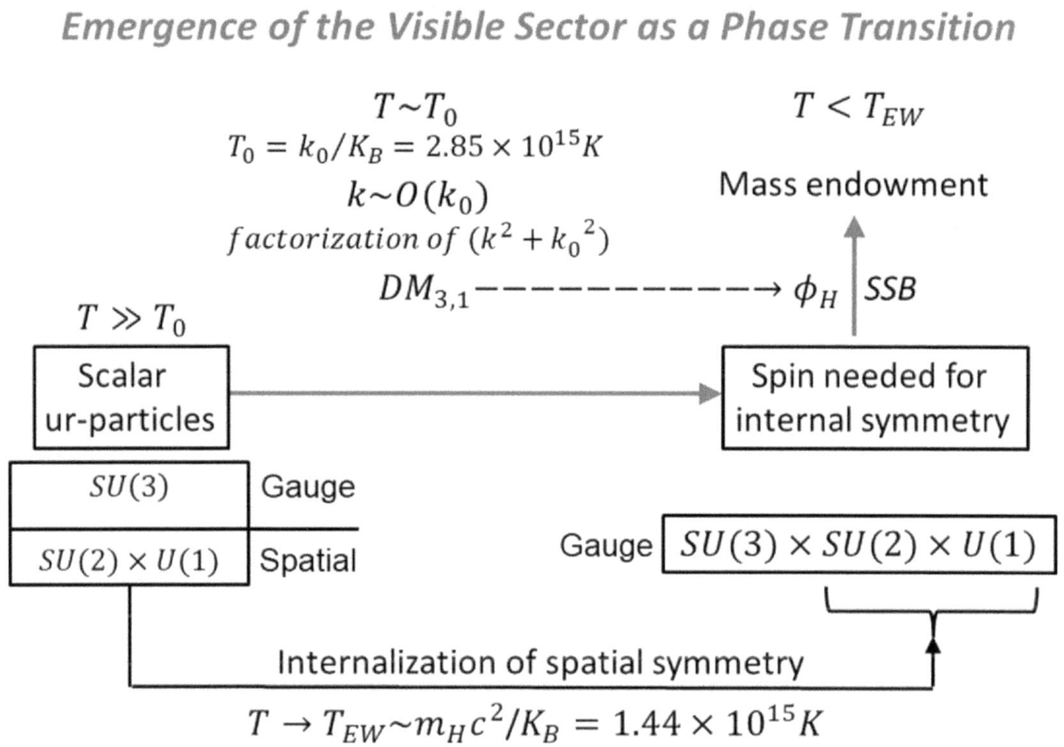

FIGURE 2.11 Emergence of the visible sector as a phase transition from ur-matter, with the internalization of the spatial symmetry in the 5D space–time. The internalization occurs as the temperature approaches the critical temperature for activation of the ur-Higgs mechanism. The scalar ur-particles are thereafter transformed into particles endowed with spin, as required to internalize the spatial symmetry as electroweak symmetry. The phase transition at $T \leq T_{EW}$ activates the Higgs field yielding the symmetry breaking associated with the true vacuum that was acquired at the higher temperature criticality where the ur-Higgs was activated.

This activation requires kinetic energies commensurate with the rest–mass energy of the Higgs boson. The scalar ur-particles in W are therefore transformed into particles endowed with spin, required to internalize the spatial symmetry of ur-matter as gauge electroweak symmetry within the visible sector. Furthermore, the phase transition at $T \leq T_{EW}$ yields the symmetry breaking associated with the true vacuum (ground state) of the Higgs field. Thus, the field excitations in the true vacuum already generated at $T_0 > T_{EW}$ yield the particle masses at $T \leq T_{EW}$ (Figure 2.11). The latent heat associated with the phase transition can therefore be assimilated to $\Delta Q_H = V(\phi_H = 0) - V_{vev}$, $V_{vev} = V({\phi_H}^{TV})$, where V is the potential energy of the Higgs field (Figure 2.4).

This scheme for mass endowment, or rather mass induction, cast in terms of critical phenomena is clearly inspired by the ideas of American physicist Philip W. Anderson on superconductivity [1,7]. In this context, the symmetry breaking associated with the formation of the organized phase below the critical Curie temperature is responsible for endowing the photon with mass, thus relenting it and precluding penetration into the superconducting material of the external magnetic field vectorially carried by the photon.

2.5 COSMOLOGICAL CONSTANT FOR A TRUE-VACUUM DOMAIN GENERATED AFTER A SYMMETRY-BREAKING PHASE TRANSITION

A phase transition at T_0 results in domains adopting the true vacuum encoded in $DM_{3,1}$, in which particles are endowed with mass through the activation of the symmetry-breaking mechanism [2]. The point of this section is to show that such domains undergo exponential expansion, thereby offering an answer to the origin of the cosmological constant problem. Our observable universe is one such domain. To compute the rate of domain expansion, we assume a single scale factor q, yielding the Hubble constant $H=\frac{\dot{q}}{q}$ (the dot indicates time derivative) and cosmological constant

$$\Lambda = 3\left(\frac{\dot{q}}{q}\right)^2 \tag{2.26}$$

The true vacuum domain is represented by the wave function

$$\psi(q) = \Upsilon(q)e^{iS(q)} \tag{2.27}$$

With action $S(q)$ defined as

$$S(q) = \frac{1}{16\pi G}\int \Upsilon(q)\sqrt{-g}d^4\boldsymbol{x} \tag{2.28}$$

The metric is given by

$$ds^2 = \frac{2G}{3\pi}\left[dt^2 - q^2(t)dV_3^2\right], \tag{2.29}$$

with 3D volume measure V_3. Heretofore, we shall assume "natural units" to simplify notation. The Lagrangian associated with Eqs. 2.28 and 2.29 is then

$$\mathcal{L} = p_q\dot{q} - \mathcal{H} \tag{2.30}$$

with scale momentum given by $p_q = -q\dot{q}$ and Hamiltonian

$$\mathcal{H} = -\frac{1}{2}\left(\frac{p_q^2}{q} + q\right) \tag{2.31}$$

Substituting quantities by operators, the domain wave function satisfies

$$\left[q^2\partial_q\left(q^{-2}\partial_q\right) - q^2\right]\psi(q) = 0 \tag{2.32}$$

Given the factorization of the wave function (Eq. 2.27), Eq. 2.32 is decomposed as

$$\partial^2_{q^2} S + 2\frac{(\partial_q \Upsilon)(\partial_q S)}{\Upsilon} - \frac{2}{q}(\partial_q S) = 0, \tag{2.33}$$

$$(\partial_q S)^2 + U(q) + U_Q(q) = 0, \tag{2.34}$$

where $U(q), U_Q(q)$ are, respectively, the classical and quantum potential given by

$$U(q) = q^2;\ U_Q(q) = -\frac{\partial^2_{q^2}\Upsilon}{\Upsilon} + \frac{2}{q}\frac{(\partial_q \Upsilon)}{\Upsilon} \tag{2.35}$$

Thus, we obtain the wave function by solving

$$\partial_{\dot{q}}\mathcal{L} = \partial_q S \quad \text{or} \quad \dot{q} = -\frac{1}{q}\partial_q S, \tag{2.36}$$

which yields

$$\psi(q) = \left[i\xi I_{\frac{3}{4}}\left(\frac{q^2}{2}\right) - K_{\frac{3}{4}}\left(\frac{q^2}{2}\right)\right] q^{\frac{3}{2}}, \tag{2.37}$$

where $I_{\frac{3}{4}}\left(\frac{q^2}{2}\right), K_{\frac{3}{4}}\left(\frac{q^2}{2}\right)$ are, respectively, modified Bessel functions of the first and second kind. The action then reads

$$S(q) = \frac{\xi}{4\Gamma\left(\frac{3}{4}\right)\Gamma\left(\frac{7}{4}\right)} q^3,\ (\xi > 0,\ q \ll 1) \tag{2.38}$$

Combining Eqs. 2.36 and 2.38, we obtain the relation $\dot{q} = \sqrt{\frac{\Lambda}{3}} q$; hence, we can readily see that the domain with true vacuum is subject to exponential expansion, with cosmological constant given by

$$\Lambda = 3\left(\frac{\dot{q}}{q}\right)^2 = \frac{27}{16}\left[\frac{\xi}{\Gamma\left(\frac{3}{4}\right)\Gamma\left(\frac{7}{4}\right)}\right]^2 \tag{2.39}$$

What is established here is that our universe constitutes one domain at true vacuum emerging at $T < T_0$ and that such a domain is subject to exponential growth in accordance with the Hubble scenario described in Chapter 1. It is unknown whether the phase transition includes other true-vacuum domains as well, i.e., a *multiverse* scenario beyond criticality. Autoencoder simulations described subsequently suggest that this is indeed the case.

2.6 DARK ENERGY AS LATENT HEAT IN THE UNIVERSE CREATION

In the physical picture put forth so far, the production of dark energy in our universe is the resultant of the phase transition leading to false-vacuum decay into the true vacuum, a transition that yields the latent heat $\Delta Q_{\mathrm{H}} = V(\phi_{\mathrm{H}} = 0) - V_{\mathrm{vev}}$. Thus, the release of latent heat is associated with the onset of symmetry breaking that occurs at $T < T_0$. At this juncture, the scalar field ϕ_{H} is endowed with its true vacuum (v.e.v.) by mixing with the dark matter scalar component $DM_{3,1}$. This true vacuum bestows mass to SM particles within the domain that we identify as our universe, a domain bound to grow exponentially with cosmological constant obtained analytically (Eq. 2.39).

An experimentally verified consequence of this analysis is that *since latent heat ΔQ_{H} scales with volume, the dark energy remains geometrically undiluted during domain growth or universe expansion.*

To derive the rate of dark energy generation, v_{DE}, we write the potential energy of the Higgs field as

$$V(\phi_{\mathrm{H}}) = -\frac{\mu_{\mathrm{H}}^2}{2}|\phi_{\mathrm{H}}|^2 + \frac{\lambda_{\mathrm{H}}^2}{4}|\phi_{\mathrm{H}}|^4 \tag{2.40}$$

This expression yields

$$\Delta Q_{\mathrm{H}} = \frac{1}{4}\left(\frac{\mu_{\mathrm{H}}^2}{\lambda_{\mathrm{H}}}\right)^2, \tag{2.41}$$

$$v_{\mathrm{DE}} = \frac{1}{4}\left(\frac{\mu_{\mathrm{H}}^2}{\lambda_{\mathrm{H}}}\right)^2 \frac{d}{dt}q^3 = \frac{\sqrt{3\Lambda}}{4}\left(\frac{\mu_{\mathrm{H}}^2}{\lambda_{\mathrm{H}}}\right)^2 q^3 \tag{2.42}$$

Solving for q as $q = q_0 \exp\left(\sqrt{\Lambda/3}t\right)$, Eq. 2.42 can be explicitly written as

$$v_{\mathrm{DE}} = \frac{\sqrt{3\Lambda}}{4} q_0^3 \left(\frac{\mu_{\mathrm{H}}^2}{\lambda_{\mathrm{H}}}\right)^2 \exp\left(\sqrt{3\Lambda}t\right) \tag{2.43}$$

The validity of the theory delineated in this section stems from the rigorous derivations from first principles, but also from numerical estimations of the relatively constant DE density notwithstanding the Hubble picture of universe expansion (Chapter 1) and contrasted with experimental observation [8], as shown in Chapter 5.

2.7 THE MULTIVERSE SCENARIO AND THE SUSTAINABILITY OF THE UNIVERSE TOPOLOGY

In consistence with the geometric dilution of gravity, its weakness relative to the other forces and the hierarchy problem of mass incommensurability with the Planck value, AI has chosen to encode mass in primeval scalar fields that incorporate an extra spatial dimension. Based on symmetry considerations one such field, the one corresponding to $DM_{3,1}$ is able to endow the scalar Higgs field with a stable vacuum that breaks gauge symmetry yielding masses of particles in the SM. Thus, a veritable phase transition occurs whereby the endowment of mass in 4D space–time $W/\sim$ promoted by the transition from the false (metastable, symmetric) vacuum to the true (stable, asymetric) vacuum correlates at the level of the 5D space–time W with a geometric dilution ($k_0 \rightarrow k_0 \cos\alpha$) of the scalar ur-Higgs field. The correlation becomes operative because the autoencoder represents a commutative diagram (Figure 2.12).

In this scheme, dark energy is characterized as the latent heat of the symmetry-breaking phase transition that yields the expanding true-vacuum domain that we came to regard as our universe. This dark energy does not get geometrically diluted because it is proportional to the domain volume itself and it is commensurate with dark matter because the latter generated it by activating the Higgs mechanism at the criticality $T \approx T_0$. A multiverse (multi-domain) scenario cannot be discarded *a priori*, and AI confirms that this is indeed the correct scenario. This is so because a significant expenditure of dark energy to fuel the transition $TV \rightarrow FV$ can be anticipated as being required to uphold the topology of space–time inherited from the early universe, an invariant in accordance with the tenets of relativity. Thus, as shown in the next section, the topology of the quotient manifold

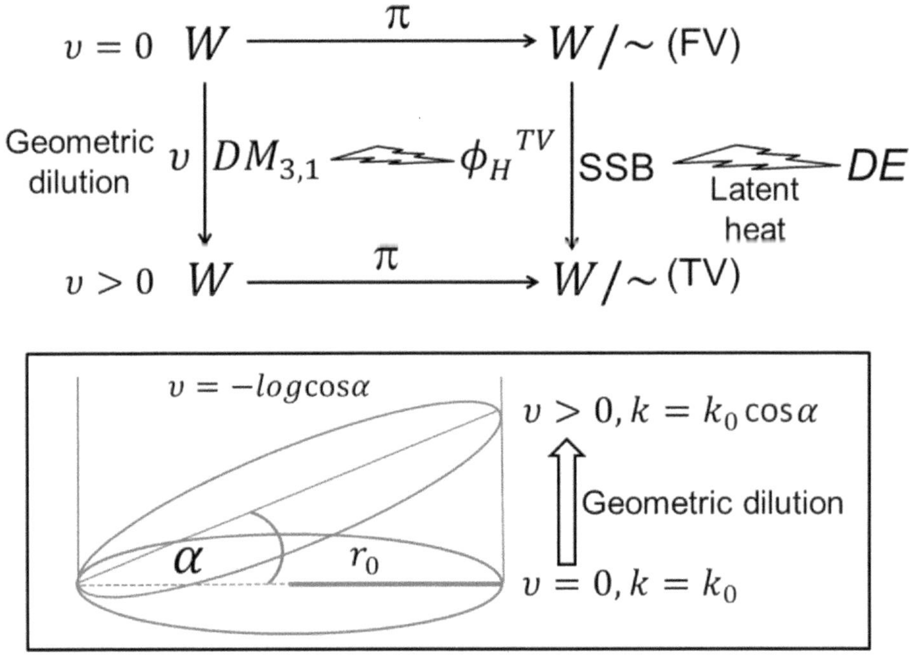

FIGURE 2.12 Autoencoder reproducing a phase transition whereby the particle endowment of mass in 4D space–time promoted by the transition from the false vacuum to the true vacuum correlates at the level of W with the geometric dilution ($k_0 \rightarrow k_0 cos\alpha$) of ur-matter. Acronyms: DM = dark matter, DE = dark energy, SSB = spontaneous symmetry breaking, TV = true vacuum, FV = false vacuum.

corresponds to a three-torus with aspect ratio 1 for any pair of circular spatial coordinates. A true vacuum domain with this topology, that is, completely covering $W/\sim$, is unsustainable because the underlying topology entails a wormhole whose existence is enabled by a false vacuum, as shown in section 2.9, and in further detail in Chapter 5. In turn, the dark energy contribution that fuels the sustainability of the false vacuum effectively addresses the problem of the "vacuum catastrophe" [8], whereby the detectable dark energy is only residual energy from an incommensurably more concentrated vacuum energy.

In fact, the primeval wormhole is enabled by the tunneling of a false-vacuum domain generated within a true vacuum and sustained through the expenditure of dark energy available as released latent heat of the phase transition at $T \approx T_0$. The autoencoder determines the dynamics of the domain walls as constrained by two classically forbidden regions. The quantum tunneling of the false-vacuum domain promotes and sustains the creation of the wormhole which in turn increases the number of instantons, so that the tunneling induces the enlargement of the wormhole throat and the increase in the mass of the primeval black hole. This "cosmological engine" required to sustain the space–time topology is thus fueled by dark energy, as outlined in this chapter and further elaborated in Chapter 5.

2.8 TOPOLOGY OF THE UNIVERSE IN QUANTUM COSMOLOGY: THE ONTOLOGICAL IMPOSSIBILITY OF BOUNDARIES

The extra spatial dimension has been incorporated to address a specific need requiring an extension of the preexisting SM. The most difficult issue we stumble upon when postulating a compact fourth spatial dimension to elucidate the nature of the quantum vacuum is that the global topology of the universe must allow for such possibility [4]. In the past, people have tried to incorporate a curled-up dimension into an Euclidean space, an idea that is *a priori* feasible but topologically meaningless. Not only such attempts failed to generate correct extensions of general relativity, but they arise from the unappealing and unwarranted assumption that one dimension would be compact, while the others are infinite [9]. Furthermore, such a ruled universe would have boundaries, which constitutes an ontological impossibility as shown below. Incorporating a compact dimension is a nuanced matter that pivotally depends on the underlying topology of the universe [4]. This was clearly not a matter of concern for Einstein because his general relativity is essentially governed by differential equations [8]. Since differential equations describe a local situation, the global topology of space–time plays no direct role. Furthermore, relativity has no concern for boundary conditions, as the universe is assumed to be infinite. In our topology, the universe is compact and multiply connected, but there are no boundary conditions simply because the existence of boundaries poses insurmountable ontological problems: The quantum vacuum would need to interface with nothingness, that is, with the absence of geometry, a clear metaphysical impossibility. To define boundaries, nothingness would need to be located beyond the boundaries but that would require extending the geometry of space beyond the boundaries, leading to a contradiction in terms.

In spite of such ontological shortcomings, Randall and Sundrum have incorporated a fourth spatial dimension as a quotient set topologically equivalent to $\mathcal{S}^1/\mathbb{Z}_2$ (Chapter 1) to address the hierarchy problem of particle masses and the weakness of gravity as arising from a geometrically diluted force [9]. Since such a quotient dimension would introduce two branes/boundaries, one to concentrate gravity and the other containing the Higgs field, their efforts, however commendable, are summarily disregarded in this book on ontological grounds.

Quantum mechanics defines the fabric of space, as determined by vacuum entanglement, but topology is not factored into the theory in any obvious way, even if entanglement is thought to be the culprit for the nonlocality of quantum phenomena. On the other hand, topology suddenly becomes relevant as one tries to reconcile general relativity with quantum physics, as it is the case with quantum gravity, as described in Section 2.13, and further on Chapter 5.

If we assume topological invariance of space throughout the universe evolution starting at the big bang, then space cannot be infinite and flat and therefore Euclidean. Provided we accept the big bang scenario, space must be compact now because it once was compact. But if space is compact, then it must be multiply connected because the dimensions are compact. In addition, as said, space cannot have a boundary because that would imply an interface with nothingness, and it is physically impossible to interface with nothingness since an interface presupposes two media described within a common geometry. This Parmenides-style analysis introduces several ontological constraints on what the topology of the universe must be. Summarizing, space must be compact, with compact dimensions, locally flat and lacking boundary. This leaves us with essentially one orientable option, and it is a multiply connected three-dimensional manifold, the three-torus or Cartesian product of three-dimensional circles: $S^1 \times S^1 \times S^1$. Hence, the incorporation of a fourth circular dimension is compatible with the topology of the space–time manifold.

As the reader may surmise at this point, there is a fundamental difference in the way the extra spatial dimension, the curled dormant dimension, is incorporated in the treatment presented in this book when compared with early attempts, especially the so-called Kaluza–Klein model [1]. These researchers incorporated the extra dimension *directly* to the Euclidean space, creating the absurdity of having one compact and three infinite spatial dimensions in space–time with no justification. This construction proved to be untenable [1,4]. Another attempt is provided by the Randall–Sundrum model [9] where, as said, the extra dimension is topologically assimilated to $S^1/\mathbb{Z}_2$, that is equivalence classes (orbifolds) on the circle modulo reflections at $y = \pi$. The extra spatial dimension has fixed points $y = 0, \pi$, taken as locations of two 3-branes that extend in the conventional x^μ directions. In this theory, the $y = 0$-brane concentrates gravity and the $y = \pi$-brane contains the Higgs field. With its three-branes as boundaries, the Randall–Sundrum construction is ontologically untenable for the reasons already given.

As we have shown, the spatial cross sections of space–time cannot be regarded as isomorphic to $\mathbb{R}^3$ but rather to the three-dimensional torus $\mathcal{T}^3 \approx \mathbb{R}^3/\mathbb{Z}^3$, that is, a quotient space of classes modulo all the spatial translations of the unit cell indexed by three integers (Figure 2.13). This implies that the space is multiply connected with all spatial dimensions being compact, hence

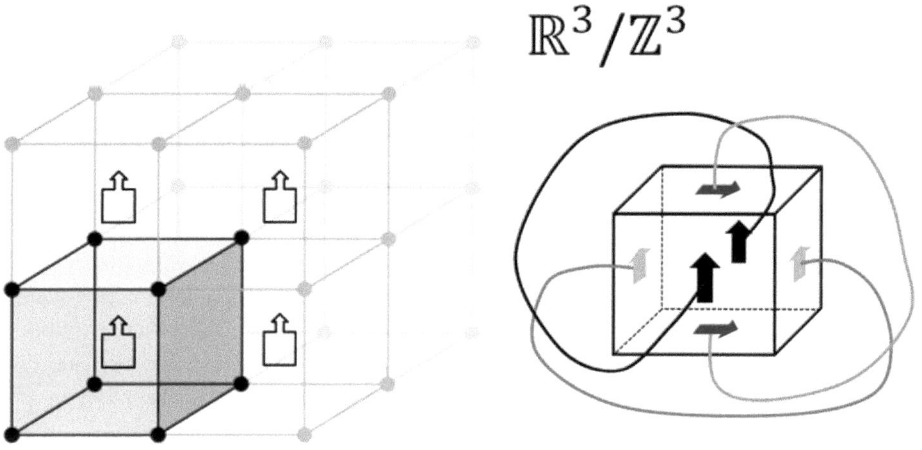

FIGURE 2.13 Generic spatial cross section of space–time with compact coordinates, as required by the topological invariance of the universe dictated by the tenets of general relativity. The compact space is isomorphic to the three-dimensional torus in accordance with the topological equivalence $\mathbb{R}^3/\mathbb{Z}^3 \approx \mathcal{T}^3$. The latter manifold is obtained by identifying opposite faces of the solid cube in three dimensions, hence featuring a wormhole upon construction with dimensional aspect ratios equal to 1 (Chapter 5).

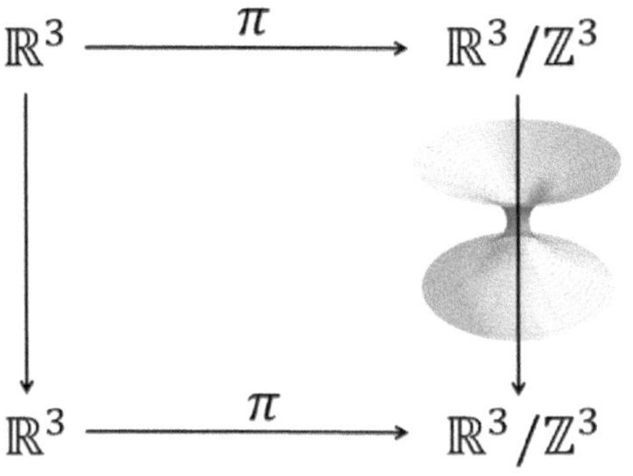

FIGURE 2.14 Commutative scheme adopted by artificial intelligence to circumvent the artefactual wormhole phenomenology associated with the universe compactness by lifting it at the level of universal covering ($\mathbb{R}^3, \pi$) of the toroidal spatial cross sections ($\mathbb{R}^3/\mathbb{Z}^3$) of space–time.

allowing for the incorporation of an the extra circular dimension ($S^1 \approx \mathbb{R}/\mathbb{Z}$) without the conceptual violence of the Kaluza–Klein or Randall–Sundrum models. A hybrid $\mathbb{R}^3 \times \mathbb{R}/\mathbb{Z}$ of three infinite and one compact spatial dimension would make no sense and is in fact not compatible with the tenets of general relativity, assuming this theory is upheld since the inception of space (i.e., once geometry is established), as topological change is forbidden.

At a fundamental level, the Poincaré group, assumed to be the group of isometries of space–time over which the whole edifice of the SM is built [8], does not really contain the proper symmetries but only approximate symmetries or rather, pseudo-symmetries, since the underlying spatial cross sections of space–time are not isomorphic to $\mathbb{R}^3$ but to $\mathbb{R}^3/\mathbb{Z}^3$.

On the other hand, a spatially compact universe implies the existence of a primeval wormhole, since the three circular dimensions are originally equivalent and therefore symmetrically isometric, as shown in Figure 2.13 and in Chapter 5. This can be visualized in a two-dimensional cross section $\mathcal{T}^2$ with dimensional aspect ratio 1. As shown in the next section and in Chapter 5, the sustainability of this wormhole is concomitant with the universe expansion. On the other hand, the very notion of universe expansion becomes meaningless if space would be assimilated with $\mathbb{R}^3$, while it makes good sense to talk about expansion of the unit cell in the quotient space $\mathbb{R}^3/\mathbb{Z}^3$. This is yet another reason to assume that the spatial cross sections of the relativistic space–time are actually isomorphic to $\mathbb{R}^3/\mathbb{Z}^3$, and not to $\mathbb{R}^3$, as assumed in the Minkowski space–time endowed with the isometries of the Poincaré group.

On the other hand, as AI adopts the commutative scheme presented in Figure 2.14, the autoencoder defines the lifting of the SM at the level of the universal covering with canonical projection $\pi: \mathbb{R}^3 \to \mathbb{R}^3/\mathbb{Z}^3$, so that the diagram presented in Figure 2.14 is commutative. This lifting has considerable advantages since it circumvents the observed universe expansion with the concomitant primeval wormhole as artefactual of a "Plato's cave physics", where the observer is placed. Even dark energy, shown in Chapter 5 to be the residual vacuum energy not invested in sustaining the primeval wormhole, becomes artefactual of the Plato's cave phenomena materializing in $\mathbb{R}^3/\mathbb{Z}^3$, the only events accessible to the quantum observer.

2.9 THE TOPOLOGY OF THE UNIVERSE AND THE VACUUM CATASTROPHE PROBLEM: UNDERSCORING THE ROLE OF DARK ENERGY

The topological invariance of 4D space–time implicit in the tenets of relativity compels us to provide a sustainable physical realization of the three-torus topology describing the throat of a wormhole. We shall show that this primeval wormhole is inducible from within a false-vacuum domain with a radial dimension z. Such a physical construct requires a metric surgically spliced into a Schwarzschild (S) component and a de Sitter (D) component [10], which may be given in polar coordinates as follows:

$$ds_S{}^2 = -A_S dt_S^2 + A_S^{-1} dr^2 + r^2\left(d\theta^2 + sin^2\theta d\phi^2\right),\ (r > z), \tag{2.44}$$

$$ds_D{}^2 = -A_D dt_D^2 + A_D^{-1} dr^2 + r^2\left(d\theta^2 + sin^2\theta d\phi^2\right),\ (r < z), \tag{2.45}$$

with $A_S = 1 - \frac{2m_{\mathrm{BH}}}{r}$ and $A_D = 1 - \frac{8}{3}\pi\Lambda r^2$, where m_{BH} denotes the mass of the black hole, equivalent to the amount of dark energy expenditure to create and sustain the false-vacuum domain. The induced metric of the false-vacuum domain wall is thus

$$ds_W{}^2 = \sigma_{00} dt_W^2 + \sigma_{11} d\theta^2 + \sigma_{22} d\phi^2 \tag{2.46}$$

The energy-momentum tensor at the domain wall, $\mathfrak{T}_{ab}$, yields the metric tensor components σ_{ab} via the relation $\mathfrak{T}_{ab} = \gamma\sigma_{ab}$, where γ denotes interfacial tension. Let g_{ab} denote the Gaussian curvature tensor depending on proper distance ζ to the domain wall, and then, we obtain

$$\left(\lim_{\zeta\to 0^+} g_{ab}\right) - \left(\lim_{\zeta\to 0^-} g_{ab}\right) = -8\pi\gamma\left[\sigma_{ab} - \frac{1}{2}\sigma\sigma_{ab}\right], \tag{2.47}$$

where σ is the induced metric trace $\sigma = \sigma^{ab}\sigma_{ab}$. Equation 2.47 yields

$$\sigma_{00} = -1,\ \sigma_{11} = z^2,\ \sigma_{22} = z^2 sin^2\theta;\ \sigma_{ab} = 0(a \neq b). \tag{2.48}$$

Since $-8\pi\left[\mathfrak{T}_{11} - \frac{1}{2}\sigma_{11}\mathfrak{T}\right] = -4\pi\gamma z^2$, $\mathfrak{T} = \sigma^{ab}\mathfrak{T}_{ab}$.

Thus, we obtain the following equation determining the dynamics of the false-vacuum domain wall:

$$-\left(1 - \frac{8}{3}\pi\Lambda z^2\right)\frac{dt_D}{dt_W} - \left(1 - \frac{2m_{\mathrm{BH}}}{z}\right)\frac{dt_S}{dt_W} = 4\pi\gamma z \tag{2.49}$$

Equation (2.49) underscores the role of dark energy in upholding both the Schwarzschild and the de Sitter contributions [10], specifically, in creating the false-vacuum black hole and upholding the cosmological constant as required to ensure the sustainability of the compact, multiply connected, and boundary-less topology of the universe. This cosmological role of dark energy analytically delineated in this section will be shown to play out in sustaining the topology of the universe, invariant since its inception, as dynamic simulations are extended with the aid of machine learning tools (Chapter 5).

2.10 QUANTUM VACUUM AS UR-MATTER: TRUE- AND FALSE-VACUUM COEXISTENCE

To truly understand what the universe was like right after the big bang would require a quantum theory of gravity, a subject that has remained for the most part unyielding to research efforts. Furthermore, a fundamental force becomes discernible only at temperatures commensurate with its critical value $T_B = m_B c^2/k_B$, where m_B is the mass of the boson that communicates the force. Consequently, as we approach the big bang with temperatures above Planck's mark, the much higher kinetic energies would render irrelevant any distinction between the fundamental forces. The curvature of space–time is determined by the concentration of energy which in the budding universe is smeared to such an extent as to lead to an uncertainty in the curvature of space–time, producing huge fluctuations in the metric: There is simply no identifiable geometry right after the big bang, while dimensions can be permuted due to complete spatial symmetry.

In this picture of the earliest universe, all we can be certain about is that the topology of space was once compact and had not boundaries (there is no geometry outside space, hence no interface with another vacuum). The topology must remain the same, since a change in topology would require a rough manifold deformation which would be forbidden according to the premises of general relativity. This has led us to propose a three-torus as the only compact orientable 3D manifold with no boundaries and with three compact dimensions. These dimensions have inflated to fit our perceived reality but we have reasons to believe there are one or perhaps more dormant dimensions.

As previously indicated, gravity is much weaker than the other fundamental forces of a quantum nature. This suggests that gravity has leaked into a higher dimension, implying that *quantum vacuum is actually ur-matter*. This is a pivotal tenet in this book. Other established facts also suggest this view, in particular the hierarchy problem, whereby the detected particle masses are incommensurably smaller than the Planck mass and the impossibility of accounting for dark matter and dark energy within the premises of the current version of the SM of particle physics.

As the issue of the need for the extra spatial dimension is settled, we faced the problem of how should the 5D space–time be modeled. The Kaluza–Klein model that does not enable geometric dilution or activate a portal into the visible sector, while the Randall–Sundrum model incorporates the extra dimension as a quotient space, a construct that is ontologically flawed since it assumes an interface with nothingness, that is, with the absence of geometry. By contrast AI resorts to the autoencoder $W \xrightarrow{\pi} W/\sim$ of the universe that we are aware of ($W/\sim$), incorporating the hidden dimension onto a 5D space–time where a scalar field spans the hidden dimension and makes the quantum vacuum behave as ur-matter (Figures 2.5 and 2.6). This representation enables the emergence of our universe as a domain $\mathfrak{D}_{3,1} \subset W/\sim$ within a symmetry-breaking phase transition induced by endowing a (3,1) portal (the Higgs field) with its v.e.v. at 246 GeV (Figure 2.11).

This domain expands exponentially according to Eq. 2.39 and is embedded in a horned three-torus representing $W/\sim$ (Figure 2.15). There may be other domains $\mathfrak{D}_{a,b} \subset W/\sim$ ($a \neq b; a,b = 1,2,3$) depending on the existence of (*a*,*b*) portals whose symmetry enables mixing with the respective ur-(*a*,*b*) particle, a form of $DM_{a,b}$. In particular, there may even be other (3,1) domains that we are not aware of. The "multiverse" is thus an untested scenario for the symmetry-breaking transition that endows particles with mass or marks the emergence of matter.

The topology of our quotient space $W/\sim$ needs to be sustainable. As shown in Section 2.9, this requires a false-vacuum (FV) domain in order to stabilize the primeval wormhole in the three-torus. Unlike the true-vacuum (TV) domain $\mathfrak{D}_{3,1}$, an FV domain does not expand spontaneously. It does so through a continuous expenditure of energy and the available energy is the latent heat ΔQ_H, generated upon the emergence of $\mathfrak{D}_{3,1}$ at $T < T_0 = T_{3,1} = 246\text{GeV}/k_B$. Furthermore, as described in detail in Chapter 5, the inflation of the FV domain is required to maintain the toroidal throat of the wormhole, in accordance with Eqs. 2.44–2.49. In summary, the energetic economy of the universe

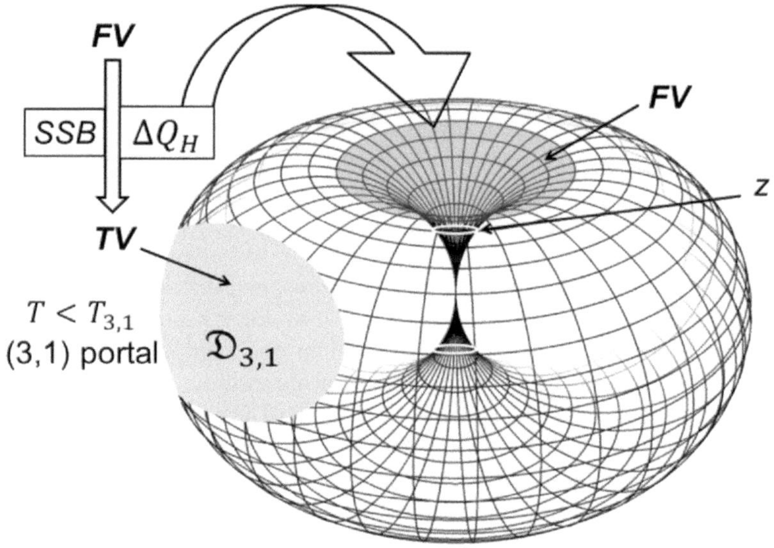

FIGURE 2.15 Flow of dark energy in the multiverse with at least two domains, one in a stable vacuum (TV) and one in a metastable vacuum (FV), with the latter sustained by a constant influx of the latent heat ΔQ_H, generated upon the emergence of the TV domain $\mathfrak{D}_{3.1}$ at $T < T_0 = T_{3,1} = 246\ GeV/k_B$. The FV domain is required to sustain the toroidal topology of the primeval wormhole.

is that of a multiverse with at least two domains, one in a stable vacuum and one in a metastable vacuum sustained by a constant influx of dark energy (Figure 2.15).

Perhaps the most subtle issue in this discussion of the emergence of the universe is the transfiguration of the symmetry of ur-matter or quantum vacuum associated with the phase transition. As noted, quantum vacuum is filled with ur-matter scalar particles with rest–mass momentum k_0 or multiples thereof. The high-dimensional factorization possibilities of the special relativity relation $k^2 + k_0^2 = \omega^2$ introduce different symmetry proposals, six shades to be precise, and only one of those shades is known to be compatible with a gauge field in the visible sector of the SM. Thus, the mixing of the ur-scalar field with a scalar boson field from the visible sector bestows the true vacuum of the visible sector, inducing a symmetry-breaking phase transition whereupon masses are bestowed onto particles. AI predicts six types of domains that may arise from phase transitions upon activation of the ur-boson (*u-B*) at $k_B T \approx m_{u-B} c^2 = k_0$. Yet, none of these domains has been detected as of yet, and the AI system (autoencoder) treats as dark matter the scalar ur-matter which differs from the (3,1) type. The five shades of purely dark matter have an abundance relative to the (3,1) shade that obeys the ratio 5:1, in close agreement to the ratio between dark and visible matter in the universe (Chapter 1). This ratio is in consonance with the six ways to factorize the energy relation for ur-matter resulting from special relativity, out of which only one (3,1) is symmetrically compatible with the gauge fields of the photon and the Higgs in the visible sector.

The autoencoder further reveals that the dark matter that behaves as ur-matter for the visible sector contains six spatial symmetries (Figures 2.7 and 2.8) that undergo transmutation into the electroweak symmetry $SU(2) \times U(1)$ upon the phase transition induced by activation of the Higgs field (Figure 2.11). No actionable gauge fields capable of inducing symmetry-breaking phase transitions for the other five symmetry schemes of ur-matter have been detected as of yet. Hence, the ratio 5:1 of dark matter to visible matter must be upheld up until the definition of "visible" is modified based on plausible extensions of the SM.

2.11 PARTICLE COSMOLOGY WITH AI: THE ORIGIN OF MATTER ENCODED IN THE QUANTUM VACUUM

Particle cosmology, the unlikely merger of particle physics and cosmology, focuses on small units of matter and their role in determining the fate and shape of the universe, while also prompting the examination of global distribution of matter and energy to make sense of local phenomena associated with mass. Thus, it constitutes a suitable platform to tackle the problem of the origin of the universe, or, in a less ambitious guise, of the origin of matter. This section delineates the implementation of an AI system suited for particle cosmology in its pursuit of the origin of the universe.

The uncanny ability to detect patterns, extract features, and distill the relevant physical picture out of algorithmic complexity in static and dynamical systems has made learning systems, especially autoencoders [4,6], the beacons of applied AI. Our goal here is more ambitious, as we strive to deploy autoencoders to reach beyond the realm of algorithmic complexity and make theoretical inferences that pertain to second-order logic [11,12]. The most crucial such cogitation by an autoencoder is the revelation that, for all its prowess [7], the Higgs field assumed to endow bosons with mass while retaining gauge symmetry is a provisional *ansatz* that needs to be revised vis-à-vis the incorporation of the extra spatial dimension. The motivation for introducing the extra dimension is amply justified in the preceding sections, where the geometric dilution of gravity and the hierarchy problem in particle masses are addressed.

An AI system set up to unravel the origin of the universe encoded in the quantum vacuum is depicted in Figure 2.16. The fifth dimension in space–time (fourth in space) is not incorporated directly, as in the Kaluza–Klein or Randall–Sundrum schemes [1,9], but within an autoencoder, represented by a commutative diagram. The autoencoder treats the quantum vacuum as the vehicle for ur-matter (UM), that is, the precursor of matter. There are six ways (shades) of encoding the scalar field spanning the fifth dimension with momentum k_0 according to the symmetries resulting from the six modes of factorization of the special relativistic relation $k^2+k_0^2=\omega^2$ (natural units are assumed). Five of these shades, namely, (1,2), (1,3), (2,3), (2,1), and (3,2), yield dark matter proper (DM), while the (3,1) shade of DM yields the ur-matter (UM) for the visible sector of the SM. This is so because the ur-field $\phi_{3,1}$ is capable of amalgamating with the Higgs scalar field ϕ_{H} since they are symmetry-compatible (Eq. 2.15), thus yielding the true vacuum (TV) of the Higgs field at 246 GeV, in accordance with the equation: $\phi_{3,1} \rhd\lhd \phi_{\mathrm{H}} = \phi_{\mathrm{H}}^{\mathrm{TV}}$ (the bowtie symbol denotes mixing). Mass in the visible sector originates through a symmetry-breaking (SB) phase transition catalyzed by the activation of $\phi_{\mathrm{H}}^{\mathrm{TV}}$ and turned on by the temperature switch $T \gg \frac{k_0}{k_{\mathrm{B}}} \rightarrow T < \frac{k_0}{k_{\mathrm{B}}}$ within an autoencoder that runs at two temperatures, one above and one below criticality. Criticality is thus parametrized by the temperature at which ur-matter acquires entity as a particle (Figure 2.16). A fermion with Lagrangian $\mathcal{L}$ first becomes symmetry-compatible with scalar field ϕ_{H} by the incorporation of a boson (B) Lagrangian, so that mass endowment in the latent 4D space–time is assimilated to the compound Lagrangian: $\mathcal{L}+\mathcal{L}_{\mathrm{B}}+\mathcal{L}_{\mathrm{H}}$. This fermion Lagrangian is lifted to $\varrho_{\alpha}^{\mathcal{L}}\left(\mathcal{L}^{3,1}\right)$, where $\mathcal{L}^{3,1}$ is the ur-matter Lagrangian and the geometric dilution operator $\varrho_{\alpha}^{\mathcal{L}}$ corresponds to the pitch angle α that yields the mass $m=k_0\cos\alpha$ for the particle with Lagrangian $\mathcal{L}$.

The autoencoder thus informs that mass only *appears* to originate from the activation of the Higgs mechanism ($\mathcal{L} \rightarrow \mathcal{L}+\mathcal{L}_{\mathrm{B}}+\mathcal{L}_{\mathrm{H}}$), but in truth, it is the geometric dilution of ur-matter as ur-particles ($\mathcal{L}^{3,1} \rightarrow \varrho_{\alpha}^{\mathcal{L}}\left(\mathcal{L}^{3,1}\right)$) that serves as precursor to the mass endowment for particles in the visible sector in accordance with the relation:

$$\pi \circ \varrho_{\alpha}^{\mathcal{L}}\left(\mathcal{L}^{3,1}\right)=\mathcal{L}+\mathcal{L}_{\mathrm{B}}+\mathcal{L}_{\mathrm{H}} \tag{2.50}$$

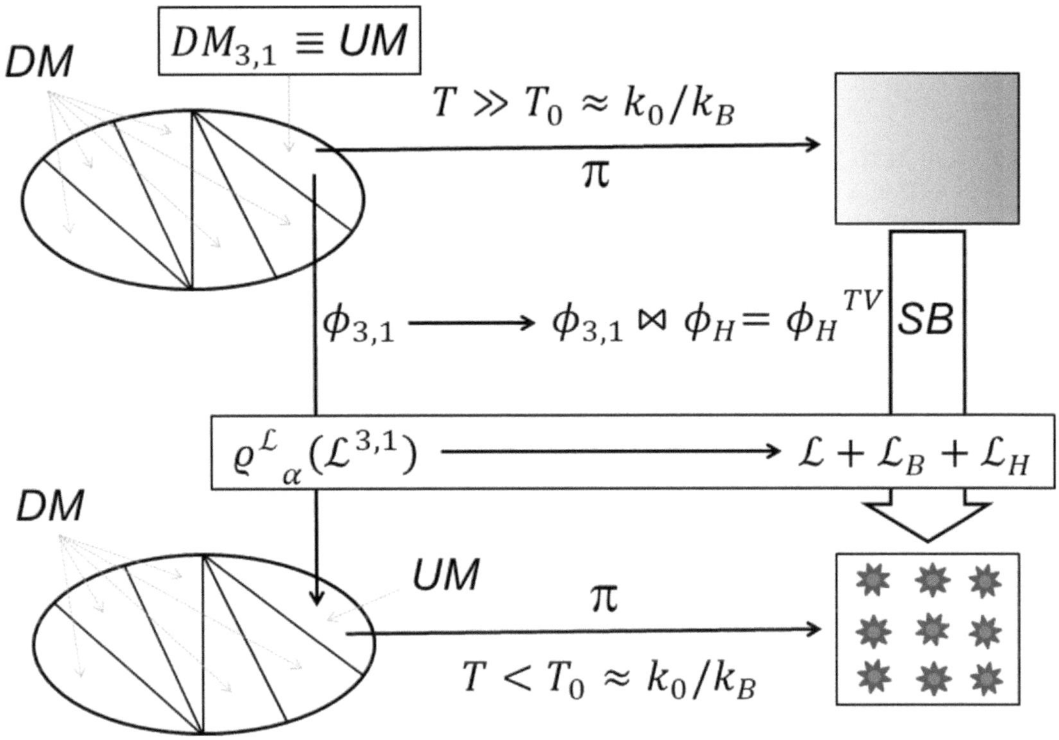

FIGURE 2.16 Artificial intelligence system set up to discover the origin of the universe encoded in the quantum vacuum. The fifth dimension is not incorporated directly but within an autoencoder, represented by a commutative arrow diagram. The autoencoder treats the quantum vacuum as ur-matter (UM), that is, as precursor of matter. There are six ways (shades) of encoding the scalar field spanning the fifth dimension with momentum k_0 according to symmetry. Five of these shades, namely, (1,2), (1,3), (2,3), (2,1), (3,2), yield dark matter (DM), while the (3,1) shade yields ur-matter for the visible sector of the SM. This is so because the ur-field $\phi_{3,1}$ is capable of amalgamating with the Higgs scalar field ϕ_H since they are symmetry-compatible (Eq. 2.15), thus yielding the true vacuum (TV) of the Higgs field, in accordance with the equation: $\phi_{3,1} \bowtie \phi_H = \phi_H^{TV}$. Mass in the visible sector originates through a symmetry-breaking (SB) phase transition catalyzed by activation of ϕ_H^{TV} and enabled by the temperature switch $T \gg \frac{k_0}{k_B} \rightarrow T < \frac{k_0}{k_B}$ within an autoencoder that runs at two temperatures, one above and one below criticality. A fermion with Lagrangian $\mathcal{L}$ first becomes symmetry-compatible with scalar field ϕ_H by incorporation of a boson (B) Lagrangian, so that mass endowment in the latent 4D space–time is assimilated to compound Lagrangian: $\mathcal{L} + \mathcal{L}_B + \mathcal{L}_H$. This Lagrangian is lifted to $\varrho_{\alpha}^{\mathcal{L}}(\mathcal{L}^{3,1})$, where $\mathcal{L}^{3,1}$ is the ur-matter Lagrangian and the geometric-dilution operator $\varrho_{\alpha}^{\mathcal{L}}$ corresponds to the pitch angle α that yields the mass $m = k_0 cos\alpha$ for the particle with Lagrangian $\mathcal{L}$. Natural units have been assumed throughout.

Such visible particles internalize the spatial symmetry of their ur-particles as gauge symmetries, thus acquiring spin, whereas the ur-particles are scalar (Figure 2.16).

Eq. 2.50 constitutes a cornerstone for a deeper understanding of quantum vacuum as ur-matter. The ramifications of this result will be pursued in the ensuing chapters. Rather than assuming the existence of a mysterious all-pervasive scalar field that endows particles with induced mass after adopting its stable true vacuum through self-interaction, we postulate that the quantum vacuum has intrinsic materiality spanning an undetectable "within" dimension. This result appears to represent a substantive way forward as it portends to reconcile the SM with the standing hierarchy problem, the cosmological constant problem, the vacuum catastrophe, and the influence of dark matter and dark energy on deep-space dynamics (Chapter 5).

2.12 AN AUTOENCODER OF THE EXTRA DIMENSION PURSUES THE GRAND UNIFICATION OF THE FUNDAMENTAL FORCES AS IT SEARCHES FOR THEIR COMMON ORIGIN

A grand unification theory (GUT) is still one of the holy grails in physics. The fundamental quantum forces respond differently to geometric dilution in the vacuum, with coupling strengths that intersect at distances approaching the Planck length $l_p = \sqrt{\hbar G/c^3} = 1.616\times10^{-35}\,\text{m}$ (Figure 2.17). The GUT primeval symmetry was proposed by Howard Georgi and Sheldon Glashow in 1974 [13]. These researchers took into account the fundamental gauge symmetry groups and assumed they are subgroups of a minimal primeval-symmetry group. Thus, since the symmetry groups *SU*(3), *SU*(2), and *U*(1) correspond, respectively, to the strong nuclear force (SNF, QCD), the weak nuclear force (WNF) and electromagnetism (EM), the primeval symmetry has been rightly assumed to be $SU(5)\approx SU(3)\times SU(2)\times U(1)$.

Given this primeval symmetry, the origin of the fundamental forces is cast by the autoencoder in terms of phase transitions corresponding to the gauging or internalization of the fundamental

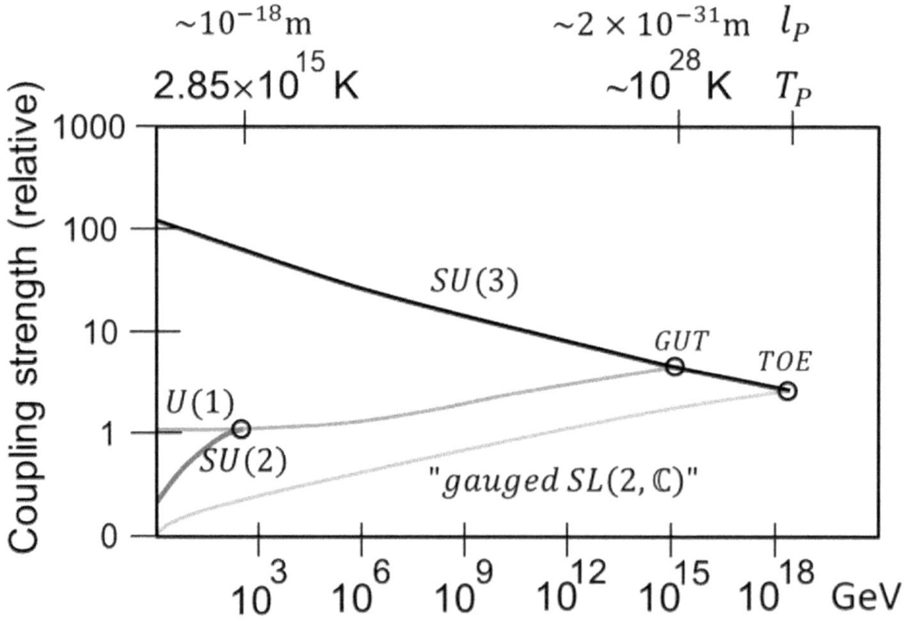

FIGURE 2.17 Gauge symmetries for the fundamental force carriers in a plot of coupling strength as a function of energy, temperature, or interaction distance. Symmetry groups *SU*(3), *SU*(2), and *U*(1) correspond, respectively, to the strong nuclear force (SNF, QCD), the weak nuclear force (WNF), and electromagnetism (EM). The circles denote the phase transitions corresponding to the gauging or internalization of the fundamental local symmetries of ur-matter. The grand unification theory (GUT) arises with the gauging of the $SU(2)\times U(1)$ symmetry of ur-matter (Figures 2.7 and 2.8) as electroweak (EW) symmetry. The theory of everything (TOE) entails a special gauging of the Lorentz $SL(2,\mathbb{C})$ symmetry to beget gravity, while *SU*(5) remains the local external symmetry of ur-matter $DM_{3,1}$. Gravity is not associated with a local gauge but with a one-form gauge symmetry.

local external symmetries of ur-matter concurrently with activation of the respective portals at critical temperatures $T = \frac{k_0}{k_B}, T = \frac{nk_0}{k_B}(n \sim 10^{13})$. Besides the $SU(2), U(1)$ symmetries depicted in Figures 2.7 and 2.8, the ur-matter possesses also an external $SU(3)$ symmetry, as shown in Figure 2.18. A generic ur-matter particle results from geometric dilution of the lightest particle in $DM_{3,1}$ with pitch angle α. The symmetry group is known to be the minimal that acts transitively over the five-sphere whose points correspond to the rotation axes of the diluted ur-matter particle. This symmetry gets gauged at the phase transition at $\sim 10^{28}$ K, concurrent with the activation of a Kaluza–Klein ur-Higgs tower with momentum nk_0, $n = \mathcal{O}(10^{13})$.

Thus, the GUT arises from the phase transition that gauges the $SU(3)$ symmetry of ur-matter with retention of spatial $SU(2) \times U(1)$ symmetry (Figures 2.7 and 2.8). The latter becomes internalized as electroweak (EW) symmetry at the next phase transition that takes place at $T = \frac{k_0}{k_B}$. On the other hand, a purported "theory of everything" (TOE) entails a special gauging of the Lorentz $SL(2, \mathbb{C})$ symmetry to beget gravity, while *SU*(5) prevails as local external symmetry for $DM_{3,1}$. It should be pointed out that gravity cannot not be really associated with a local gauge but with a one-form generalized symmetry, as discussed subsequently.

As stated, the unifications (GUT and TOE) are suggested by the differences in response to the quantum vacuum marking convergence of the fundamental forces (Figure 2.19). This convergence (Figures 2.17 and 2.19) *hints that all fundamental forces are actually one and the same that propagates under different symmetries providing different levels of geometric dilution as they emerge from phase transitions in early stages of the universe* (Figure 2.20).

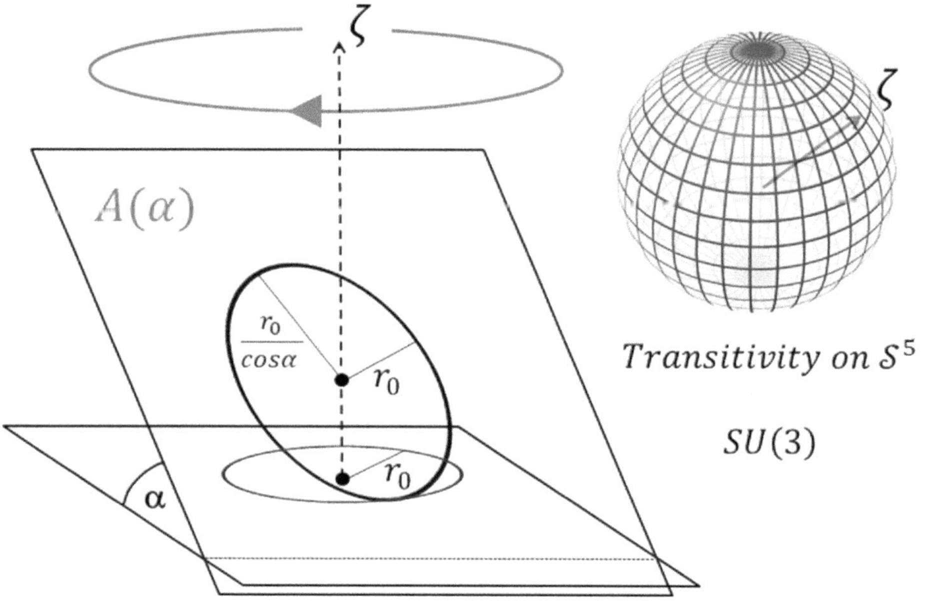

FIGURE 2.18 External local *SU*(3)-symmetry of ur-matter. The generic ur-matter particle results from geometric dilution of the lightest $DM_{3,1}$ particle parametrized by pitch angle α. The symmetry group is known to be the minimal that acts transitively over the five-sphere whose points correspond to the rotation axes of the diluted ur-matter particle. This symmetry gets gauged at the phase transition at $\sim 10^{28}$ K, concurrent with the activation of a Kaluza–Klein ur-Higgs with momentum nk_0, $n = \mathcal{O}(10^{13})$.

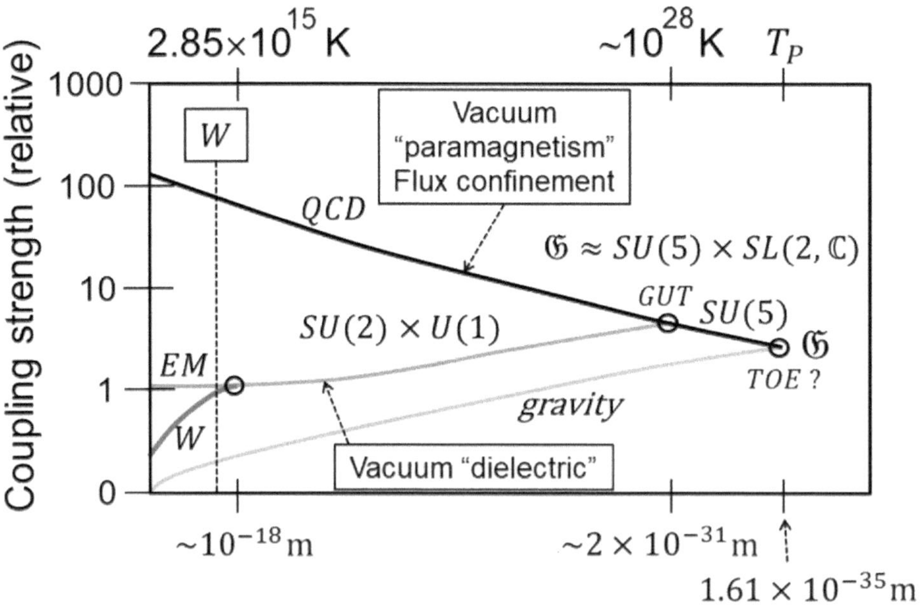

FIGURE 2.19 Response to the quantum vacuum by the fundamental forces. The SNF grows monotonically with distance as per the scale indicated in the figure. The SNF is significantly enhanced at distances larger than $\sim 10^{-18}$ m due to vacuum paramagnetism brought about by confinement of quark-gluon field lines that make the strength of the coupling behave as $\propto R$, rather than $\propto R^{-2}$. By contrast, the transient (e^-, e^+) – pair creation in the vacuum introduces polarization enfeebling electric field lines in a kind of vacuum dielectric, and hence, the response of EM to quantum vacuum opposes that of SNF. On the other hand, the WNF gets significantly enfeebled at distances larger than $\sim 10^{-18}$ m because it is conveyed by massive boson W, so its range is $r_W \sim \frac{\hbar}{m_W c} \sim 10^{-18}$ m. The different responses of the fundamental forces to the quantum vacuum enable line crossing at points indicative of force unification as interaction distances decrease and approach Planck's critical value. Thus, at GUT, the external $SU(5)$ symmetry of ur-matter is partially internalized into gauged symmetry $SU(3)$, while the external local symmetry $SU(2) \times U(1)$ is retained by ur-matter, respectively, begetting the SNF and electroweak symmetry. At temperatures higher that the Planck temperature, ur-matter retains its full symmetry under the group $\mathfrak{G} \approx SU(5) \times SL(2,\mathbb{C})$. At the Planck temperature, the relativistic component $SL(2,\mathbb{C})$ gets internalized as one-form-gauged Lorentz symmetry, begetting gravity, while ur-matter retains the full $SU(5)$ – symmetry up until the GUT point.

The SNF grows monotonically with distance as per the scale indicated in Figure 2.19. Thus, this force is significantly enhanced at distances larger than $\sim 10^{-18}$ m due to vacuum paramagnetism [14] brought about by area confinement of quark-gluon field lines that make the strength of the field behave as $\propto R$, rather than $\propto R^{-2}$. By contrast, the transient (e^-, e^+)-pair creation in the vacuum introduces polarization enfeebling electric field lines in a kind of vacuum "dielectric" that weakens EM in such a way that its response to quantum vacuum opposes that of SNF (Figure 2.19). On the other hand, the WNF gets significantly enfeebled at distances larger than $\sim 10^{-18}$ m because the force is conveyed by massive boson W ($W^{\pm}$), so its range is $r_W \sim \frac{\hbar}{m_W c} \sim 10^{-18}$ m.

Thus, the different responses of the fundamental forces to the quantum vacuum eventually result in intersections of their plots of coupling strength as functions of the energy or distance. As interaction distances approach Planck's critical value, the points of intersection suggest a unification of

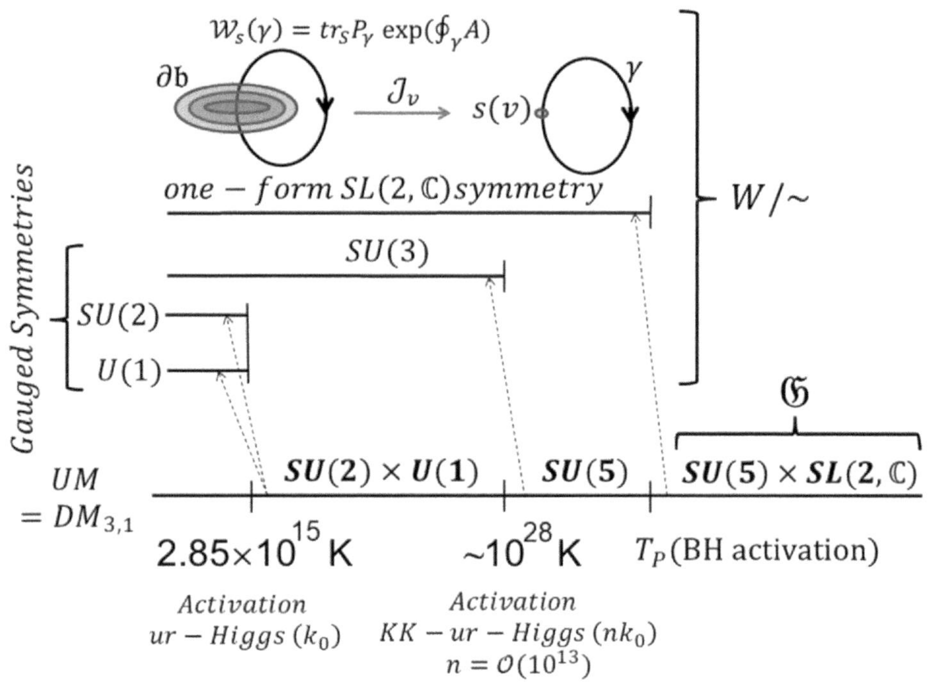

FIGURE 2.20 Phase transitions of ur-matter into symmetry components of the visible sector of the SM with concurrent gauging of the symmetry components. The symmetry bifurcation begetting gravity does not yield a local gauge symmetry but a more generalized Yang–Mills one-form gauge symmetry identified with the subgroup $Z(SL(2,\mathbb{C}))$, the center of the local Lorentz group [14].

forces. Thus, at GUT, a phase transition takes place where the external $SU(5)$ symmetry of ur-matter is partially internalized into the gauged $SU(3)$ symmetry ($SU(3)$ is a subgroup of $SU(5)$), while the residual external local symmetry $SU(2)\times U(1)$ of ur-matter becomes the electroweak symmetry that bifurcates after being gauged at the next phase transition.

At temperatures higher that the Planck temperature, ur-matter retains its full symmetry under the group $\mathfrak{G} \approx SU(5)\times SL(2,\mathbb{C})$. At the Planck temperature (T_P), the primeval phase transition takes place and the relativistic component $SL(2,\mathbb{C})$ of the full symmetry of ur-matter gets gauged as one-form Lorentz symmetry to beget gravity, while ur-matter retains the full $SU(5)$ symmetry up until the GUT point (Figures 2.19 and 2.20). Figure 2.20 depicts the phase transitions of ur-matter into symmetry components of the visible sector of the SM with progressive and concurrent gauging of the symmetry. The internalization of relativistic ur-symmetry at the phase transition that begets gravity does not yield a local gauge symmetry but a more generalized Yang–Mills one-form generalized symmetry identified with the subgroup $Z(SL(2,\mathbb{C}))$, the center of the local Lorentz group [15]. This implies that quantum gravity cannot be cast in terms of standard gauge theory but requires further elaboration, as provided in the next section.

2.13 AI DELINEATES THE ONSET OF QUANTUM GRAVITY AS A PRIMEVAL PHASE TRANSITION

So far, the picture that has emerged from the AI's autoencoder of the fifth dimension describes the origin of matter and of the fundamental forces of quantum nature as outcomes of phase transitions from an ur-matter-endowed quantum vacuum. This vacuum is endowed with a primeval $SU(5)\times SL(2,\mathbb{C})$ symmetry, and this symmetry may be further simplified, as shown in Section 2.14. Yet, the picture is not complete unless it can encompass the "theory of everything" (TOE) critical point (Figures 2.17, 2.19, and 2.20), where gravity emerges below the critical Planck temperature concurrent with the gauging of its symmetry enshrined in the relativistic Lorentz group.

As profusely discussed in this chapter, symmetry has become a cornerstone of modeling efforts in particle physics. Historically, the efforts have focused on symmetries that act on local operators, with the SM exhibiting symmetries of this type, for example, relating to charge in electromagnetism and chiral symmetry associated with the strong nuclear force. To cast gravity in terms of a gauge symmetry, however, the autoencoder needs to extend the concept of symmetry significantly. This is an ambitious undertaking where AI encompasses higher-form symmetries acting on extended objects described by nonlocal operators supported on lines or branes. These higher-form symmetries act trivially on the local operators of standard gauge theory, which are regarded by the autoencoder as "zero-form symmetry".

Thus, a generalized symmetry is required to cast the onset of gravity as a phase transition in quantum field theory, on a par with the phase transitions previously described in this chapter. This is purportedly the "autoencoder version of quantum gravity", where a field theory of gravity requires a gauged generalized symmetry. We now outline the gravitational one-form symmetry gauged by the autoencoder at the phase transition putatively taking place in the earliest universe at $T < T_P = \sqrt{\dfrac{\hbar c^5}{G k_B^2}} \sim 10^{32}$ K to decode gravity (Figure 2.20) concurrently with the activation of the ur-black hole with mass $\dfrac{n k_0}{c}, n = \left[\sqrt{\hbar c^3 / k_0^2 G}\right]\left([.] = \text{integer part}\right)$.

To identify the generalized symmetry associated with gravity, the autoencoder casts general relativity as a special sort of gauge theory of the local Lorentz group $SL(2,\mathbb{C})$. Thus, gravity is associated not with a local gauge as in the other forces in the SM, but with a one-form symmetry. This generalized symmetry is enshrined in an operator $\mathcal{J}_v$ determined by an element from the center of the local Lorentz group: $v \in Z(SL(2,\mathbb{C}))$. The corresponding charge-line operator $\mathcal{W}_s$ is the holonomy in a spin representation s, which is topologically linked to $\mathcal{J}_v$ [15]. Via its associated parallel-transport map, the holonomy captures the curvature of space–time as a measure of gravity. Specifically, $\mathcal{J}_v$ generates a chiral cosmic string defect in space–time whose deficit angle is estimated by $\mathcal{W}_s$, as can be directly verified within an Anti-de-Sitter-Schwarzschild surgical background, similar to the one described in Section 2.9 [15]. Thus, the topological linking of the line and the one-form symmetry operators ($\mathcal{W}_s$ and $\mathcal{J}_v$, respectively) corresponds to the measurement of a quantized conical deficit angle by the spin holonomy.

The novelty in this AI approach is that the autoencoder treats the onset of gravity at criticality in gauge-theoretic terms *albeit* not within the usual meaning of gauge symmetry adopted in the SM. In the standard picture, gravity is a theory of diffeomorphisms realized by a self-interacting massless spin-two field. In an abuse of language, this symmetry is often labelled "gauge symmetry", but it truly is not. Gravity may be alternatively described by a gauge theory of local Lorentz transformations for the tetrad and spin connection [15]. These equivalent descriptions are appropriated by the autoencoder to internalize a generalized symmetry for gravity at the phase transition corresponding to the critical Planck temperature. The "tetradic gravity" exhibits the one-form symmetry introduced by $\mathcal{J}_v$ and thereby is charged by $Z(SL(2,\mathbb{C})) \approx \mathbb{Z}_2$. The one-form symmetry operator $\mathcal{J}_v$ acts on $\mathcal{W}_s$, which is essentially the spin connection computed in the spin representation s along a given contour [15]. More specifically, $\mathcal{J}_v$ creates a chiral defect in space–time that serves as gravitational analog of the Dirac string (Chapter 3) and is directly gauged by the spin holonomy $\mathcal{W}_s$. In essence, the symmetry operator inserts the point $s(v)$ into the trace of the Wilson loop associated with $\mathcal{W}_s$.

In accordance with the tenets of quantum gravity, a global one-form symmetry can be implemented as a field transformation that takes the form of a local Lorentz transformation determined by the center of the Lorentz group. The object charged under this symmetry is the spin holonomy $\mathcal{W}_s(\gamma)$ for the irreducible representation s defined over a one-dimensional contour γ by

$$\mathcal{W}_s(\gamma) = tr_S P_\gamma \exp\left(\oint_\gamma A \cdot ds\right), \tag{2.51}$$

where P_γ is the path-ordering operator associated with the spin connection and defined over the tangent bundle for the space–time Riemannian manifold. The parallel transport map is associated with A, the one-form EM gauge connection defined by the vector potential: $A^a = A^a_\mu dx^\mu$. Thus, the curvature of space–time is factored into the parallel transport map for the gauge connection and as such becomes the determinant of gravity which is identified with the charged holonomy. The relativistic duality gravity curvature is thus enshrined in the extended object charged under the gauged Lorentz group.

On the other hand, the one-form symmetry transformation is implemented by the "twisted Lorentz transformation" $\mathcal{J}_v(\mathcal{S})$ supported by the codimension-two surface $\mathcal{S} = \partial\mathfrak{b}$, where the coboundary $\mathfrak{b}$ determines the homotopy for shrinking $\mathcal{S}$ down to a point on γ (Figure 2.20). In this way, the operator $\mathcal{J}_v(\mathcal{S})$ may be regarded as the inducer of a chiral defect carrying the gravitational charge.

In consonance with the three fundamental forces of a quantum nature, we have shown that gravity emerges from a phase transition concurrently with the gauging of a generalized $SL(2,\mathbb{C})$ symmetry of ur-matter, as the primeval ur-black hole gets activated below Planck's temperature.

2.14 SYMMETRY OF THE UNIVERSE 10^{-43} S AFTER THE BIG BANG AND THE ONSET OF QUANTUM GRAVITY

An AI system in the guise of an autoencoder of the dark dimension has cast the origin of the visible universe as made up of three phase transitions corresponding to the progressive unraveling of the four fundamental forces (Figures 2.19 and 2.20). Thus, the earliest universe became endowed with gravity, once the Lorentz $SL(2,\mathbb{C})$ symmetry component of the primeval $\mathfrak{G}$ symmetry of $DM_{3,1}$ becomes gauged at a phase transition at the Planck temperature into the special one-form $SL(2,\mathbb{C})$ symmetry defined in the previous section. This special gauging implies that the 4D space–time is assumed to be tractable as a Riemann manifold with a well-defined curvature (cf. Eq. 2.51), and hence, it must occur concomitantly with the activation of the primeval black hole at the Planck temperature T_P, that is, at time $t = \hbar/(k_\mathrm{B}T_\mathrm{P}) \sim 10^{-43}\,\mathrm{s}$ after the big bang. The other two phase transitions follow a similar pattern: gauging of specific components of the inherent symmetry of ur-matter $DM_{3,1}$ with the concurrent activation of the respective particle portals (Figures 2.16 and 2.20).

Thus, the autoencoder casts the grand unification theory (GUT) in terms of two phase transitions of ur-matter as described in Figure 2.20. One such transition gauges the $SU(3)$ symmetry and the other gauges the $SU(2)\times U(1)$ symmetry of the (3,1) shade of dark matter (ur-matter), described in Figure 2.18 and Figures 2.7, 2.8, respectively. However, as the autoencoder was trained with the Lie-group symmetries under which the fundamental bosons of the SM are charged, it became apparent that the incorporation of the relativistic symmetry group $SL(2,\mathbb{C})$ introduced redundancies when

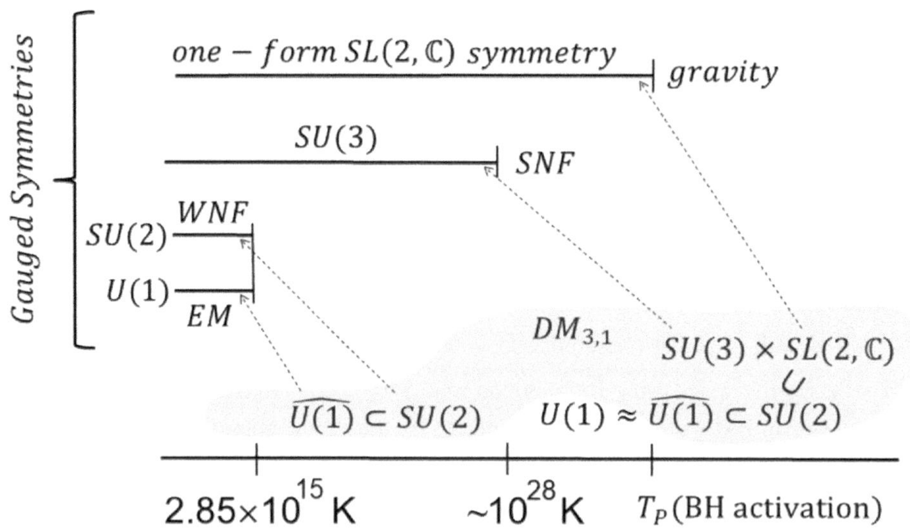

FIGURE 2.21 Gauging of ur-matter symmetries at the three phase transitions that span the origin of the visible universe in an irreducible version of the primeval symmetry of ur-matter at $t = 10^{-43}$ s after the big bang.

inferring the primeval symmetry group $\mathfrak{G}$ for the quantum vacuum at the onset of quantum gravity that takes place as a phase transition at the criticality defined by the Planck temperature.

It is obvious that $SU(2)$, the group of unitary complex 2×2 matrices with determinant 1 is a subgroup of the $SL(2,\mathbb{C})$, the group of complex 2×2 matrices with determinant 1. On the other hand, $U(1)$ is essentially contained in $SU(2)$ (and thereby in the Lorentz group) via the isomorphism:

$$U(1) \approx \widehat{U(1)} \subset SU(2) \tag{2.52}$$

given explicitly as

$$U(1) \ni e^{i\Theta} \rightarrow \begin{pmatrix} e^{i\Theta} & 0 \\ 0 & e^{-i\Theta} \end{pmatrix} \in \widehat{U(1)}, \Theta \in [0,\ 2\pi] \tag{2.53}$$

So the ur-matter symmetries that need to be gauged to trigger the onset of the electroweak unification are subsumed as subgroups of the relativistic Lorentz group (Figures 2.21 and 2.22):

$$U(1) \approx \widehat{U(1)} \subset SU(2) \subset SL(2,\mathbb{C}) \tag{2.54}$$

Once space–time turns into a Riemannian manifold and curvature becomes a feature of space–time, we may assert that gravity emerges charged under the one-form $SL(2,\mathbb{C})$ symmetry (Eq. 2.51) gauged at the phase transition at Planck's temperature. In accordance with the emergence of the other three forces of quantum origin, gravity is charged under a gauged symmetry component of the ur-matter $\mathfrak{G}$-symmetry. The key problem is to identify the *simplest* primeval symmetry of the universe at the time of the earliest phase transition that yielded gravity. Since AI treats the precursor symmetry groups at the onset of the electroweak force as subgroups of the Lorentz group (Eqs. 2.53

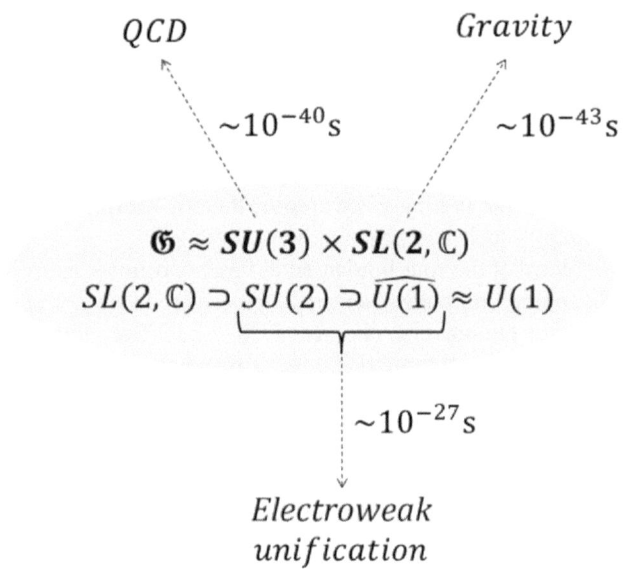

FIGURE 2.22 Primeval symmetry of the ur-matter endowed quantum vacuum giving rise to the fundamental forces through one-form (gravity) or zero-form (QCD, EW) gauging at critical points in the development of the universe (Figures 2.19 and 2.20).

and 2.54; Figures 2.21 and 2.22), we may assert that the irreducible primeval symmetry group of the universe is

$$\mathfrak{G} \approx SU(3) \times SL(2, \mathbb{C}) \tag{2.55}$$

Just like the electroweak synthesis is implied by the fact that QED group $U(1)$ is a subgroup of $SU(2)$ under the isomorphism given by Eqs. 2.52 and 2.53, a synthesis encompassing gravity and the electroweak unification is readily implied by Eq. 2.54. These inferences are of course made within the frame of the autoencoder that portrays the onset of the fundamental forces as phase transitions dictated by the gauging of the symmetries of ur-particles that serve as precursors to the force carriers.

The synthesis implied by Eq. 2.54 suggests the AI answer to the quest for quantum gravity, as the electroweak field is charged under a zero-form gauging of subgroups of the relativistic symmetry group of ur-matter, while a one-form gauging of the full relativistic group begets gravity concurrently with the activation of the primeval black hole at $T \lesssim T_{\mathrm{P}}$.

2.15 AI'S QUEST FOR QUANTUM GRAVITY AND AN ULTRA-UNIFICATION SUBSTANTIATING THE MASS–GRAVITY DUALITY

This chapter has covered vast territory in an effort to describe how an AI system known as autoencoder distills quantum reality from relativistic symmetries susceptible of being gauged at particular junctures that constitute veritable phase transitions. In this way, AI approaches the core problems in particle cosmology in a quest to gain perspectives on the origin of the universe. If not complete solutions, this chapter provides meaningful and original insights on the entire plexus of interrelated problems and as such constitutes a significant contribution to the field of particle cosmology.

Of all the problems that impinge on the conundrum of the origin of the universe, none is more relevant than quantum gravity, the much needed description of the earliest universe defined by Planck scales. In so far as geometry can be assumed at criticality, so that space–time may be associated with a manifold with definable local curvature and the relativistic symmetry group becomes operational, AI informs that the problem of quantum gravity is amenable to a solution. This solution implies that gravity may be quantized if the relativistic symmetry group, the Lorentz group, can be gauged by the earliest phase transition after the big bang. Of course, there is no standard gauge theory for gravity that would summarily qualify as leading to quantum gravity.

The AI model postulates that the four fundamental forces originate from ur-matter symmetries of the primeval quantum vacuum internalized as gauge generalized symmetries at the phase transitions encompassed by the birth of the universe (Figures 2.20–2.22). The gauging of the relativistic symmetry internalizes it not as local symmetry but as a one-form generalized symmetry, where the purported curvature of the Riemannian manifold is factored into the extended object charged under the gravity-associated symmetry (Eq. 2.51). This object is the holonomy of a spin connection, only meaningful if geometry is upheld.

Intriguingly, the primeval symmetry of space–time can be simplified because the symmetry groups associated with the electroweak unification are isomorphic to subgroups of the relativistic Lorentz group (Eqs. 2.53 and 2.54). This observation prompts us to postulate that gravity may be treated in a quantized ultra-unification since its ur-symmetry is inclusive of that of the weak force and electromagnetism. In other words, the Lorentz ur-symmetry of gravity subsumes the ur-symmetry of the lightest particle in $DM_{3,1}$, the shade of ur-matter whose scalar field mixes and thereby activates the Higgs field by endowing it with the true vacuum (Eq. 2.15).

The symmetry relatedness, linking gravity and the Higgs-induced mass with retention of electroweak symmetry, paves the way for a quantum gravity ultra-unification under a gauge generalized symmetry. It would be highly commendable if future research efforts in particle cosmology took up this challenge since this ultra-unification would substantiate the gravity–mass duality of general relativity. Furthermore, the ultra-unification suggests that the fundamental forces in nature stemmed from one primeval force present at the Planck epoch. This force got diluted differently into the geometry of the universe, following the gauge symmetries that charged the fundamental force carriers.

2.16 EVOLVING DARK DIMENSION AND THE ORIGIN OF THE PRIMEVAL FORCE

The inflation along the dark dimension may be regarded as a proxy for the universe evolution in so far as the radius of the circular extra dimension determines the masses of the ur-bosons whose symmetries become gauged at the phase transitions. This fact prompts the autoencoder to introduce the possibility that the ur-bosons that get gauged to yield the quantum force carriers are actually relics of the holonomic boson, the gravity carrier that measures space–time curvature through the parallel transport map associated with the spin connection. The generic contour in this primeval boson projects onto the dark dimension with radius $r_0 = \sqrt{\hbar G/c^3}$, which is precisely the curvature radius of the space–time computed by the holonomy defined by the contour. In this sense, we may state that *gravity is stored in the dark dimension* and its geometric dilution into the visible dimensions may explain its incommensurably low coupling strength relative to the other forces (Figure 2.17).

This observation bears a significant analogy with the storage of the true vacuum for the Higgs field in the dark dimension. In this case, the ur-Higgs boson bestows the true vacuum on the Higgs field as a result of the symmetry compatibility between the (3,1) dark matter and the Higgs field (Eq. 2.15). The Higgs boson with its geometric dilution of the true vacuum is therefore treated by the autoencoder as a relic of the cosmic string contour that defines the primeval holonomic boson or carrier of the gravity stored in the dark dimension (Figure 2.23). This morphing of one boson into the other is reinforced by two facts: (a) The internal symmetry of the Higgs field is subsumed by the

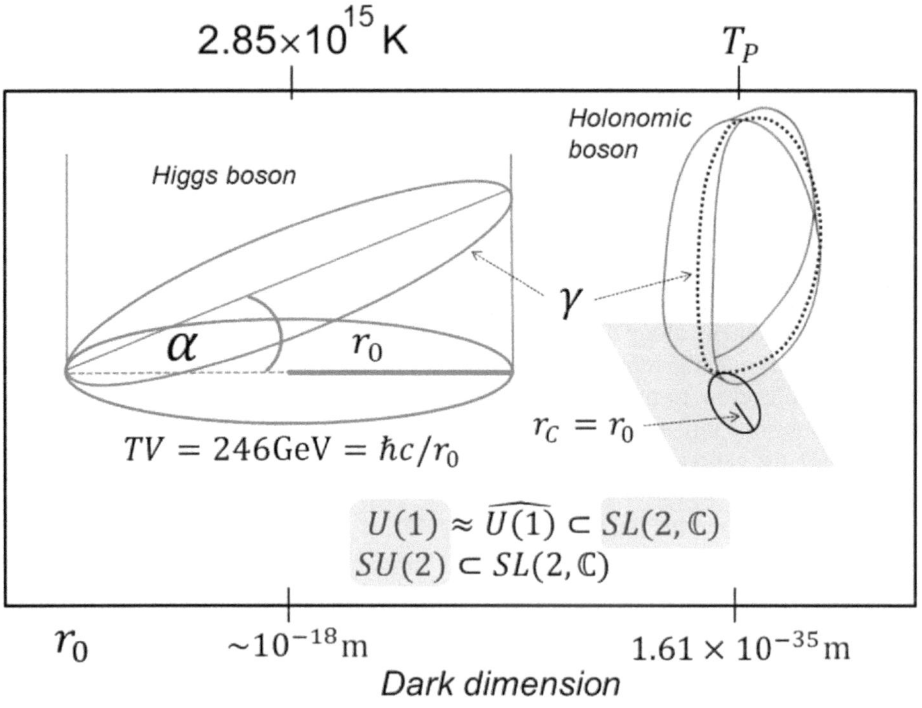

FIGURE 2.23 Morphing of the holonomic boson into the Higgs boson realizing the mass-gravity duality as the universe transitions from the Planck epoch to the electroweak epoch concurrently with the inflation in the dark dimension and inclusion of the respective gauge symmetries.

generalized symmetry of the holonomic boson and (b) the mass-gravity duality of general relativity naturally links the Higgs boson as endower of mass and the holonomic boson as carrier of gravity.

In the autoencoder approach, all four forces are manifestations of a primeval force under different symmetries, and therefore, all bosons must morph into the holonomic boson as the dark dimension functions as proxy for the universe evolution.

2.17 LARGE LANGUAGE MODEL FOR THE WARPED GRAVITON

2.17.1 Detectable Warped Gravitons in an Ultraviolet Completion of Gravity

A quest for the graviton as an excitation of a quantized gravitational field requires a gauge representation of the Lorentz transformations, reconciling the curved manifold of general relativity with the Euclidean support of its purported quantum wavefunction. A way forward builds upon the tetrad frame fields introduced to construct a gauge theory of gravity [16–18]. In the weak field limit (WFL), a perturbative graviton with Minkowski background yields a massless spin-2 particle with $U(1)$ gauge symmetry, consistent with the infinite range of gravity communicated at the speed of light. This section addresses the uncharted strong field limit (SFL), obviously not amenable to a perturbative treatment and requiring generalized gauge representations to account for massive gravitons.

Modeling efforts geared at describing massive Kaluza–Klein (KK) gravitons, purportedly produced in the LHC or other colliders, required a framework where standard model (SM) fields propagate within a warped extra dimension enabling the geometric dilution of gravity [19]. Although much of the experimental parameter space remains unexplored, such efforts have so far proven unyielding to experimental validation [20,21]. Thus, massive warped gravitons have eluded detection in instruments such as the ATLAS detector at the CERN Large Hadron Collider (LHC) [20,21]. Specifically, a search for neutral heavy spin-2 resonances in the 300 GeV to 4 TeV range with dominant $W+W-$ or ZZ decay channels and generated via gluon fusion or quark–antiquark annihilation has not proven successful under current constraints at 139 fb^{-1} integrated luminosity and 13 TeV center-of-mass hadronic energy ($\sqrt{S}$) [21,22].

While the WFL appears tractable by regarding the graviton tensor field as a perturbation of the Minkowski tensor, the SFL is uncharted territory for gauge theories of gravity. This section deals precisely with this regime using a large language models (LLMs) within a functional language program (Lean, Appendix). This section is inspired by prospects to identify dark matter (DM) as an excitation of a KK graviton tower decaying via gravitational interaction into lighter particles with wavelength inhomogeneity along the dark dimension [23,24]. These ideas are grounded in available experimental evidence on the relics of cosmological evolution. Furthermore, this section shows that self-interaction is inherent to the graviton, leading to the conclusion that at least a portion of DM in the Universe must be self-interacting. Such self-interactions would allow energy and momentum transport through the galaxy halos. Thus, this section may provide a new vantage point to interpret the diversity of galactic dynamic patterns and other astrophysical phenomenology that cannot be reconciled with the current picture of collisionless DM.

The section is geared at predicting warped gravitons in an ultraviolet completion of gravity amenable of detection in a collider. This requires an LLM for a fiber-bundle formulation of a gauge theory pivoting on the spin connection that determines the tangent bundle of a space-time endowed with an extra warped dimension [25,26] (Appendix). In this way, this section provides a predictive differential geometry model of the graviton as transducer of space–time curvature into energy stored in a warped dimension. The curvature-energy transduction becomes operative as the boson begets curvature in the dualistic relativistic framework, implying that the stress-energy tensor must be associated with a scalar describing self-interacting DM within a ϕ^4 - model.

In the WFL, the graviton is expected to be massless since gravity presumably propagates at the speed of light and has infinite range. On the other hand, in the SFL, the compatible model predicts massive warped gravitons that bear on the collider setting, where the curvature associated with the emission cross section gets transduced into warped rest–mass energy that conveys gravity. Thus, the theory yields testable predictions for the LHC [20-22] or future colliders. Furthermore, it shows a compatibility between the WFL and SFL depictions, as shown subsequently.

A warped spatial dimension needs to be incorporated to address the problematic geometric dilution of mass and gravity [19,26]. This dilution provides an explanation of the perplexingly small coupling strength of gravity relative to the other fundamental forces of established quantum origin. Thus, an extra circular dimension is incorporated with radius $r_0 = 0.802695 \times 10^{-18}$ m within the range of the smallest material scale, the effective quark diameter established from parton models of inelastic $e^{\pm}p$ scattering [27]. Significantly, the minimum rest–mass energy stored in this warped dimension is $E_0 = k_0 c = \hbar c / r_0 = 246$ GeV, the true vacuum of the Higgs field [28]. This choice of the warped dimension becomes crucial for our intended purpose. Thus, the stress-energy tensor in a fiber-bundle gauge theory of the graviton will be shown to introduce a mass-endowing mechanism via the curvature-mass relativistic duality modeled by a self-coupling potential energy term. This ϕ^4 -model yields massive gravitons vis-à-vis the vacuum expectation value (v.e.v.) of a Wilson loop representation in the SFL.

At a variance with alternative approaches [19,23], in our model, the matter wave quantized within the warped dimension admits geometric dilution into visible dimensions. This dilution is parametrized by a pitch angle α (Figure 2.24), so that the angled cylindrical cross section is elliptical, with major axis $r_0/\cos\alpha$ and minor axis r_0. The geometric dilution is defined as $\upsilon = -\log\cos\alpha$ [28]. This implies that gravitons become massless at infinite dilution, that is, with $\alpha = \dfrac{\pi}{2}$, thus probing a locally flat space–time. They may also probe any value of space–time curvature since the projection

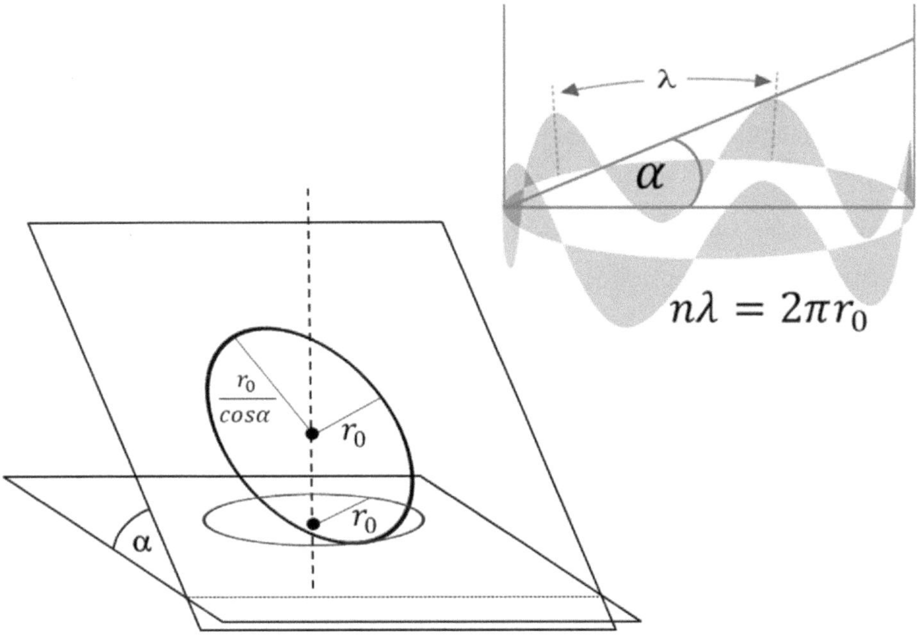

FIGURE 2.24 Geometric dilution of a matter-wave enshrined in the warped dimension. Geometric dilution onto a generic observable dimension is parametrized by a single pitch angle α [28].

onto the warped dimension is quantized by $\lambda = \frac{2\pi r_0}{n}$ (Figure 2.24) while the pitch angle is not subject to quantum constraints.

The geometric dilution parameter υ plays a role akin to the scale and amplitude of the de Broglie wavelength inhomogeneity in the dark dimension in models representing cosmological gravitons as dark matter [23]. Thus, the decay of KK gravitons into lighter gravitons, a hallmark of cosmological evolution in warped reality [23,29], plays out in a similar way in this work, as $\alpha \to \frac{\pi}{2}$ or, equivalently, $\upsilon \to \infty$, reckoning that geometric dilution serves as proxy of time in cosmological evolution [28].

The approach to incorporate an extra spatial dimension differs fundamentally from other procedures that incorporate it within a quotient topology of a circle modulo the transformation $\varphi \to \varphi + \pi$; thus, generating two branes $\varphi = 0, \varphi = \pi$ identified with manifold boundaries [19]. This poses an ontological problem since there is no obvious meaning to "boundary" as there is no geometry to describe "outside": The signifier "$\varphi > \pi$" becomes "inside"; that is, it is included in the quotient space that constitutes reality [28]. In any case, the KK gravitons predicted by such an approach to exist in the bulk within the branes [29] remain unyielding to experimental validation [20-22].

2.17.2 Fiber-Bundle Formulation of a Gauge Theory of Gravity within a Functional Language Program

To account for quantum gravity in the SFL, a fiber-bundle formulation of its gauge theory is implemented within a functional language program operating a LLM (Appendix). Thus, spinors are incorporated into the general relativity framework provided by the space–time curved manifold

$\mathcal{M}$. This procedure is enabled by first tending the spin connection across the tangent bundle $T\mathcal{M}$. This connection is implemented via a frame field required to piece together the atlas $\{T_x\mathcal{M}\}_{x\in\mathcal{M}}$ of local tangent spaces and relates to the geometry of the curved manifold in such a way that a tangent space Lorentz transformation translates onto a gauge transformation of the spinor field. To enable interconversion between the metric tensor $g_{\mu\nu}$ of $\mathcal{M}$ and the flat (Minkowski) metric tensor η_{ab} of $T\mathcal{M}$, a pentad frame field $\{e_\mu^a\}$, (Greek indices for curved space tensor, Latin isospin indices for flat tangent space Lorentz indices) is introduced as uniquely defined by the relations:

$$g_{\mu\nu} = e_\mu^a e_\nu^b \eta_{ab},\ e_a^\mu e_b^\nu g_{\mu\nu} = \eta_{ab},\ e_\mu^a e_b^\mu = \delta_b^a,\ e_\mu^a e_a^\nu = \delta_\mu^\nu,\ e^{\mu a} = g^{\mu\nu} e_\nu^{\ a},\ e_{\mu a} = e_\mu^{\ b}\eta_{ba} \tag{2.56}$$

Equipped with the pentad frame field relations, the spin connection w_μ^{ab} is defined as follows:

$$w_\mu^{ab} = \frac{1}{2}e^{\nu a}\left(\partial_\mu e_\nu^b - \partial_\nu e_\mu^b\right) - \frac{1}{2}e^{\nu b}\left(\partial_\mu e_\nu^a - \partial_\nu e_\mu^a\right) - \frac{1}{2}e^{\lambda a}e^{\sigma b}\left(\partial_\lambda e_{\sigma c} - \partial_\sigma e_{\lambda c}\right)e_\mu^c. \tag{2.57}$$

The spin connection enables the implementation of the covariant derivative by introducing the gamma matrices $\{\breve{\gamma}^a\}$ that fulfill the Dirac–Clifford algebra with anticommutator relations: $\{\breve{\gamma}^a, \breve{\gamma}^b\} = 2\eta^{ab}$, with the extended 5×5 matrices defined as $\breve{\gamma}^a = \begin{pmatrix} \gamma^a & \mathbf{0}^T \\ \mathbf{0} & 1 \end{pmatrix}$:

$$\nabla_\mu \psi = \left(\partial_\mu - \frac{i}{8}w_\mu^{ab}\sigma_{ab}\right)\psi,\ \sigma^{ab} = \frac{i}{2}[\breve{\gamma}^a, \breve{\gamma}^b] \tag{2.58}$$

Reciprocally, the curved Dirac–Clifford algebra $\{\breve{\gamma}^\mu = e_a^\mu \breve{\gamma}^a\}$ is defined by the anticommutators $\{\breve{\gamma}^\mu, \breve{\gamma}^\nu\} = 2g^{\mu\nu}$ with generators $\frac{i}{2}[\breve{\gamma}^\mu, \breve{\gamma}^\nu]$. On vectors, the covariant derivative yields

$$\nabla_\mu V_a = \partial_\mu V_a + w_{\mu a}^{\ \ \ b} V_b = \partial_\mu V_a + w_\mu^{\ cb}\eta_{ca}V_b \tag{2.59}$$

Hence, the curvature $\Omega_{\mu\nu}^{ab}$ associated with the spin connection is defined via the commutator $[\nabla_\mu, \nabla_\nu]$:

$$\Omega_{\mu\nu}^{\ \ \ ab} = \Omega_{\mu\nu c}^{\ \ \ \ b}\eta^{ca};\ \Omega_{\mu\nu a}^{\ \ \ \ b}A_b = [\nabla_\mu, \nabla_\nu]A_a, \tag{2.60}$$

which enables us to relate the spin curvature to the Riemannian curvature tensor as: $R_{\nu\mu\rho}^{\ \ \ \ \lambda}A_\lambda = e_\rho^a \Omega_{\mu\nu a}^{\ \ \ \ b} e_b^\lambda$. This relation holds from the relations

$$R_{\nu\mu\rho}^{\ \ \ \ \lambda}A_\lambda = [\nabla_\mu, \nabla_\nu]A_\rho = [\nabla_\mu, \nabla_\nu]e_\rho^a A_a = e_\rho^a \Omega_{\mu\nu a}^{\ \ \ \ b}A_b = e_\rho^a \Omega_{\mu\nu a}^{\ \ \ \ b} e_b^\lambda A_\lambda \tag{2.61}$$

Hence, denoting $e = \det\left(e^{\mu}_{a}\right)$, with $k = 8\pi G c^{-4}$ (Einstein's gravitational constant), the Lagrangian for fermions coupled with gravity on $\mathcal{M}$ reads in natural units:

$$\mathcal{L} = \frac{e}{2k}\Omega_{\mu\nu}{}^{ab} e^{\mu}_{a} e^{\nu}_{b} + e\overline{\psi}\left(i\breve{\gamma}^{\mu}\nabla_{\mu} - k_{0}\cos\alpha\right)\psi \tag{2.62}$$

A fiber-bundle formulation of the gauge theory of gravity in the SFL will be now formalized within an AI system implemented upon the pentadic framework defined by Eqs. 2.56–2.62. The intent is to model the warped graviton as conveyor of the gravitational force within the inferential framework generated by relations 2.56–2.62. In the SFL, this graviton behaves as an holonomic transducer of space–time curvature [30] and is charged under a generalized gauge Lorentz symmetry [31-34].

2.17.3 Autoencoder LLM for the Graviton within a Fiber-Bundle Formulation of Quantum Gravity

An AI-empowered inferential scheme to model the graviton within a fiber-bundle gauge formulation requires a differential geometry formalization supporting two interplaying components: (a) a theorem/proposition prover such as Lean 4 [35,36] with an intensional logic under the guise of a dependent type theory [37] spanned by the differential geometry propositional kernel [38] determined by Eqs. 2.56–2.62 (Appendix), and (b) a generalized autoencoder LLM [28,39] serving as the supporting inferential framework within which the formal prover operates by maintaining compatibility between general relativity and its gauged version on the tangent bundle of space–time, as described by the commutative scheme given in Eq. 2.63.

$$\begin{array}{ccccc}
 & & \text{boson emission} & & \\
\mathcal{M} & \xrightarrow[\{e^{\mu}_{a}\}]{\mathcal{E}} & T\mathcal{M} & \xrightarrow{\pi} & T\mathcal{M}/\sim \\
\mathcal{F}\Big\downarrow & & w^{ab}_{\mu}\ \Big\downarrow \mathcal{W}\ \ \sigma_{ab} & & \Big\downarrow \mathcal{K}_{SM} \\
\mathcal{M} & \xrightarrow[\mathcal{E}]{\{e^{\mu}_{b}\}} & T\mathcal{M} & \xrightarrow[\pi]{} & T\mathcal{M}/\sim \\
 & & \text{boson decay} & &
\end{array} \tag{2.63}$$

Equations of motion (EOM) defined by the maps $\mathcal{F}$, $\mathcal{W}$, $\mathcal{K}_{SM}$ are mutually compatible and represent the process [boson emission $\rightarrow$ boson decay] with the graviton cast as a Wilson loop [40] in three different and interrelating settings, namely, $\mathcal{M}$, $T\mathcal{M}$, $T\mathcal{M}/\sim$, with the coarser level $T\mathcal{M}/\sim$ supporting the SM. These settings indicate respectively the 5D space–time manifold, its tangent bundle, and the quotient space of the tangent bundle modulo the equivalence relation "$\sim$". Under this equivalence, points are identified when they possess identical coordinate representation in the tangent bundle for the underlying standard 4D space–time manifold. The maps $\mathcal{E}$, π, enabling transitions between the three different representations in the LLM are, respectively, the pentadic frame field map or "encoder" and the canonical projection or "decoder" [28,39]. The latter associates each point in the 5D tangent bundle with its equivalence class modulo "$\sim$". The map $\mathcal{W}: T\mathcal{M} \to T\mathcal{M}$, representing the transition between boson emission (i.e., within the birth cross section of gluon fusion) and boson decay (i.e., a decay ZZ channel) [20-22]. The map is determined by the spin connection w_μ^{ab} tending the atlas $\{T_x\mathcal{M}\}_{x\in\mathcal{M}}$, and enabling the charge of the boson under a generalized symmetry [31-34]. The compatibility of the three representations is then ensured by the commutativity of the diagram in Eq. 2.63 (Appendix), whereby any two paths designated by consecutive arrows with identical origin and endpoint yield the same result:

$$\mathcal{W} \circ \mathcal{E} = \mathcal{E} \circ \mathcal{F},\ \mathcal{K}_{\text{SM}} \circ \pi = \pi \circ \mathcal{W} \tag{2.64}$$

These two commutative relations defining the autoencoder LLM must be invariably fulfilled for compatibility, thereby becoming constraints in all formal propositional inferences, ultimately enabling the computation of the warped graviton mass in the SFL. This compatibility implies that the relations in Eq. 2.64 generate all other commutative relations in the diagram in Eq. 2.63, such as $\mathcal{K}_{\text{SM}} \circ \pi \circ \mathcal{E} = \pi \circ \mathcal{E} \circ \mathcal{F}$, signaling the constraining equivalence between the gauge theory of general relativity and its fiber-bundle formulation within the LLM.

The equations defining the gauge theory of gravity are automatically extrapolated by the LLM to hold even near a black hole in strong gravity, as the tangent-bundle connections are implemented in the formal autoencoder enshrined in the LLM (Appendix).

2.17.4 Encoding the Warped Graviton in the Weak and Strong Field Limits

Within the commutative scheme (Eq. 2.63), compatibility of relativistic and fiber-bundle descriptions implies that the Einstein–Hilbert action

$$S_{E-H} = \frac{1}{2\kappa}\int \sqrt{-g}R_{\mu\nu\rho\sigma}g^{\mu\rho}g^{\nu\sigma}d^5x = \frac{1}{2\kappa}\int \sqrt{-g}Rd^5x, \tag{2.65}$$

obtained by fully contracting the Riemann curvature tensor $R_{\mu\nu\rho\sigma}$, is identified by the autoencoder LLM [28,39] as a pentadic action obtained by fully contracting the spin curvature via the shift tensor identified as the pentad:

$$S = \frac{1}{2\kappa}\int e\Omega_{\mu\nu}{}^{ab}e_a^\mu e_b^\nu d^5x = \frac{1}{2\kappa}\int e\Omega d^5x \tag{2.66}$$

The equivalence of both actions within the commutative scheme in Eq. 2.63 follows readily from the relations:

$$g^{\xi\lambda}R_{\nu\xi\mu\rho}A_\lambda = R_{\nu\ \mu\rho}^{\ \lambda}A_\lambda = e_\rho^a\Omega_{\mu\nu a}{}^bA_b = e_\rho^a\Omega_{\mu\nu a}{}^be_b^\lambda A_\lambda = e_\rho^a\Omega_{\mu\nu}{}^{cb}\eta_{ca}e_b^\lambda A_\lambda \tag{2.67}$$

Thus, the Riemann and Ricci curvature tensors may be given in terms of the spin connection curvature:

$$R_{\nu\xi\mu\rho} = g_{\xi\lambda}e_\rho^a\Omega_{\mu\nu}{}^{cb}\eta_{ca}e_b^\lambda,\ R_{\mu\nu} = \frac{e}{\sqrt{-g}}\Omega_{\mu\nu}{}^{ab}\eta_{ab}, \tag{2.68}$$

thereby asserting the equivalence of Eqs. 2.65 and 2.66 within the autoencoder defined by Eq. 2.63. To obtain a warped graviton storing the quantized projection of its rest mass in the warped dimension of radius $r_0 = \hbar c(246\text{ GeV})^{-1}$, it becomes essential to generate a pentadic relativistic equation where the stress-energy tensor is associated with a scalar potential $V(\phi)$ [40]. This potential yields the vacuum expectation value (v.e.v.) for the graviton field and endows the geometrically diluted warped graviton with mass. Specifically, the potential is generated by the autoencoder [28,39] as associated with a Wilson loop [40] intended to represent the graviton when constrained by the shift tensor equivalence between general relativity in the curved manifold $\mathcal{M}$ and "relativity with spinors" on $T\mathcal{M}$. Thus, we define the curvature tensor $\widehat{\Omega}_{\mu\nu}^{ab}$ and its curvature scalar $\widehat{\Omega} = \widehat{\Omega}_{\mu\nu}^{ab} e_a^{\mu} e_b^{\nu}$ by the relativistic relations:

$$
\begin{aligned}
\widehat{\Omega}_{\mu\nu}{}^{ab}\eta_{ab} - \frac{1}{2}\widehat{\Omega} g_{\mu\nu} &= \frac{\kappa\sqrt{-g}}{2e} T_{\mu\nu}(\phi) \\
&= \frac{\kappa\sqrt{-g}}{2e}\left[\left(g^{\rho\alpha}g^{\sigma\beta} + g^{\rho\beta}g^{\sigma\alpha} - g^{\rho\sigma}g^{\alpha\beta}\right)\partial_\alpha\bar{\phi}\partial_\beta\phi - g^{\rho\sigma}V(\phi)\right]g_{\rho\mu}g_{\sigma\nu} \\
&= \frac{\kappa\sqrt{-g}}{e}\left\{\partial_\mu\bar{\phi}\partial_\nu\phi - g_{\mu\nu}\left[\frac{1}{2}g^{\alpha\beta}\partial_\alpha\bar{\phi}\partial_\beta\phi + V(\phi)\right]\right\},
\end{aligned}
\tag{2.69}
$$

$$
V(\phi) = -\frac{1}{2}m_\phi^2\bar{\phi}\phi + \frac{\lambda}{4}(\bar{\phi}\phi)^2 \tag{2.70}
$$

The potential given by Eq. 2.70 endows the warped graviton with mass in the SFL. This is so because, as shown below, the graviton field is represented by a Wilson loop $\mathcal{W} = \mathcal{Y}^{\mu\nu}g_{\mu\nu}$ with vacuum expectation value:

$$
\langle\mathcal{W}\rangle = \frac{m_\phi}{\sqrt{\lambda}} \tag{2.71}
$$

In the WFL, the warped graviton tensor field $h_{\mu\nu}$ is the one assigned to a massless spin-2 particle with $U(1)$ gauge symmetry ($h_{\mu\nu} \rightarrow h_{\mu\nu} + \partial_\mu a_\nu + \partial_\nu a_\mu$). This graviton is treated perturbatively on a Minkowski background, so that $g_{\mu\nu} = \eta_{\mu\nu} + h_{\mu\nu}$. In the WFL, the graviton Lagrangian becomes

$$
\mathcal{L}_{WF} = \epsilon^{\mu\nu\rho\sigma}\left(\epsilon^{\alpha\beta\gamma}{}_{\sigma}\right)\partial_\mu h_{\nu\beta}\partial_\alpha h_{\rho\gamma}, \tag{2.72}
$$

where the ϵ's are Levi–Civita's permutation symbols. The nonlinear completion of its action $S = \int\mathcal{L}_{WF}d^5x$ becomes $S = \frac{1}{2\kappa}\int\sqrt{-g}Rd^5x$, or, equivalently, $S = \frac{1}{2\kappa}\int e\Omega_{\mu\nu}{}^{ab}e_a^{\mu}e_b^{\nu}d^5x$. This last assertion provides the clues to construct the ultraviolet (SFL) completion of the warped graviton, a task now undertaken by running the LLM within the Lean environment (Appendix).

Massive spin-2 particles are known to arise from 5D theoretical models that address both the hierarchy problem and the geometric dilution of gravity [19,29]. Such models predict the LHC production of such particles mainly relying on a gluon fusion mechanism and the particles are in principle detectable within parameter ranges so far not probed at the LHC or other colliders [20,21]. As shown subsequently, the scenario proposed in this letter for graviton creation and the tuning of the emission/detection parameters differ fundamentally from those stemming from the exotic theories (i.e., outside the SM) that have been tested by the ATLAS and CMS experiments based on 139 fb^{-1} (ATLAS) and 137 fb^{-1} (CMS) integrated luminosities for proton-proton collisions at $\sqrt{s}$ = 13 TeV collected during LHC Run-2. No evidence for the graviton signal was found in such experiments [22]. Nevertheless, much of the parameter space remains to be probed, so the search for exotic resonances at the LHC is likely to continue. The results that follow invite such an endeavor, requiring the exploration of other parameter regions.

To identify the warped graviton tensor field $\mathcal{Y}_{\mu\nu}$ in the strong field limit, the LLM represents the curvature with Clifford algebra generators that transform under the spinor winding "electromagnetic" $U(1)$-symmetry elements $U_\xi, \xi \in [0, 2\pi]$, just like the terms $w_\mu^{ab}\sigma_{ab}$ in Eq. 2.59, transform in the manner of the electromagnetic vector potential A_μ. The warped graviton field is implemented in such a way that $\mathcal{Y}_{\mu\nu}$ may be charged under a symmetry transformation $U_\zeta, \zeta = \pi$, associated with the nontrivial idempotent element $U_\zeta \in Z(SL(2,\mathbb{C})) \approx \mathbb{Z}_2$ in the center of the Lorentz group. This chiral-defect-inducing symmetry effectively mounts the contour of the line operator onto a Moebius strip creating a chiral defect that amounts to a $\pi-$ winding deficit in the spinors attached to space–time (Eq. 2.63), while two such deficits created by the U_ζ iteration $U_\zeta^2 = U_{2\pi}$ amount to the identity acting on the parallel transport map [41]. Thus, the need to represent the nontrivial center element U_ζ arises from the invariance under the chiral winding symmetry transformation:

$$U_\zeta^{-1}\mathcal{Y}_{\mu\nu}U_\zeta = U_\zeta^2\mathcal{Y}_{\mu\nu} = U_{2\pi}\mathcal{Y}_{\mu\nu} = \mathcal{Y}_{\mu\nu}, \tag{2.73}$$

From this argument, it becomes necessary to construct the Wilson loop so that space–time curvature may be holonomically transduced as rest mass as implied by Eq. 2.69. Thus, the following compatibility relation is established, implying that the massive warped graviton indeed belongs to the ultraviolet completion of gravity, as verified by Lean (Appendix):

$$\lim_{\varnothing\gamma\to 0}\mathcal{Y}_{\mu\nu} = h_{\mu\nu} \tag{2.74}$$

for $\gamma:[0,1]\to\mathcal{M}$, a generic closed map upon which the holonomy is defined. Eq. 2.74 is valid because the following three actions are equal:

$$\frac{1}{2\kappa}\int e\Omega\, d^5x = \frac{1}{2\kappa}\int\sqrt{-g}R\,d^5x = -\frac{1}{3}\int\sqrt{-g}T(\phi)\,d^5x \tag{2.75}$$

The r.h.s. equality in Eq. 2.75 follows from the contraction of Einstein's equation of general relativity ($D = 5$) yielding

$$-\frac{3}{2}R = \kappa T(\phi), \left(T(\phi) = T_{\mu\nu}(\phi)g^{\mu\nu}\right). \tag{2.76}$$

The warped graviton tensor field in the SFL is identified by the fiber-bundle autoencoder of the gauge theory as

$$\mathcal{Y}^{\mu\nu} = \mathcal{P}_\gamma \exp\left(\oint_\gamma \hat{w}_\lambda^{\ ab} e_a^\mu e_b^\nu dx^\lambda\right), \tag{2.77}$$

where $\mathcal{P}_\gamma$ is the path-ordering operator associated with the closed map $\gamma:[0,1]\to\mathcal{M}$ and $\hat{w}_\mu^{\ ab}$ is the spin connection associated with the curvature $\widehat{\Omega}^{ab}_{\mu\nu}$. The associated action for the tensor field $\mathcal{Y}_{\mu\nu} = \mathcal{Y}^{\lambda\sigma} g_{\lambda\mu} g_{\sigma\nu}$ in the strong field limit measures the field strength of the spin connection:

$$\int \mathcal{L}_{SF} d^5x = \frac{1}{4\kappa}\int \epsilon_{abcd} e^a \wedge e^b \wedge \left(\hat{w}^k{}_d + \hat{w}^k{}_f \wedge \hat{w}^f{}_d\right) \eta_{kc} d^5x = -\frac{1}{3}\int \sqrt{-g} T(\phi) d^5x, \qquad (2.78)$$

yielding the warped mass-endowing potential (Eqs. 2.70 and 2.71). The self-interaction term is introduced because the curvature–rest–mass–energy transduction materializes on the curvature bestowed on the space–time manifold by m_ϕ via the relativistic duality. Thus, Eqs. 2.69 and 2.70 provide the relativistic framework for the massive warped graviton represented as a ϕ^4 -self-coupled transducer of curvature in the ultraviolet completion of gravity. Furthermore, the $U(1)$ gauge symmetry in the WFL translates into the generalized gauge symmetry concomitantly with the transition to the SFL.

2.17.5 Experimental Validation of the Warped Graviton LLM Prediction for the Strong Field Limit

A warped graviton is expected to be generated and detected at the LHC via the high luminosity gluon fusion (gg) mechanism [20-22]. This is expected since the cross section $\sigma_{gg\to G}$ results from effective g-G-coupling contributions from the graviton KK tower $\left\{m_{G_n} = \frac{n\hbar c}{r_0}\right\}$, with the nth version of an undiluted graviton holonomically transducing curvature with radius $\frac{r_0}{n}$ and energetically storing it in the warped dimension with de Broglie wavelength $\lambda = 2\pi r_0/n$:

$$\sigma_{gg\to G} = \sum_{n=1}^{\infty}\left(\int G^{\mu\nu}\mathcal{Y}^{(n)}_{\mu\nu} d^5x\right) / \sqrt{\left(\int G^{\zeta\vartheta} G_{\zeta\vartheta} \mathcal{Y}^{(n)\varrho\xi} \mathcal{Y}^{(n)}_{\rho\xi} d^5x\right)} = \pi C_{g-G}\sum_{n=1}^{\infty}\left(\frac{r_0}{n}\right)^2$$
$$= C_{g-G}\pi\frac{\pi^2}{6}r_0^2, \qquad (2.79)$$

where $\frac{\pi^3}{6}r_0^2 = 33.89pb$ and $G^{\mu\nu} = \frac{i}{g_s} g^{\mu\alpha} g^{\nu\beta}\left[D_\alpha, D_\beta\right]$ is the gluon field stress tensor with covariant derivatives $D_\mu = \partial_\mu - ig_s\mathcal{A}_\mu$, where $g_S = \sqrt{4\pi\alpha_S(Q)}$ is the strong coupling constant for momentum transfer Q, and $\mathcal{A}_\mu$, the gluon field. This gives

$$C_{g-G} = \left(\frac{\hbar c}{r_0\sqrt{S}}\right)^2 \lim_{\varnothing\gamma\to 2r_0} \mathrm{tr}\mathcal{P}_\gamma \int g^{\mu\alpha} g^{\nu\beta}\left(\partial_\alpha\mathcal{A}_\beta - \partial_\beta\mathcal{A}_\alpha + ig_S\left[\mathcal{A}_\alpha, \mathcal{A}_\beta\right]\right)$$
$$\exp\left(\oint_\gamma \widehat{\Omega}^{ab}_{\mu\nu}\sigma_{ab} ds\right) d^5x, \qquad (2.80)$$

with $\alpha_S = \alpha_S\left(\frac{\hbar c}{r_0}\right) = 0.12 \pm 0.01$ [20] and hadronic center-of-mass energy $\sqrt{S} = 13$ TeV. Expanding the exponential to first-order in-spin connection curvature, we obtain $\sigma_{gg \to G} = 14.92 \pm 0.52$ fb. Since the graviton-line operator is a curvature–energy transducer, this hadronic cross section yields the warped graviton mass 9.050 ± 0.055 TeV. Assuming a graviton ZZ-decay channel, with one Z-boson decaying hadronically and the other, leptonically, we expect to leading order a detectable excess peaking at 9.05 TeV in the m_{ZZ} distribution over the SM background at $\sqrt{S} = 13$ TeV.

2.17.6 The LLM-Inferred Warped Graviton Is Self-Interacting Dark Matter and Can Be Created in a Collider

To identify the warped graviton as a spin holonomy transducer of space–time curvature, a fiber-bundle autoencoder LLM [28,39] of general relativity builds upon a pentadic formalism that casts gravity in gauge theoretic terms (Appendix). The graviton is thus identified in the uncharted SFL by a Wilson loop [40] with a v.e.v. defined by a self-interacting dark matter scalar incorporated in the stress–energy tensor. The warped graviton is thus massive and charged under a symmetry represented by inducing a chiral defect in the guise of a winding deficiency in the spinors attached to the relativistic framework. In the WFL, the graviton becomes massless and treated as a perturbation of the Minkowski background. On the other hand, in the SFL, it may be produced and detected in the laboratory as a warped boson that transduces the space–time curvature at its emission cross section into a geometrically diluted warped dimension that stores rest–mass energy via relativistically operational self-interaction (Eqs. 69 and 70). The curvature-transducing mechanism materializes within a Wilson loop. The generation and detection of the warped graviton invites probing regions in parameter space hitherto unexplored at the LHC or future colliders [2-22].

Despite their potential impact to come to grips with quantum gravity, effective theories of the graviton have not yielded to experimental test so far. This work takes a novel path by describing a gauge theory of nonlocal operators built upon a fiber-bundle framework that enabled quantization of the gravitational field wherein a self-coupling mass-endowing mechanism becomes operational via the stress–energy tensor. In the SFL, the theory predicts warped gravitons that convey gravity as Wilson loops that holonomically transduce space–time curvature, becoming amenable of experimental detection.

Since gravitons are likely DM candidates [23], and self-interactivity is evidently inherent to the graviton in the SFL, we can expect that at least a portion of DM in the universe should be self-interacting. Such self-interactions would allow energy and momentum transport through the galaxy halos, creating structural and dynamical patterns fundamentally different from those produced by collisionless DM. Thus, this work may provide a new vantage point to interpret the diversity of galactic patterns and other astrophysical phenomenology associated with DM.

REFERENCES

1. Weinberg S (2005) *The Quantum Theory of Fields, Volume 1: Foundations*, Cambridge University Press, Cambridge.
2. Higgs PW (1964) Broken symmetries and the masses of gauge bosons. *Phys Rev Lett* 13: 508–509.
3. Perelman G (2002) The Entropy Formula for the Ricci Flow and Its Geometric Applications. arXiv:math.DG/0211159.
4. Fernández A (2024) *Artificial Intelligence Models for the Dark Universe: Forays in Mathematical Cosmology*. CRC Press, Boca Raton, FL.
5. Abramowiczy H, Abtt I, Adamczykh L, Adamusae M, Antonelli S, et al. Zeus Collaboration (2016) Limits on the effective quark radius from inclusive ep scattering at HERA. *Phys Lett B* 757: 468–472.

6. Kutz JN, Stranton SL (2022) *Data-Driven Science and Engineering: Machine Learning, Dynamical Systems, and Control*, 2nd Edition. Cambridge University Press.
7. Randall L (2013) *Higgs Discovery: The Power of Empty Space*. Ecco, HarperCollins Publishers, New York.
8. Weinberg S (2008) *Cosmology*. Oxford University Press.
9. Randall L, Sundrum R (1999) Large mass hierarchy from a small extra dimension. *Phys Rev Lett* 83: 3370–3373.
10. Visser M (1989) Traversable wormholes from surgically modified Schwarzschild spacetimes. *Nucl Phys B* 328: 203–212.
11. He Y-H (2021) *The Calabi–Yau Landscape: From Geometry, to Physics, to Machine Learning* (Lecture Notes in Mathematics). Springer, Berlin.
12. He Y-H (2021) *Machine Learning in Pure Mathematics and Theoretical Physics* (Yan-Hui He, Editor). World Scientific Publishing Co., Singapore.
13. Georgi H, Glashow SL (1974) Unity of all elementary-particle forces. *Phys Rev Lett* 32: 438–441.
14. Wilczek F (1982) Quantum chromodynamics: The modern theory of the strong interaction. *Ann Rev Nucl Particle Sci* 32: 177–209.
15. Brennan TD, Hong S (2023) Introduction to Generalized Global Symmetries in QFT and Particle Physics. arXiv:2306.00912.
16. Palatini A (1919) Deduzione invariantiva delle equazioni gravitazionali dal principio di Hamilton. *Rend Circ Mat Palermo* 43: 203–217.
17. Ashtekar A (1986) New variables for classical and quantum gravity. *Phys Rev Lett* 57: 2244.
18. Capovilla R, Jacobson T, Dell J (1991) A pure spin connection formulation of gravity. *Class Quant Grav* 8: 59–78.
19. Randall L, Sundrum R (1999) An alternative to compactification. *Phys Rev Lett* 83: 4690–4693.
20. The ATLAS Collaboration (2022) Combination of searches for heavy resonances using 139 fb^{-1} of proton–proton collision data at $\sqrt{S} = 13$ TeV with the ATLAS detector. ATLAS-CONF-2022-028.
21. The ATLAS Collaboration (2022) Search for heavy resonances in the decay channel $W^+W^- \rightarrow e\nu\ \mu\nu$ in pp collisions at $\sqrt{S} = 13$ TeV using 139 fb^{-1} of data with the ATLAS detector. ATLAS-CONF-2022-066.
22. Klimek P (2021) Searches for exotic resonances at LHC. CERN ATL-PHYS-Proc-2021-015.
23. Obied G, Dvorkin C, Gonzalo E, Vafa C (2024) Dark dimension and decaying dark matter gravitons. *Phys Rev D* 109: 063540.
24. Dienes KR, Thomas B (2012) Dynamic dark matter. I. Theoretical overview. *Phys Rev D* 85: 083523.
25. Eguchi T, Gilkey PB, Hanson AJ (1980) Gravitation, gauge theories and differential geometry. *Phys Rep* 66: 213–393.
26. Arkani-Hamed N, Dimopoulos S, Dvali G (1998) The Hierarchy problem and new dimensions at a millimeter. *Phys Lett B* 429. 263.
27. Abramowiczy H, Abtt I, Adamczykh L, Adamusae M, Antonelli S, et al. Zeus Collaboration (2016) Limits on the effective quark radius from inclusive EP scattering at HERA. *Phys Lett* B 757: 468.
28. Fernández A (2024) *Artificial Intelligence Models for the Dark Universe: Forays in Mathematical Cosmology*. CRC Press, Boca Raton, FL.
29. Dillon BM, Han C, Lee HM, Park M (2017) KK graviton resonance and cascade decays in warped gravity. *Int J Mod Phys A* 32: 1745006.
30. Jacobson T, Romano JD (1993) The spin holonomy group in general relativity. *Commun Math Phys* 155: 261.
31. Brennan TD, Hong S (2023) Introduction to Generalized Global Symmetries in QFT and Particle Physics. arXiv:2306.00912.
32. Gaiotto G, Kapustin A, Seiberg N, Willett B (2015) Generalized global symmetries. *J High En Phys* 2015: 172.
33. Benedetti V, Bueno P, Magan JM (2023) Generalized symmetries for generalized gravitons. *Phys Rev Lett* 131: 111603.
34. Casini H, Magan JM (2021) On completeness and generalized symmetries in quantum field theory. *Mod Phys Lett A* 36: 2130025.
35. Nawrocki W, Ayers EW, Ebner G (2023) An Extensible User Interface for Lean 4. In 14th International Conference on Interactive Theorem Proving (ITP 2023). Leibniz International Proceedings in Informatics (LIPIcs), 268: 24:1–24:20.

36. Moura L, Ullrich S (2021) The Lean 4 Theorem Prover and Programming Language. Conference on Automatic Deduction. CADE 28. Platzer A and Sutcliffe G (Eds.), pp. 625–635.
37. Streicher T (1993) Investigations into intensional type theory. Habilitiation Thesis, Ludwig Maximilian Universitaet.
38. Cavalleri N (2020). mathlib3/src/geometry/manifold/algebra/left_invariant_derivation.lean at master · leanprover-community/mathlib3 · GitHub
39. Fernández A (2022) *Topological Dynamics in Metamodel Discovery with Artificial Intelligence: From Biomedical to Cosmological Technologies*. CRC Press, Boca Raton, FL.
40. Giles R (1981) Reconstruction of gauge potentials from Wilson loops. *Phys Rev D* 24: 2160–2168.
41. Modanese G (1993) Geodesic round trips by parallel transport in quantum gravity. *Phys Rev D* 47: 502.

3 Methods
Formal Autoencoders for Quantum Cosmology

"The physical world is only made of information, energy and matter are incidentals".

John A. Wheeler

This chapter introduces AI systems and broadly delineates their possibilities for model discovery. Two avenues of model discovery are pursued: (A) processing of big data and (B) formal mathematics incorporating large language models and interactive proof assistants. Part A of the presentation deals mostly with deep learning (DL), autoencoders, and other more specialized architectures and is tailored to researchers seeking to unravel physical models distilled from big data. With the leveraging of artificial intelligence (AI), dynamical systems have found a fertile ground for development. Machine learning identifies parsimonious models providing physical underpinnings of time-series data. However, such heavily parametrized models hardly yield physical laws. The problem becomes daunting as we turn to multi-scale complexities. This chapter addresses these imperatives as it introduces topological methods that enable metamodel discovery and the proper computational tools to decode the metamodel as an inferential framework. The methods advance model discovery, enabling reverse engineering of big data arising in the cosmological context, where metamodels with emergent quantum behavior are crucial to provide physical underpinnings of quantum gravity (QG).

This chapter also addresses the problem of quantum gravity as an emergent property in the physics of machine learning. To that effect, the chapter explores the possibility of an AI-based construction of a quantum holographic autoencoder (HAE), which requires that we first deal with the physics of machine learning and specifically enquire whether emergent quantum behavior can arise in a neural network (NN). By emergent quantum mechanics, we mean a formulation within a framework of nonlocal equilibrated hidden variables, as in the Bohm scheme. Once an emergent quantum behavior is shown to become possible within the machine learning system equilibrated on the nontrainable – i.e. hidden – variables, we address the question of developing a relativistic string gravitational scheme on the hidden variables adopted. Thus, the network with equilibrated nontrainable variables becomes in effect a quantum gravity autoencoder for the network exhibiting emergent gravity in the nonequilibrium regime prior to the equilibration of the nontrainable variables. In this way, we build a quantum metamodel for gravity that fulfills at least in part a major imperative for physicists seeking a unified field theory. Additionally, by coupling two physical embodiments of quantum-gravity autoencoders, the chapter implements a cosmic technology for universe reproduction through quantum tunneling across the interface between the two autoencoders. The technology is supplemented with a space–time rendering of the quantum

DOI: 10.1201/9781003632924-3

entanglement of the two primeval wormholes associated with the creation of a progeny universe through quantum tunneling.

We may summarize part A by stating that it encompasses the identification of parsimonious metamodels of dynamical systems that describe complex contexts represented by big data.

The discovery avenue B is tailored to develop a gauge theory of gravity, deemed in Chapters 2 and 6 as a crucial step to develop an AI-based "Theory of Everything". Such a theory pivots on a differential geometry formalization carried out by an autoencoder that reconciles quantum mechanics (QM) with general relativity (GR) and the classical theory of gravity. The differential geometry approach is instrumental because the geometries that support GR and QM are different: While GR is built upon a curved manifold that represents space–time, the wave function in QM is supported by an Euclidean space. The goal here is to implement an interactive theorem prover within the framework of an autoencoder to generate new results within the frame-field formalization rules made up of a set of declaratory propositions that constitutes an inductive type. The interactive proof assistant pivots on the proposition-as-type paradigm. From a constructive perspective, proofs are mathematical objects that are denoted by suitable expressions generated by the autoencoder in dependent type theory.

3.1 PRIMER ON NEURAL NETWORKS FOR AI-EMPOWERED MODEL BUILDING

The leveraging of artificial intelligence (AI) for model discovery in dynamical systems is revolutionizing both disciplines, leading to a mutually beneficial development. The cases of interest in this book are not amenable of model discovery in the sense of yielding a sparse system of differential equations that entrains the full dynamical system. Rather, we are seeking for something more elementary and subtle: a metamodel or topological representation of the dynamics. With the implementation of topological methods, AI-empowered metamodel discovery is able to focus on levels of system complexity and multi-scale hierarchies considered off limits for current AI technologies. The information on time series is encoded at the maximum level of coarse graining, hence greatly simplifying the computations while enabling a decoding of the information generated at the level of a topological description.

In dealing with dynamical systems using AI-based approaches, we address the following core question: What constitutes an insightful parsimonious model? The standard answer is as follows: "a sparse system of differential equations on latent coordinates". As argued in this chapter, this is not necessarily the format chosen by AI, given the "dimensionality curse" associated with the ultracomplex realities we chose to work on. The deployment of AI requires a paradigm shift, where dynamic information gets encoded in what would be termed a "topological metamodel". The metamodel is essentially pattern-based, where AI-interpretable topological patterns encode physics laws. These methods are likely to advance model discovery as they enable the reverse engineering of time series stemming from vastly complex realities hitherto inaccessible to other AI methods.

Artificial intelligence (AI) refers to machines capable of exhibiting behavioral traits that humans regard as indicators of intelligence, such as learning and problem solving [1]. Within this protean subject, machine learning (ML) refers to the ability to learn without being explicitly instructed to do so, while deep learning (DL) refers to an automated extraction of features, patterns, and ultimately models from arrayed data that is sequentially represented within an abstraction hierarchy organized as a multi-layered neural network (NN) [2-4].

DL has been shown to be highly efficacious at identifying features that are in principle discoverable from the data [2,5]. As in face recognition, features are hierarchically organized, so that large-scale patterns (eyes, noses, and face shapes) emerge after several layers of abstraction from simpler or more rudimentary patterns (lines, curves, and shades). The beauty and power of DL resides in the fact that the feature extraction process may be carried out in an unsupervised manner: The features emerge from the training of the system without human input or bias and enable the network to make accurate inferences. In this era of big data, we may state that there are several compelling reasons for implementing DL approaches:

- Fields like biology, particle physics, and cosmology are generating vast amounts of data and time series that can be easily stored and interpreted to achieve a conceptual unification within overarching models,
- we have the right hardware, i.e., graphics processing units (GPUs) that are massively parallelizable, and
- we have adequate software such as TensorFlow (TF) that enables suitable modular coding if the data can be pixelized or voxelized into a tensorial array, be it a vector, a matrix or a tensor proper [3,4].

At the most basic level, the building block of a NN is the neuron, referred to as *perceptron* [5]. The perceptron enables forward propagation of information encoded in an array of inputs $x_1, x_2, \ldots, x_n$ weighted by parameters $w_1, w_2, \ldots, w_n$ to generate an output of the form $y = f\left(\sum_{i=1}^{n} x_i w_i + w_0\right)$, where w_0 may be regarded as a bias term and f is a nonlinear activation function, often a sigmoid or sigmoid shaped, as shown in Figure 3.1. The bias term enables to shift the activation function. In vector representation, we often write $\boldsymbol{y} = \boldsymbol{f}(z)$, $z = \boldsymbol{x}.\boldsymbol{w} + \boldsymbol{w}_0$, where $\boldsymbol{x}$, $\boldsymbol{w}$ are input and weight vector, respectively. One can subsequently build a fully connected layer of perceptrons indexed by *j*, whereby $z_j = \boldsymbol{x}.\boldsymbol{w}_j + w_{0,j}$. The dense layer can be readily implemented in TF code,

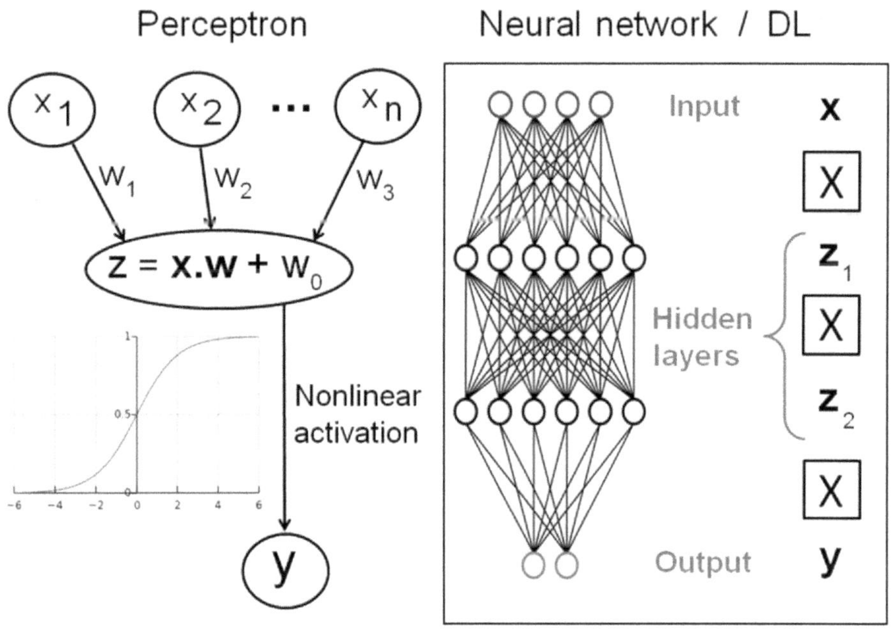

FIGURE 3.1 Scheme of the perceptron or neuron activation by linear transformation of input stimulus ($\boldsymbol{x}$) followed by nonlinear signal transmission as output *y*. The panels on the right show the organization of NNs with one and multiple hidden layers. The "boxed X" indicates full (dense) node connectivity between consecutive layers.

by simply specifying number of outputs/perceptrons [3,4]. We usually refer to the vector of linearly transformed inputs z as "hidden layer", as it does not explicitly describe observables [5].

Hidden layers may be stacked as in DL architectures. Thus, for the jth perceptron in the kth hidden layer, we get $z_j^{(k)} = \sum_{i=1}^{n_{k-1}} y_i^{(k-1)} w_{ij}^{(k)} + w_{oj}^{(k)} = \sum_{i=1}^{n_{k-1}} f(z_i^{(k-1)}) w_{ij}^{(k)} + w_{oj}^{(k)}$. The sequential composition of the network by stacking hidden layers has a standardized script in TF that generates the propagation of information as specified by the equation above [3,4]. In a NN with K hidden layers, we may regard the output $y_j^{(K)} = f\left(\sum_{i=1}^{n_{K-1}} y_i^{(K-1)} w_{ij}^{(K)} + w_{oj}^{(K)}\right)$ as an "inference" made by the DL system.

The accuracy of the DL inference represents the level of optimization of network performance and may be assessed vis-à-vis a training set of input/output paired data points. This assessment is often referred to as the *loss* of the network [2,5]. The loss is parametrically dependent on the full weight tensor $\boldsymbol{W} = \left[w_{i_{k-1}j_k}^{(k)}\right]_{k=1,\dots,K}$ which may contain a huge number of weights, in the thousands if not millions, depending on the size of the network. Thus, for DL NN with K hidden layers, the loss function or empirical risk $J(\boldsymbol{W})$ becomes $J(\boldsymbol{W}) = |\mathfrak{J}|^{-1} \sum_{\xi\in\mathfrak{J}} \mathcal{L}\left(\boldsymbol{y}_\xi^{(K)}(\boldsymbol{W}), \boldsymbol{y}_\xi\right)$, where $\mathfrak{J}$ is the training set with number of elements $|\mathfrak{J}|$, $y_\xi^{(K)}(\boldsymbol{W})$ is the predicted output vector for input vector $\boldsymbol{x}_\xi$, $\xi \in \mathfrak{J}$, $\boldsymbol{y}_\xi$ is the actual output vector, and $\mathcal{L}\left(\boldsymbol{y}_\xi^{(K)}(\boldsymbol{W}), \boldsymbol{y}_\xi\right)$ measures the discrepancy between actual and predicted output. In regression problems, where the output is a numerical vector, it is often convenient to adopt $\mathcal{L}\left(\boldsymbol{y}_\xi^{(K)}, \boldsymbol{y}_\xi\right) = \left\|\boldsymbol{y}_\xi^{(K)}(\boldsymbol{W}) - \boldsymbol{y}_\xi\right\|^2$. In such cases, the optimization of the NN through training becomes a problem of least squares. To optimize, the NN is tantamount to minimize the loss $J(\boldsymbol{W})$, which requires a careful fine tuning of the size of the training set vis-à-vis the size of the weight parametrization. In principle, the optimal network is defined as follows: $\boldsymbol{W} = \boldsymbol{W}^* = Arg\min J(\boldsymbol{W})$.

An insufficient number of training input/output pairs relative to the size of the weight tensor would give rise to *overfitting*, requiring special techniques, generically known as *regularization*, in order to trim the network, i.e., randomly remove connections, without compromising predicting efficacy [2,5].

Optimizing the network involves the laborious and costly computation of the minus gradient $-\frac{\partial J(\boldsymbol{W})}{\partial \boldsymbol{W}}$, which locally indicates the direction of steepest descent in the multi-dimensional surface $J = J(\boldsymbol{W})$. An iterative gradient descent computation generating a fine-tuned weight updating $\boldsymbol{W} \rightarrow \boldsymbol{W} - h\frac{\partial J(\boldsymbol{W})}{\partial \boldsymbol{W}}$ should eventually lead to convergence to a local minimum of $J = J(\boldsymbol{W})$ when adopting a suitable learning step h. This parameter should be tuned to effectively escape local minima while avoiding overshooting in trying to reach the global minimum. Most gradient descent algorithms use an adaptive learning step during training, in accordance with the constraints indicated. In practice, the gradient problem is approached by what is called the *stochastic gradient descent* method, whereby not all datapoints (input/output pairs) in the training set are used in each minimization step, but the gradient is approximated by an average over randomly chosen batches of datapoints in a trade-off between accuracy and computational efficiency. Obviously, batch size and learning rate are correlated, so the more accurate the gradient estimation, the larger the learning step may be (a token of computational confidence). To achieve significant speed, the stochastic gradient descent computation may be massively parallelized by splitting up batches into multiple GPUs.

3.2 NEURAL NETWORKS FOR DYNAMICAL SYSTEMS

In this section, we shall be concerned with data organized as a time series arrayed as $\{x(t_0), x(t_0+\tau), x(t_0+2\tau), \ldots, x(t_0+L\tau)\}$, where $x(t)$ is the vector of observables at time t and τ is the interval that determines the time coarse graining inherent to the sequential detection registered in the vector x. The time series enables a training of the NN such that the output vector $y = y(x(t), W)$ should approximate $x(t+\tau)$ when the input is $x(t)$, and this correspondence is carried over all t in the training time series. Thus, for a given network architecture, the optimal network is the one that realizes the minimum of the loss function:

$$J(W) = \sum_{n=1}^{L} \left\| x(t_0 + n\tau) - y(x(t_0 + (n-1)\tau), f, W) \right\|^2 \tag{3.1}$$

Obviously, for a fixed activation function, a model for the time series may be given simply by $Arg\min J(W)$, but such a model would lack universality as it would be extremely parametrized, most likely overparametrized, and would not prove insightful in the sense that it is not parsimonious. The discussion prompts us to enquire what truly constitutes a model. This question will be addressed subsequently.

The most common time series that humanity has collected since time immemorial stems from astronomical observation. For about two millennia until the time of Copernicus, and Newton later on, humanity had been striving to find a suitable model that would fit and explain the data, i.e., the recorded sequential positions of a set of celestial bodies. In today's more general context, biology, particle physics, and cosmology are generating dynamical data at a staggering rate, while models that fit and explain the data are sorely lacking or hopelessly inconsequential. It is expected that the advent of AI will dramatically impact this sort of model discovery.

In essence, we seek for what is known as autoencoder, an intermediate output with a dimensionality reduction and a simplified discerning physical picture that should therefore prove insightful to make sense of the patterns enshrined in the dynamics, enabling meaningful output inferences. These autoencoders and the models they give rise to will be studied in detail in the subsequent chapter.

3.3 DEEP LEARNING AND FEATURE EXTRACTION

The huge output of biomedical and biostructural data has become the hallmark of the post-genomic era, while chemical combinatorial possibilities make it forbiddingly difficult to parse chemical space in search for suitable leads for targeted therapy [6]. In this scenario, pharmaceutical researchers have turned to DL for guidance in drug discovery and development and target validation [7-10]. Thus, pharmacoinformatics has benefited immensely from the advent of DL systems trained to pair chemical compounds with molecular attributes likely to have therapeutic impact. These computational and informational techniques enable the evaluation of chemical compounds for specific properties, including target affinity, affinity screening profiles, structural features of drug-target docking, and ADMET (absorption, distribution, metabolism, excretion, and toxicity) profiling [8].

A major challenge in implementing DL models for pharmacoinformatics in accordance with the generic scheme outlined stems from the need to represent the chemical structure as a tensor array (vector, matrix, or tensor proper) of pixelated or voxelated inputs that may be subsequently interrogated geometrically across the hidden layers of NN in search for features that are indicative of the molecular properties indicated. There are a number of representations of chemical space amenable to TF encoding. The most obvious one is to order the atoms in the compound on a 1D array (following, for example, the IUPAC numerical labeling convention) and represent the chemical structure of the molecule as a covalent bond matrix pairing atoms in row and column in accordance with their covalent linkages, including single, multiple, and resonant (aromatic) bonds. The matrix is subsequently transformed into a topological descriptor that describes the invariants arising from different atom ordering. The input describing chemical structure is paired within a training set against the molecular attributes of therapeutic relevance that the network is meant to infer. Then, compounds that need to be evaluated/profiled are inputted as the array of pixels/voxels, transformed into a topological representation, and feature extraction is achieved through the sequential activation of hidden layers at increasing levels of abstraction, eventually leading to the profile inference which is subsumed in the output layer.

Feature extraction through the NN often requires particular architectures known as convolutional NNs (CNNs) [2-5]. The idea is to pixelize or voxelize the input data in a matrix or 3D tensor array and then scan (convolve) the array with a filter associated with a specific pattern to generate feature map. For the sake of the argument, let us consider a 2D-array input $\boldsymbol{M}$. The filter $\boldsymbol{F} = (w_{ij})$ is an m×m matrix actually representing a convolutional kernel, so that a neuron in the F-associated hidden layer $\boldsymbol{M}_F$ only senses the pixels in an m×m patch (receptive field), and the layer becomes the feature map

$$\boldsymbol{M}_F = \boldsymbol{M} * \boldsymbol{F} = \left(\sum_{j=1}^{m} \sum_{i=1}^{m} w_{ij} x_{i+ap\ j+bp} + w_0 \right)_{ab}, \tag{3.2}$$

where p, usually set at $p = 2$, is the stride adopted as the filter slides along the input matrix array, with dummy integers a, b indicating the patch location. Essentially, the CNN is a NN where the set of weights in each mxm patch of inputs are always the same as the filter slides along the input array to generate the hidden layer that constitutes the feature map. The convolution operation becomes the entry-by-entry (Frobenius) matrix inner product as the filter slides along the input matrix with a given stride. Successive filters may be applied reducing progressively the size of the feature maps as higher and higher levels of abstraction are achieved in the representation of the data (Figure 3.2). Thus, the parametrization required for feature extraction in CNNs is relatively small, as all the neurons in a hidden layer share the same connectivity parameters that define the filter that generated that layer.

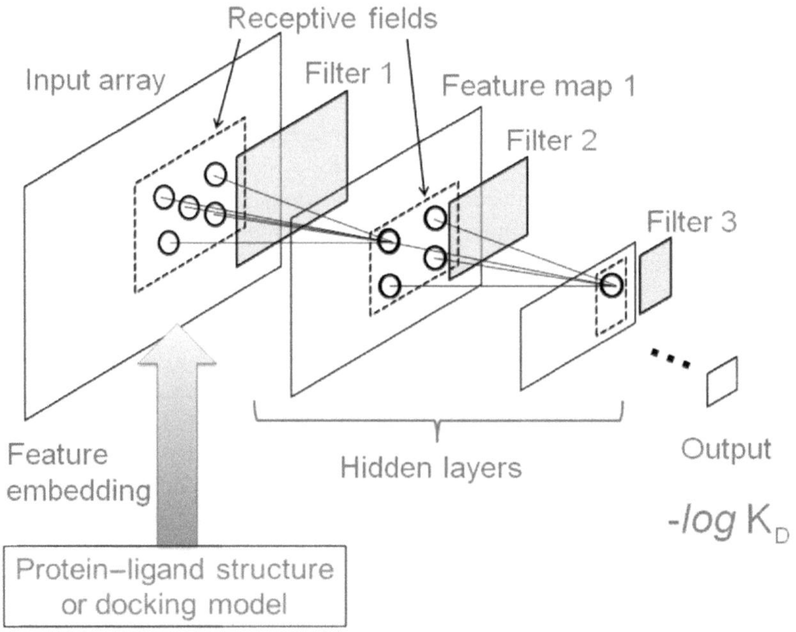

FIGURE 3.2 Scheme of a CNN used for inference of drug–target affinity through sequential application of filters that become optimized through minimization of the network loss.

In CNNs, the filters are not specified *a priori*, i.e., their weights are not fixed through human intervention. The filter parametrization is automatically determined by the training/optimization process without introducing any assumption, other than the size of the receptive fields for each convolution operation and the overall number of filters to be applied. Thus, feature extraction in a CNN is carried out in an unsupervised manner and only requires that we script (in TF coding) the number of hidden layers or feature maps and the size of the filters to be applied to generate each hidden layer. The features themselves emerge as the network is trained.

Thus, CNNs often constitute AI-empowered platforms for drug discovery, where the structure of a protein–ligand interface that serves as precursor for a predicted drug–target interaction is pixelated in a feature embedding process as a 3D spatial array of protein–drug atom pairs deemed to be interacting across the interface [11]. The inference of target affinity for a given drug is assessed through a sequence of feature extractions using convolution filters until the output feature map becomes a number directly associated with the affinity $pK_D = -logK_D$, where K_D is the dissociation constant for the drug–target complex inputted via feature embedding (Figure 3.2). The training of the network is carried out by minimizing the network loss or empirical risk over a set of drug–target complexes whose structure and affinity are both known (preferably, experimentally determined). The proteins in the complexes of the training set are typically homologs of the one whose affinity for a specific drug we seek to infer, so that the features that enable the affinity inference may emerge from structural alignment.

In other more complex applications of CNNs, where the output is not simply a numerical parameter but rather a numerical array, it is often convenient to adopt a variation of the CNN architecture in which the patches that yield the pixels with the highest weights in the feature map are reconstituted into features perfectly stenciled by the filter, while other patches that do not yield discernible features are left invariant in nonoverlapping regions. Thus, with the application of each filter, a feature-enriched reconstitution of the input layer is carried out prior to the application of the next filter (Figure 3.3).

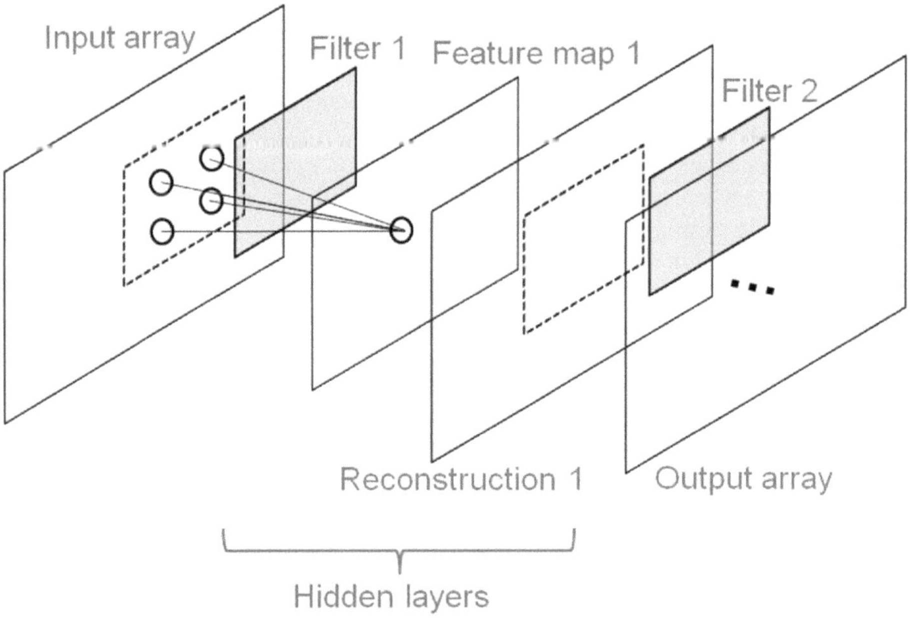

FIGURE 3.3 Scheme of a CNN with a particular type of architecture in which the patches that yield the pixels with the highest weights in the feature map are recreated as filter-stenciled "perfect" features. Thus, a feature-enriched reconstitution of each layer is carried out prior to the application of the next filter.

3.4 AUTOENCODERS DISTILLING THE LATENT DYNAMICS FROM A HIGHER DIMENSIONAL SPACE

It is widely expected that AI and DL in particular will become a major player in model discovery for data-driven research [12]. Within this vast array of possibilities, the focus of this book is model discovery for dynamical systems that underlie big time-series data from vast areas as distant as biomedicine and cosmology [13-15]. The biggest hurdle we stumble upon is that the level of complexity of the data generated in fields like biology and cosmology is so extreme that it challenges the notion of model itself. As shown subsequently, the model itself can seldom be cast as a system of differential equations. Thus, much of the ensuing discussion deals with the question of what constitutes a meaningful model with predictive value when dealing with the ultracomplex realities represented in the contexts of biology or cosmology.

Encoding has proven to be a necessary category in AI [12-15]. The type of AI we are mostly concerned with involves deep learning, which requires an encoding of the raw information that needs to be acquired and processed further to make meaningful inferences. Just like with human intelligence, the encoding problem stems from the core question: What is essential and what it superfluous? The encoding problem becomes solvable when the system under scrutiny is hierarchical and the hierarchical structure is fairly obvious or at least discoverable. In the cases treated in this book, the structure of the data is always hierarchical, implying that the learning process admits a reductive approach represented by the encoding. When the network architecture is such that this process is automatically generated, we name the NN *autoencoder*.

We shall deal mostly with time-dependent data representing physical or biophysical systems, where detailed fast motions may be systematically averaged out, so the relevant information may be stored in a coarse grained representation. Rigorous mathematical constructs will be introduced to implement the hierarchical encoding materialized by the autoencoder. Some examples of hierarchical encoding of physical or biophysical processes that needs to be taken into account when designing the autoencoder architecture are as follows:

1. The adiabatic approximation, where fast-relaxing or fast-evolving enslaved degrees of freedom are averaged out, or thermalized or equilibrated when incorporated into a model for the time evolution of a dynamical system [16]. In molecular physics, examples of such degrees of freedom are vibrational hard modes that evolve on timescales of the order of picoseconds to nanoseconds, while soft modes evolve on longer timescales ranging from the submicrosecond to seconds.
2. In atomic physics, the Born–Oppenheimer approximation represents regions of the potential energy surface where an adiabatic regime holds, so the motion of electrons is enslaved or entrained by the slower motion of the atomic nuclei [17].
3. The latent manifold in dynamical systems [12], where the system is entrained or enslaved in the long-time limit by the evolution in a manifold of lower dimension than the original space. The latent manifold is often referred to as center manifold [16], especially in the context of dissipative systems, and contains the attractors of the system (Figure 3.4).
4. In the biophysical context of functionalized soluble proteins (enzymes), the quantum mechanics for specific chemical processes take place at the *epistructure* (solvent organization around the protein structure), where water is chemically functionalized [18]. A rigorous treatment of this problem would require that we solve a time-dependent Schrödinger equation. In this context, an AI approach would need to be incorporated to learn to infer the nodal structure of molecular orbitals within a voxelated 3D grid. With current technologies, this AI approach is plausible only under the adiabatic regime given by the Born–Oppenheimer approximation.
5. In clonal population dynamics, cancer phenotypes that become selected under therapeutic pressure [19]. The dynamics are complex but hierarchical, and the mathematical procedure to "encode what is essential" in this context is based on the center manifold reduction (Figure 3.4). This reduction discussed in (3) is actually a projection of the dynamical system onto a

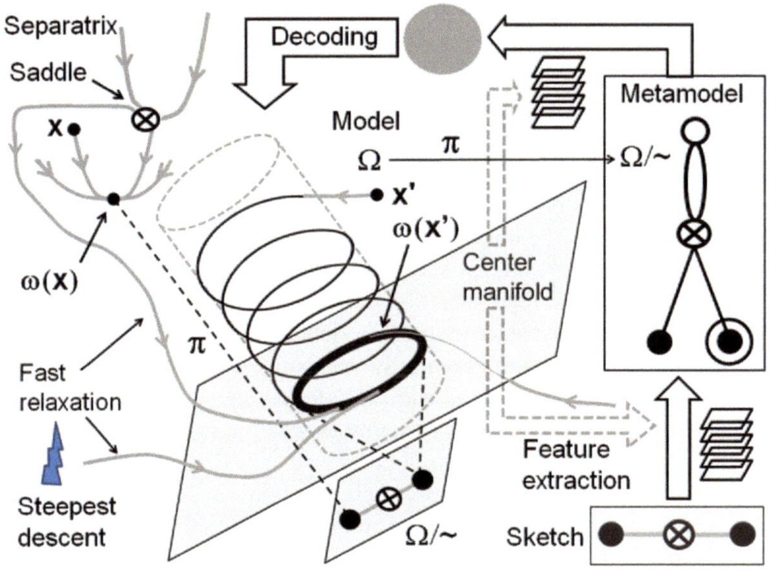

FIGURE 3.4 Schematics of the topological "metamodel" encoding of a dynamical system that contains a center manifold. The center manifold enslaves or entrains the dynamics for timescales associated with the adiabatic elimination of fast-relaxing and thermalized degrees of freedom and therefore constitutes a latent manifold (Ω) within which a model (differential equations on latent coordinates) may be identified by an autoencoder. The dynamic information may be encoded further by a second autoencoder at a higher level of abstraction, where the dynamics are represented more coarsely as "modulo-basin" transitions. The modulo-basin dynamics is mapped on a quotient space, $\Omega/\sim$, where two states x, y are regarded as equivalent, $x \sim y$, if they have the same destiny state ($\omega(x) = \omega(y)$). The encoding processes are symbolized by dashed lines, and the enslaving center-manifold dynamics is highlighted by a thick dark circle. Thus, the second autoencoder materializes the projection $\pi: \Omega \to \Omega/\sim$ and represents the dynamics as a walk in a graph whose vertices are the critical points (minima, saddles of different indices, maxima) and attractors of the system, and the edges represent connections along pathways of steepest descent. This topological metamodel may be subsequently decoded back to flesh out the dynamical system using learning technology.

lower dimensional system that entrains it. The autoencoder may discover the center manifold by interrogating vast amounts of time-dependent data, as it seeks to meaningfully reduce dimensionality in such hierarchical systems.

6. Chapter 4 introduces the *quotient space*, a fundamental mathematical construct to take advantage of dynamic hierarchy in order to encode information as required to implement DL systems [20]. The quotient space is built upon an underlying dynamics and may be equated with the orbit space, i.e., points in the same trajectory are regarded as equivalent, and the quotient space is the set of equivalence classes with a topology inherited from the underlying space where the dynamical system is mapped (Figure 3.4). In rigorous terms, two points–states (x, x') are equivalent ($x \sim x'$) or belong to the same equivalence class ($x' \in \overline{x}$, $\overline{x'} = \overline{x}$) if and only if they share the same destiny state ($\omega(x) = \omega(x')$) vis-à-vis the trajectory to which they both belong. Thus, we simplify the space by lumping microstates into basins of attraction (of destiny states) in the potential energy surface. The modulo-basin dynamics constitutes a coarse model named *metamodel*, which is far easier to encode as the system learns data generated by atomistic molecular dynamics simulations. In this way, AI learns to propagate dynamics in quotient space, discovering a metamodel to cover physically realistic timescales usually inaccessible to detailed atomistic computations.

While the center manifold encoding is suited for dissipative dynamics, where the attractor may be nontrivial (cf. Figure 3.4), the quotient space simplification is better suited for Hamiltonian systems. Both levels of encoding converge as free-energy dissipation tends to zero. In fact, as the AI system encodes time-dependent raw data, it implicitly composes the two levels of encoding, with the center-manifold reduction averaging out of fast modes, followed by projection onto quotient space (Figure 3.4).

In all cases dealt with in this book, the hierarchy of the data that make it amenable to encoding is either AI-discoverable, or it may be unraveled through a rigorous mathematical construct that needs to be incorporated to the learning code, as shown below.

The modulo-basin hierarchical representation of the dynamics in quotient space enables the construction of metamodels, that is, coarse-grained models that represent transitions between equivalence classes of states of the system, essentially reproducing the topological dynamics within a graph representation (Figure 3.5). As an illustration, let us consider the "keto" ←→ "enol" interconversion of acetone (Figure 3.5). This is a chemical reaction involving an intramolecular proton migration that can be modeled by the potential energy surface (PES) representing the ground-level electronic energy sheet under the Born–Oppenheimer approximation. From quantum mechanics calculation, we know that this PES has three minima, representing the less stable "enol" form (1), the more stable "keto" form (2) and a state where the jumping proton is completely dissociated from the rest of the molecule (3). The minima are separated by four saddle points at the top of the path of steepest descent joining pairs of minima on both sides of the saddle along the direction of negative curvature. This direction is actually the eigendirection associated with the eigenvalue of the Hessian matrix (Jacobian of the gradient vector field) with negative real part computed at the saddle point. In turn, the lines of steepest descent joining the four saddles with the three maxima become the separatrices of the basins of attraction of the minima. Thus, all lines of steepest descent may be represented as a graph (metamodel II, Figure 3.5) joining critical points that are in turn organized in

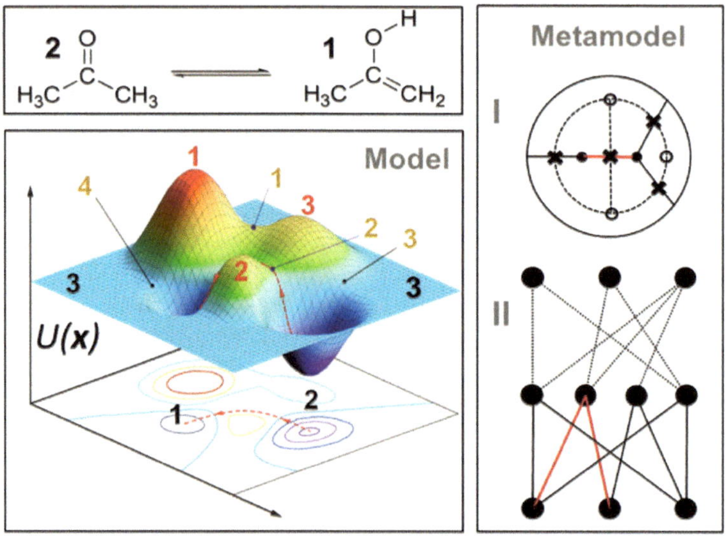

FIGURE 3.5 Three levels of abstraction in the modeling of the chemical dynamics for the isomerization "keto" ⇔ "enol" of acetone, representing the intramolecular migration of a proton. The potential energy surface (PES) represents the model and the topological relations between critical points, together with the "modulo-basin" transitions along the paths of steepest descent joining critical points represent the metamodel. On metamodel I, the modulo-basin topology is mapped onto the latent manifold, while on metamodel II, the modulo-basin topology is abstracted further, and displayed as a graph.

tiers, where the lowest tier corresponds to points where all eigendirections have positive curvature (minima) and the next tier includes all critical points with only one eigendirection with negative curvature (saddles of index 1), and the next tier is associated with two eigendirections of negative curvature (maxima in the case of Figure 3.5). The edges on the graph represent paths of steepest descent joining critical points in adjacent tiers. In this way, the topological dynamics of the PES may be encoded as a metamodel represented in graph form, while the reversible chemical reaction pathway becomes a walk in the graph.

As argued subsequently, metamodels enable the discovery of hierarchical dynamical systems underlying processes that unravel in realities of high multi-scale complexity. The implementation of a metamodel factorization of the dynamics requires the concerted participation of several components operating in a coordinated manner within an AI platform. First, we need to introduce the so-called autoencoders that constitute the deep NN systems that encode the dynamics on the "latent manifold" Ω. This manifold entails a significant dimensionality reduction relative to state space W and is spanned by the internal coordinates that label the orbits of the symmetry group $\mathcal{G}$ inherent to the system and acting on W: $\Omega \approx W/\mathcal{G}$. Traditionally, it is expected that the first autoencoder, hereby denoted AE1, that generates the latent manifold is jointly optimized to generate also the most sparse or minimal set of differential equations on the manifold that can be decoded back onto the dynamical system defined on W. This is what is usually meant by "model discovery". In practice, the level of multi-scale complexity of the processes dealt with in this book does not make models amenable to discovery at the geometric level. As previously argued, in such cases, another level of abstraction $\pi : \Omega \to \Omega/\sim$ needs to be introduced so that the coarse-graining of time within $\Omega/\sim$ reflects equilibration within the basins of attraction in the latent dynamics. This hierarchical escalation in the level of abstraction requires a second autoencoder, AE2, capable of encoding the topological features of the latent vector field.

At this stage, a different sort of NN architecture is required to propagate the metadynamics on $\Omega/\sim$. To properly delineate the architecture of the NN required for metadynamic propagation, we limit the discussion to the case where the dynamics is generated by a smooth (i.e., C^1) potential energy function $U : W \to \mathbb{R}$ invariant upon the isometries (distance-preserving transformations) of W. Furthermore, $\mathcal{G}$ is an Euclidean group, so that Ω is compact, and hence, $\Omega/\sim$ becomes a discrete set of basins, and a basin assignment represents a coarse state of the system. Then, the metadynamics may be encoded as "evolving text" representing a time series of basin transitions. The textual processing requires the implementation of a particular type of DL architecture known as *transformer*, while the transformer-based propagation of the metadynamics constitutes a Markovian process (Figure 3.6). Through AE2, this metadynamics is decoded as latent dynamics on Ω and validated by contrasting the latent dynamics against the hidden Markov process upheld under the adiabatic conditions described.

Thus, to implement a metamodel within an AI platform, it is essential that the two autoencoders and the transformer are optimized to work concertedly with complete compatibility (Figure 3.6). This means that an input state yields the same destiny state regardless of the pathway chosen provided the pathways have identical endpoint spaces in the commutative diagram presented in Figure 3.6.

3.5 REVERSE ENGINEERING OF THE STANDARD MODEL WITH A REVERSE AUTODECODER

The notion of diagram commutativity (Figure 3.6) is central to the encoding of a dynamical system into its latent dynamics and, reciprocally, to the decoding of the latent dynamics onto the full dynamics. Thus, the task of decoding the standard model, canonically defined on the four-dimensional space–time $W/\sim$, onto a five-dimensional compact multiply connected manifold W requires an autoencoder operating in reverse. This reverse autoencoder lifts the flow $\mathcal{F} : W/\sim \to W/\sim$ to a flow $\tilde{\mathcal{F}} : W \to W$ that must be compatible in accordance with the commutativity condition: $\mathcal{F} \circ \pi = \pi \circ \tilde{\mathcal{F}}$, with

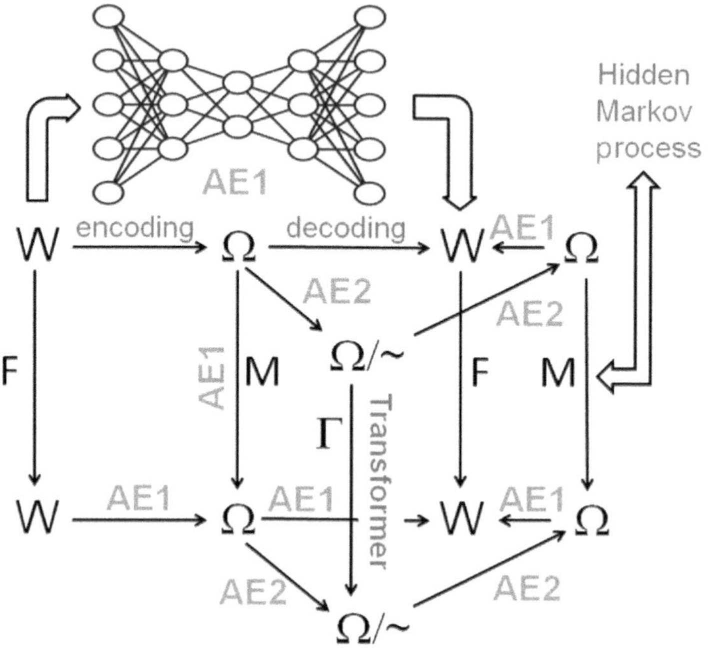

FIGURE 3.6 Commutative diagram representing two coupled autoencoders operating sequentially in tandem and representing two levels of abstraction of a dynamical system. The first autoencoder, labeled AE1, projects the dynamics onto the latent manifold, while autoencoder AE2 projects the latent dynamics onto a discretized "modulo-basin" version where a coarse state of the system is represented in textual form and the "modulo-basin" dynamics may be learned and propagated within a special type of autoencoding architecture known as "transformer".

$\pi: W \rightarrow W/\sim$ denoting the canonical projection that assigns each point in W to its equivalence class in $W/\sim$. The key issue with this formulation is that the standard model *per se* does not represent a dynamical system in any obvious way, so what sort of flow are we actually discussing in this context? This question is crucial because we are introducing AI technology that is tailored for dynamical systems.

Fortunately, we may treat the interaction processes described by the standard model as transformations created by a time-dependent propagator. For each elementary process, this operator has a time step associated with it, and this time step is exactly the lifetime of the boson that communicates the force in the process of particle transformation, as obtained from Heisenberg's uncertainty principle: $\sim \frac{\hbar}{2M_B c^2}$, where M_B is the rest mass of the respective gauge boson. There may be more than one boson involved, as in the case of creation and decay of the Higgs boson through gluon fusion, in which case the time step is the sum of the respective lifetimes. The dedicated AI technology of variational autoencoders may be thus leveraged to reverse engineer the standard model, a process tantamount to identify the lift flow $\tilde{\mathcal{F}}: W \rightarrow W$ for a propagator $\mathcal{F}: W/\sim \longrightarrow W/\sim$ generically determining an elementary process in dynamic form. In this setting, "reverse engineering" is equated with identifying the lift propagator that would make the diagram in Figure 3.7 commutative.

The AI technology described in this chapter will be applied in reverse in Chapters 4 and 5 in order to identify dark matter and dark energy in five-dimensional space–time. This is tantamount to reverse engineer the standard model by incorporating the dormant dimension shown to store gravity.

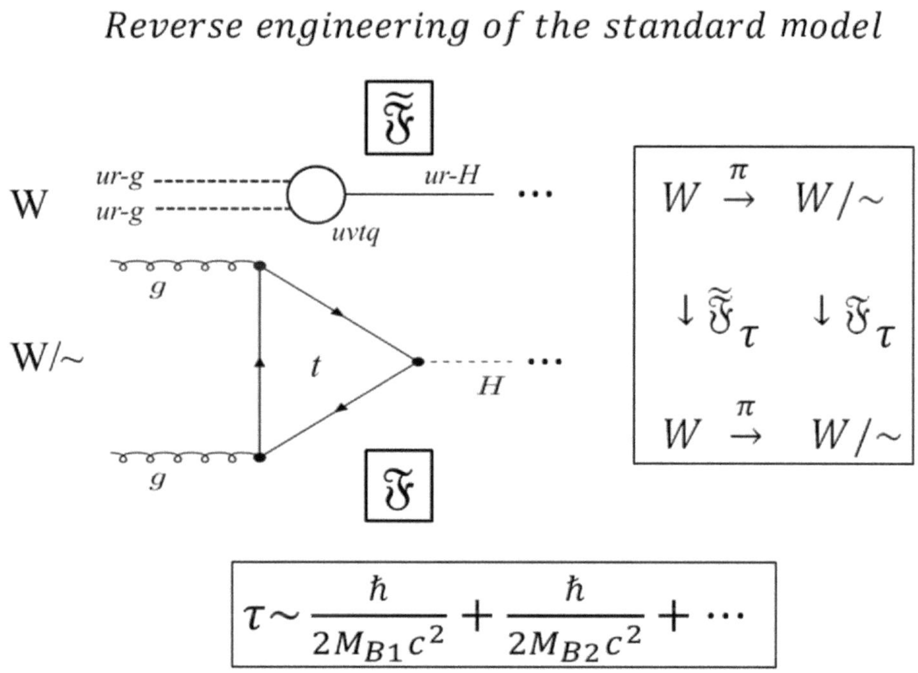

FIGURE 3.7 Diagram commutativity fulfilled by the autoencoder that carries out the reverse engineering of the standard model.

3.6 AUTOENCODERS FOR DYNAMICAL SYSTEMS

With the leveraging of artificial intelligence (AI) [1] and in particular, machine learning approaches [2,3], dynamical systems have found a new fertile ground for further development [12,13]. Showcase problems in applied mathematics, including the Lorenz strange attractor, reaction–diffusion spatio-temporal systems, and fluid dynamic flows captured by the Navier–Stokes equation, are being examined in a new guise as autoencoders identify parsimonious models with reduced dimensionality [14]. Such models are meant to provide the physical underpinnings of the phenomena enshrined in time-series data or generated by systems of raw differential equations.

Machine learning or more broadly, AI, is being leveraged for model discovery of dynamical systems underlying data represented as a time series. The data regression system, which in this context is a neural network predicting future behavior, is trained by the time series and regarded as the model itself. However, this model is heavily parametrized and hence too "fragile" to allow for extrapolation [12]. In other words, such models are not really amenable to yield physical laws, the way other data-regression approaches are [15,21]. This statement has been voiced repeatedly and hints at some level of dissatisfaction: Machine learning and AI in general are very efficient at providing predictive models when trained on a sufficiently long time series but often do a poor job at providing physical insights regarding the underlying dynamical system.

This problem gets significantly amplified as we turn to biological or, more broadly, biomedical matter [19]. It is widely felt that, when examined in their multi-scale richness and complex heterogeneity, dynamical systems in biology or biomedicine cannot reach the level of maturation required to be subsumed into applied mathematics. This statement should be interpreted in the sense that we lack sparse enough models that provide physical underpinnings of biological/biomedical phenomena and are suitable for extrapolation. Will the leveraging of alternative AI-based approaches change the *status quo*? This chapter portends to address this problem and provide insights that will be methodologically fleshed out in the subsequent chapters to enable metamodel discovery by reverse engineering time series stemming from highly complex multi-scale realities.

A key question in fostering the mathematical maturation of model discovery in biology may be cast as follows: What constitutes a parsimonious model that provides physical underpinnings of biological phenomena? A standard answer with broad bearing on most problems considered tractable in applied mathematics is as follows: "a sparse system of differential equations on a smooth manifold of latent coordinates" [14,21]. As this chapter argues, this may be simply too much to ask for in the context of biological matter. Furthermore, this is not necessarily the format or framework that AI would typically choose for model discovery, given the "dimensionality curse" associated with the molecular reality of biological systems [19]. In principle, a single autoencoder that optimizes for sparsity in the discovery of the latent manifold might not provide a satisfactory solution to the modeling problem. Molecular reality *in vivo* typically has well over a million coordinates required to specify the state of the system, and the extent of connectivity parametrization for an autoencoder capable of handling such level of complexity would be simply enormous, implying that the training and variational optimization of the neural network would be off limits, at least with current computational capabilities [22].

This chapter squarely addresses this matter. To do so, it leverages AI methods to circumvent the difficulties associated with model discovery for time-dependent phenomena arising in soft or biological matter. In essence, as we shall show, AI dwells on a paradigm of what constitutes a parsimonious metamodel that is significantly different from the one adopted by applied mathematicians [19]. Thus, to identify the most economic yet faithful metamodel, AI will be shown to use two or more tandem autoencoders instead of one, as it is typically done in model discovery [14]. The autoencoders are coupled and become fully compatible with each other at the completion of the parameter optimization process, as defined precisely in the subsequent sections. In the simpler cases, where two autoencoders are required, the second autoencoder translates the dynamic information embossed in the latent manifold, turning it into a topological dynamics metamodel [23] which can be decoded and enables significant propagation of the dynamics into the future. This property is essential for coverage of realistic timescales relevant to the level of state extrapolation required. Crucially, the topological dynamics metamodel constructed by AI is essentially a pattern-based model, not a system of differential equations, as it would be expected for standard model discovery in dynamical systems. This does not mean that AI is discovering laws without equations, but simply that AI adopts a different way of casting models susceptible of extrapolation as recognizable physics laws. Thus, AI may not straightforwardly give us the equations that govern *in vivo* protein folding, but the underlying physics discovered may in all likelihood be cast in terms of AI-interpretable topological patterns that signal the commitment of the chain to fold into a steady conformation [20].

These topological dynamics methods are likely to advance the field of AI-based model discovery as they enable the reverse engineering of time-series data stemming from vastly complex hierarchical realities. Such contexts are illustrated for example by the cellular setting that assists and expedites molecular processes, which have been hitherto considered off limits to machine learning approaches to dynamical systems.

At this stage of development, the topological methods introduced yield AI-recognizable patterns but do not beget latent differential equations that have traditionally been the hallmarks of model discovery. This may be viewed as a limitation in some sense, but we argue that that assertion is perhaps a reflection of narrow-mindedness. Synergistic efforts involving AI are likely to dominate future human endeavor in science, and AI systems are very much attuned to encode and process patterns in metamodels such as those produced by the topological approaches introduced in this book.

The impact of the topological methodology is likely to be broad, as it would render tractable problems in dynamical model discovery that have been hitherto considered off limits by applied mathematicians that are currently incorporating machine learning in their toolbox. Thus, ultra-complex realities recreating cellular environments that influence and steer molecular dynamics are likely to be within reach as topological methods are incorporated to the AI-empowered metamodel discovery.

Since the dawn of modern science, Western civilization upheld the belief that understanding the workings of the universe pivots on finding fundamental equations that govern physical processes. In the case of a dynamical system, the underlying differential equations represent the basic model assumed to enshrine the physical laws that underpin the process described. The specific constraints, conservation principles, and symmetries of the system must all be taken into account when positing the differential equations. After centuries of work within this paradigm, it would be interesting to see how the leveraging of AI in synergy with the human endeavor will affect the choice of the format within which the physical laws are encoded and how this choice impacts the field of dynamic modeling. With the exponential development of computer technologies [22], we may soon be witnessing a paradigm revision. Equation-based modeling may or may not remain the dominant paradigm. Other types of dynamic descriptors are already making strides and contributing to a new understanding of the universe, or at least, of the dynamic multi-scale complexities of reality [19]. These descriptors are certainly different and arguably more pliable than what biological humans managed to achieve so far.

The discovery of physical laws distilled from sequential data representing a time series remains a major challenge as well as an imperative to enhance our understanding and control of physical processes. In recent times, this type of data-driven model discovery has been fueled by significant breakthroughs [15,19,21,24]. Yet, we also live in an era of big data, especially stemming from fields like biology and astrophysics, where the multi-scale complexity of the data organization makes model discovery particularly daunting. It is unclear whether the enormous richness of the data describing *in vivo* contexts at the molecular level is amenable to the kind of parsimonious models based on differential equations that mankind holds dear and has sought since time immemorial. More than in any other field, in biology, it is likely that a willy-nilly application of Occam's razor may lead to self-inflicted wounds.

Deep learning (DL) approaches realized through autoencoder architectures have proven particularly valuable for the discovery of data-driven models represented by differential equations framed on "essential coordinates". The later, often referred to as "latent coordinates" [21], span the so-called center manifold in dynamical systems [19,16]. This reduction entails a significant dimensionality reduction and usually identifies the enslaving "slow" process that dynamically subordinates fast relaxing modes [12,16] and serves to encode the physical process we seek to model. Latent coordinates need to be selected very carefully, not only by weighing the extent of dimensionality reduction and the compactness and smoothness of the latent manifold, but also the economy of the resulting model. This economy is typically assessed by the complexity of the differential equations in latent space, quantified by the number of nonlinear terms [14].

In generic terms, an autoencoder seeks to identify latent coordinates **x** as output of a feed-forward neural network (NN) with multiple hidden layers. The NN is inputted observable vectors, denoted generically as **z**, that serve to train the network, as the decoding of **z** from the latent coordinates must reproduce the inputted **z**-value (Figure 3.8). The autoencoder is optimized variationally, meaning that the activation weight parametrization of the multi-layered encoding and decoding functions denoted, respectively, γ and μ, minimizes the loss function $\mathfrak{L}(\gamma,\mu)$ that measures the efficacy in the recovery of the input vectors:

$$\mathfrak{L}(\gamma,\mu) = Q^{-1}\sum_{z\in\mathfrak{F}} \left\|z-(\mu\circ\gamma)z\right\|^2 \; ; \; \gamma,\mu = \arg min_{\gamma,\mu}\mathfrak{L} \tag{3.3}$$

where $\mathfrak{F}$ is the training set and $Q = |\mathfrak{F}|$ is its cardinal, for now regarded as a hyperparameter fixed to avoid overfitting [2,3,12,13] in accordance with the dimensions of the NN.

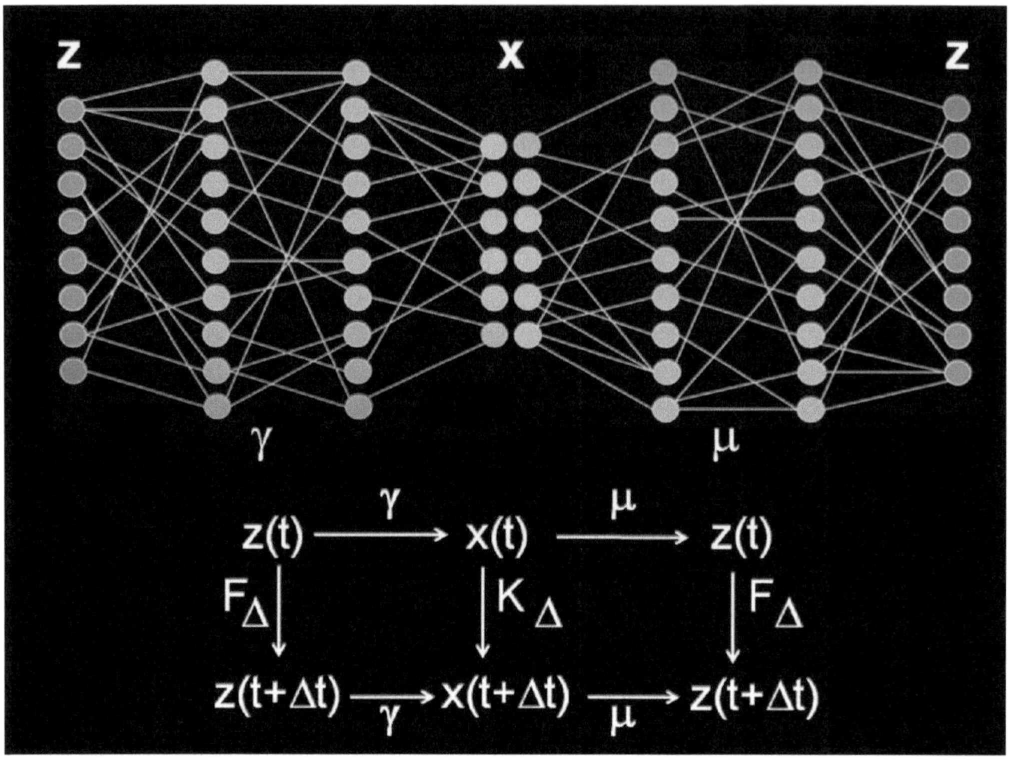

FIGURE 3.8 Generic scheme of the neural network (NN) architecture for autoencoder (γ,μ). The vector $z \in W$ represents the state of the system and $x \in \Omega$ represents an encoded latent state that is decoded back into z. The encoding process entails a dimensionality reduction with $\dim\Omega < \dim W$ and, concomitantly, a coarse graining of time in multiples of a time step Δ associated with propagation of the dynamics on the latent manifold Ω. Gray disks represent nodes in hidden layers for encoder $\gamma = \gamma(\Theta)$ and decoder $\mu = \mu(\Theta)$, where Θ denotes the weights of node connections that are variationally optimized according to a loss function $\mathfrak{L} = \mathfrak{L}(\Theta)$. The weights realizing the variational mimimum $(\Theta = \arg\min \mathfrak{L})$ in the training of the NN are optimal at making the diagram commutative.

The autoencoder is functionally operative to model the dynamical system that underlies the given time series $\left\{z(t_0 + n\Delta t), z(t_0 + (n+1)\Delta t)\right\}_{n=0,1,2,\ldots,L}$ if and only if the following relations hold for $t = t_0 + n\Delta t$ with $n = 0, 1, 2, \ldots, L$ ($L \gg 1$) and any initial time t_0:

$$\begin{aligned}(K_\Delta \circ \gamma)z(t) = \gamma z(t+\Delta t) = (\gamma \circ F_\Delta)z(t); (F_\Delta \circ \mu \circ \gamma)z(t) \\ = (\mu \circ K_\Delta \circ \gamma)z(t) = z(t+\Delta t),\end{aligned} \tag{3.4}$$

where F_Δ, K_Δ are the infinitesimal time maps in state space W and latent manifold Ω, respectively (cf. Figure 3.9). In other words, as the autoencoder becomes variationally optimized, the functional relations $K_\Delta \circ \gamma = \gamma \circ F_\Delta$; $F_\Delta \circ \mu = \mu \circ K_\Delta$ hold.

Thus, diagram commutativity (Figure 3.8) becomes the key property that enables us to assert that the choice of latent coordinates $x \in \Omega$ was the "right" one in the sense that it captures the entrainment of the dynamics on W by the reduced dynamics on the latent manifold Ω. The commutativity rules ensuring the compatibility of raw and latent dynamics may be cast in terms of derivatives using the chain rule [14] as Δt is taken to the infinitesimal limit dt.

3.7 MODEL DISCOVERY WITH DEEP LEARNING

Model discovery in dynamical systems has been maturing for some time, to the extent that it has been integrated to the corpus of applied mathematics. Furthermore, many applied mathematicians have incorporated DL to their toolbox as they make forays in reverse engineering of dynamical systems [12]. The overarching goal is to develop regression methods that enable extraction of parsimonious dynamics from large data organized as time series $\{\mathbf{z}(t)\}_t$ with t given as multiples of a fixed time step [12,14]. With the advent of DL, autoencoder architectures have successfully identified latent (essential) coordinate frames with significant dimensionality reduction ($\dim\Omega < \dim W$). Ultimately, the parsimonious description should become solvable in the form of a differential system $\dot{\boldsymbol{x}} = \boldsymbol{K}(\boldsymbol{x})$ which spans a differential system $\dot{z} = \boldsymbol{F}(z)$ for the time-series vector z. Yet, a mere stark reduction in dimensionality does not necessarily ensure a parsimonious description [14]. The variational optimization of the autoencoder should simultaneously consider both the dimensionality of the coordinate frame for the latent manifold and the sparsity of the equations in the latent manifold, with the latter measured by the number of terms in the flow (vector field) that determines $\dot{z}$.

The autoencoder architecture and the required dual diagram commutativity between the latent flow **K** and the given time-series flow **F** are jointly represented in Figure 3.9. We adopt a time step τ as the time-series hyperparameter to discretize the flows by coarse graining time resolution. The loss function $\mathfrak{L}(\gamma,\mu,\boldsymbol{K})$ now includes five terms: $\mathfrak{L}_r(\gamma,\mu)$, previously introduced, weights the efficacy of the recovery of the **z**-value upon encoding and decoding, $\mathfrak{L}_s(\boldsymbol{K})$ accounts for the sparsity of the latent flow **K**, and the terms $\mathfrak{L}_C(\gamma,\boldsymbol{K}), \mathfrak{L}_C(\gamma,\mu), \mathfrak{L}_C(\gamma,\mu,\boldsymbol{K})$ represent the penalties associated

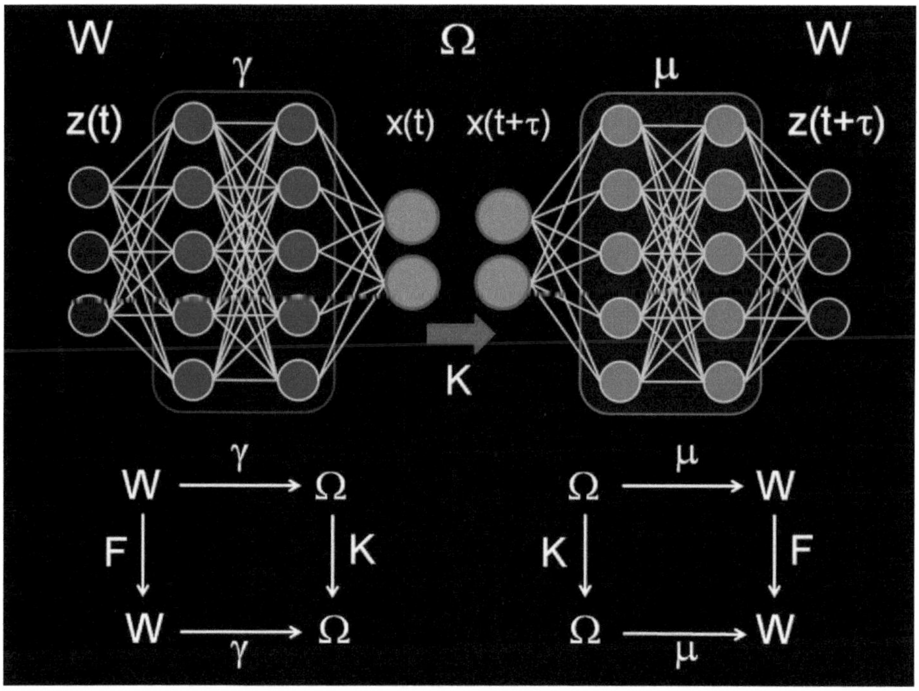

FIGURE 3.9 Scheme of autoencoder for latent dynamics optimizing simultaneously the dimensionality reduction and the parsimony of latent dynamical system governed by the flow $\boldsymbol{K}$. The optimization is extended further relative to the architecture displayed in Figure 3.8 by adding the extra term $\mathfrak{L}_s(\boldsymbol{K})$ to yield the loss functional $\mathfrak{L} = \mathfrak{L}(\Theta,\boldsymbol{K}) = \mathfrak{L}(\Theta) + \mathfrak{L}_s(\boldsymbol{K})$. The optimal $\Theta,\boldsymbol{K}$ pair realizing the variational minimum is precisely the one that makes the bottom diagrams commutative.

with imperfect diagram commutativity that need to be imposed to guarantee the dynamic compatibility of the **z** and **x** descriptions of the system. Thus, we get

$$\mathfrak{L}(\gamma,\mu,\boldsymbol{K})=\mathfrak{L}_r(\gamma,\mu)+\vartheta_s\mathfrak{L}_s(\boldsymbol{K})+\vartheta_C\left[\mathfrak{L}_C(\gamma,\boldsymbol{K})+\mathfrak{L}_C(\gamma,\mu)+\mathfrak{L}_C(\gamma,\mu,\boldsymbol{K})\right], \tag{3.5}$$

where the relative weights ϑ_s, ϑ_C are hyperparameters and

$$\mathfrak{L}_r(\gamma,\mu)=Q^{-1}\sum_{z\in\mathfrak{F}}\left\|z-(\mu\circ\gamma)z\right\|^2, \tag{3.6}$$

$$\mathfrak{L}_C(\gamma,\boldsymbol{K})=Q^{-1}\sum_{z\in\mathfrak{F}}\left\|(K\circ\gamma)z-(\gamma\circ F)z\right\|^2, \tag{3.7}$$

$$\mathfrak{L}_C(\gamma,\mu)=Q^{-1}\sum_{z\in\mathfrak{F}}\left\|(F\circ\mu\circ\gamma)z-(\mu\circ\gamma\circ F)z\right\|^2, \tag{3.8}$$

$$\mathfrak{L}_C(\gamma,\mu,\boldsymbol{K})=Q^{-1}\sum_{z\in\mathfrak{F}}\left\|(\mu\circ K\circ\gamma)z-(F\circ\mu\circ\gamma)z\right\|^2, \tag{3.9}$$

with $Q=|\mathfrak{F}|$ indicating the cardinal of the training set.

To determine the sparsity term $\mathfrak{L}_s(\boldsymbol{K})$, we first take the limit case where τ becomes the time infinitesimal dt and assume that $\dot{\boldsymbol{x}}=\boldsymbol{K}(\boldsymbol{x})=\boldsymbol{AP}(\boldsymbol{x})$, where $\mathbf{A}=[A_{ij}]$ is a $m\times p$ matrix ($m=\dim\Omega$) whose coefficients are variationally optimized jointly with the parametrization of the autoencoder maps γ and μ, and $\mathbf{P}(\mathbf{x})$ is a p-vector given by

$$\boldsymbol{P}(\boldsymbol{x})^T=\left(1\ x_1\ x_2\ x_3\ x_1^2\ x_2^2\ x_3^2\ x_1x_2\ x_2x_3\ x_1x_3\ x_1x_2x_3\ldots\right), \tag{3.10}$$

consisting of p functional terms. The terms do not need be of polynomial form, and they may be chosen in accordance with the type of symbolic regression. Thus, the variational optimization of $\boldsymbol{A}$ is tantamount to a symbolic regression in the latent manifold Ω. To ensure the sparsity of the model, we define the loss term as $\mathfrak{L}_s(\boldsymbol{K})=\|\boldsymbol{A}\|_F$, where $\|.\|_F$ is the Frobenius norm given by

$$\|\boldsymbol{A}\|_F=\left[\sum_{i=1}^{m}\sum_{j=1}^{p}A_{ij}^{\ 2}\right]^{1/2} \tag{3.11}$$

The incorporation of this term into the loss function given by Eq. 3.5 ensures that the latent manifold is selected to provide a parsimonious model with reduced dimensionality governed by the simplest possible set of differential equations in consonance with the time series provided. A different but equivalent formulation of the autoencoder has been given elsewhere [14] and uses the chain rule of derivation rather than diagram commutativity to infer the proper loss functional.

3.8 EMPOWERING MOLECULAR DYNAMICS WITH AI

As applied mathematicians make their forays combining dynamical systems with machine learning, they adopt specific systems that have traditionally represented showcases for model building.

Examples of such systems are the Lorenz attractor generated by three coupled ordinary differential equations, the spatio-temporal reaction–diffusion systems and the hydrodynamics and turbulence models governed by the Navier–Stokes equation [14,24]. Seldom, if ever, we had a chance to evaluate how autoencoder techniques pan out in the realm of molecular dynamics for ultra-complex multi-scale many-body systems. We are specifically referring to the discovery of models that distill the cooperative collective motion of ensembles of atoms, atomic groups, molecules, or molecular assemblages in condensed phases characterized as clusters, liquids, glasses, crystals, polymers, etc., under overarching categories such as soft matter and biological matter. The foundational approach to model discovery for such systems has been traditionally provided by statistical mechanics. These methods have met considerable success, except in the realm of biological matter, where problems such as the discovery of protein folding pathways, or the role of in vivo/cellular contexts in expediting the folding process cannot be even remotely addressed due to their sheer complexity and heterogeneity. Interacting molecular units are simply too diverse and the cellular environment is too complex at multiple scales and heterogeneous for statistical mechanics to make successful forays in biology [19,25,26]. This is precisely the context where AI may empower dynamical systems by leveraging highly specialized autoencoders and even batteries of autoencoders, as described subsequently.

To furnish a general framework, we may start by noting that we are dealing with N particles that may be free or tethered through covalent linkages forming assemblages such as biopolymers that may interact with one other through ephemeral or permanent associations that do not involve covalent bonds. A protein chain embedded in an aqueous environment and interacting with other biomolecular entities such as protein enzymes, chaperones, or ribosomes represents the quintessential situation that we wish to address in this book as we learn to leverage AI methods and integrate them on a model discovery platform.

Keeping for now the discussion at its most generic level, let us consider an ensemble of N particles that has associated with it an internal energy $\mathcal{U} = \mathcal{U}(\hat{z}), \hat{z} \in \mathbb{R}^{3N}$ only dependent on interparticle distances and hence invariant upon isometries – distance-preserving maps – of $\mathbb{R}^3$. Then, the state or configuration of the system may be represented as a point z in the quotient space $W = \mathbb{R}^{3N}/E^+(3)$, where $E^+(3)$ is the special Euclidean group of isometries of $\mathbb{R}^3$ including only rigid-body translations and rotations but excluding reflections [27]. The latter are excluded because they do not preserve the chirality (handedness) of asymmetric tetrahedral carbon groups, which constitutes a key constraint: Chirality is known to be strictly preserved in biology. Thus, W is (3N-6)-dimensional and its coordinates represent all the internal degrees of freedom of the system. In addition to the reduction modulo isometries, another quotient is required to represent the latent manifold Ω. This further dimensionality reduction depends on the constraints brought about by the covalent linkages that tether specific units in the system. For example, if we are studying the dynamics of protein folding, we note that high-frequency vibrational motions involving covalently paired atoms may be averaged out as their associated timescales, in the femtosecond (fs) range [28], are incommensurably shorter than those associated with soft modes represented by dihedral torsions of the polymer backbone (Figure 3.10), typically in the nanosecond (ns) or sub-ns range. A similar reduction applies to planar angular vibrations, with frequencies in the order of $(\text{fs})^{-1}$ to $(10\ \text{fs})^{-1}$ [19,28]. These simplifications point to a parsimonious model representing an adiabatic system that incorporates only soft modes of the chain. Thus, as protein folding is represented as a dynamical system with the protein chain searching in conformation space in an *in vitro* setting, Ω becomes a compact manifold in the form of a 2M-torus, where M is the number of amino acid units in the chain [20]. This latent manifold is deduced assuming that the water molecules surrounding the protein chain that explores conformation space are treated implicitly; that is, their influence is energetically subsumed in the potential energy function U only dependent on distances between the protein chain subunits and the local configurations that result thereof.

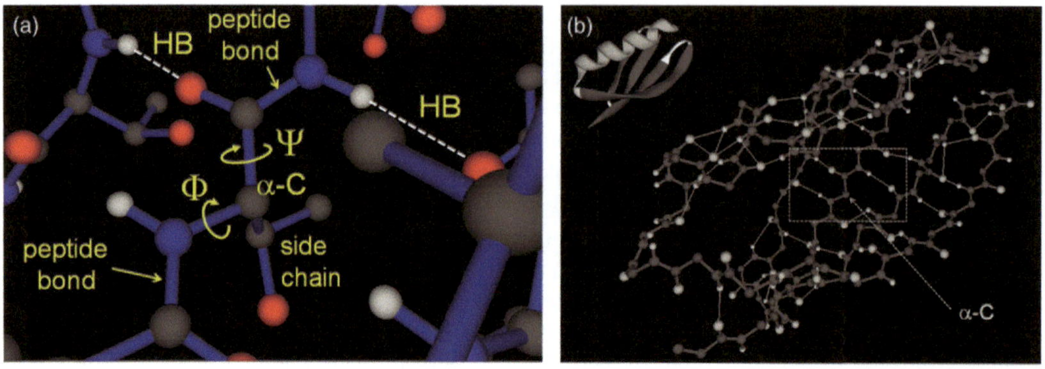

FIGURE 3.10 Latent coordinates for a folding protein chain. (a) Individual residue unit within a protein chain attached to an adjacent residue through a torsionally rigid linkage known as peptide bond. For the sake of illustration, a particular protein has been selected, namely, the thermophilic variant of the B1 domain of protein G from *Streptococcus*, whose native 3D structure is found at the entry 1GB4 in the protein data bank (PDB). In the latent coordinate frame, the state of each residue is represented by a pair of coordinates (Φ, Ψ) representing the torsional dihedral degrees of freedom of the protein backbone. The amino acid type for the residue, describing the local chemical composition of the chain, is identified by the side-chain group that is covalently linked to the alpha-carbon in the protein backbone. In turn, polar groups in the backbone and side chain can be engaged in orientationally and distance-dependent noncovalent linkages known as hydrogen bonds (HB, dashed lines), whose cumulative folding-stabilizing effect becomes a determinant of the protein 3D structure. (b) Zoom out of the detail shown in (a), revealing the backbone HB pattern and topology (ribbon rendering) of the native structure. Backbone HBs exposed to water are indicated by thin green lines (buried ones in gray) and they signal structural deficiencies, since the structure may get locally disrupted when the polar groups paired by intramolecular HBs get hydrated, that is, interact with surrounding molecules of the aqueous solvent.

The dynamics in W obeys the basic physical law:

$$\dot{z} = v; \; \dot{v} = -\nabla_{\mathbf{z}} U. \tag{3.12}$$

To properly define $U = U(z)$, we introduce the following.

Theorem 3.1. There exists a map $U : \mathbb{R}^{3N}/E^{+}(3) \to \mathbb{R}$ that makes the following diagram commutative:

$$\begin{array}{ccc} \mathbb{R}^{3N} & \xrightarrow{\mathcal{U}} & R \\ \pi \downarrow & \nearrow U & \\ \mathbb{R}^{3N}/E^{+}(3) & & \end{array} \Rightarrow \mathrm{U} \circ \pi = \mathcal{U}, \tag{3.13}$$

where $\pi : \mathbb{R}^{3N} \to \mathbb{R}^{3N}/E^{+}(3)$ is the canonical projection associating each point in $\mathbb{R}^{3N}$, specifying the 3D coordinates of each of the N particles of the system, with the set of points in $\mathbb{R}^{3N}$ resulting from rotations and translations of the N-particle system treated as a rigid body that is preserving all inter-particle distances. Thus, the projection associates a state of the system with its class in quotient space, that is, with the collection of all points contained in the group orbit generated by the action of the Euclidean group on the point in the domain.

To prove this "factorization" result, it suffices to note that the potential energy $\mathcal{U} : \mathbb{R}^{3N} \to \mathbb{R}$, $\mathcal{U} = \mathcal{U}(\hat{z}), \hat{z} \epsilon \, \mathbb{R}^{3N}$ is invariant on the orbits of $E^{+}(3)$ in $\mathbb{R}^{3N}$.

An autoencoder (γ, μ) yielding a parsimonious model in Ω would require that the autoencoder constructs $\tilde{U}(x) = U(\mu(x))$, so the dynamic equations in the latent manifold become

$$\dot{x} = \tilde{v};\ \dot{\tilde{v}} = -\nabla_x \tilde{U}(x). \tag{3.14}$$

This implies that in the infinitesimal limit $\tau = dt$, the potential energy function may be obtained by noting that the sparse map $\boldsymbol{K}$ defined by the autoencoder obeys

$$\dot{x}(t) = \boldsymbol{K}(x(t)) = -\int_0^t \nabla_x \tilde{U}(x(t'))dt' + \dot{x}(0) \tag{3.15}$$

On the other hand, if the time series used to train the encoder is generated using molecular dynamics governed by potential energy function $U = U(\mathbf{z})$, then an additional loss term in $\dot{x}$ of the form

$$\mathfrak{L}_U(\mu, \boldsymbol{K}) = \left\| \boldsymbol{K}(x(t)) - \nabla_z x(t) \left[-\int_0^t \nabla_z U(z(t'))dt' + \dot{z}(0) \right]_{z(t')=\mu(x(t'))} \right\|^2 \tag{3.16}$$

needs to be incorporated to optimize the autoencoder and its associated propagator $\mathbf{K}$.

3.9 PHYSICS ON LATENT MANIFOLDS

The sheer complexity of the molecular dynamics arising in soft and biological matter, especially in *in vivo* settings where $N \sim 10^6$–10^7 including solvent molecules, is unlikely to ultimately allow for the type of reductive approach that standard autoencoders usually provide (Figure 3.9). A case in point is the discovery of the physical underpinnings of protein folding assisted by an *in vivo* context that enhances the expediency of the process (Figure 3.11) [29]. The space is often anisotropic, the system itself is highly heterogeneous, and its components are too diverse, with potentials or force fields that cannot fully account for the complexities and many-body effects enshrined in the time series. It is doubtful that the latent compact manifold spanning soft mode coordinates (backbone torsional dihedrals in the case of the folding protein, see Figure 3.10) will be amenable to the type of model discovery that is usually cast in terms of sparse differential equations. For example, generating a minimal set of 2M (M~100) coupled differential equations that govern the backbone torsional dynamics underlying the protein folding process with implicit treatment of the environment is out of reach given current capabilities in deep learning.

Other many-body systems share similar problems, as their wanton complexity is off limits for state-of-the-art autoencoders seeking to identify parsimonious models with differential equations. Yet, as we shall now show, a generic topological understanding of the latent dynamics may yield a way of learning dynamic data that enables suitable propagation of the time series into the future, endowing the autoencoder with predictive value. Thus, topological methods will be readily incorporated in AI-based model discovery for systems with multi-scale complexity.

The approach entails a simplification based on the topological dynamics in the latent manifold that is on the dynamics modulo the basins of attraction of the generic singular points of the map $\boldsymbol{K} : \Omega \to T\Omega$, where $T\Omega$ denotes the tangent bundle of Ω and Ω itself is assumed to be C^1-differentiable and compact [30], as it is the case in the previously discussed example.

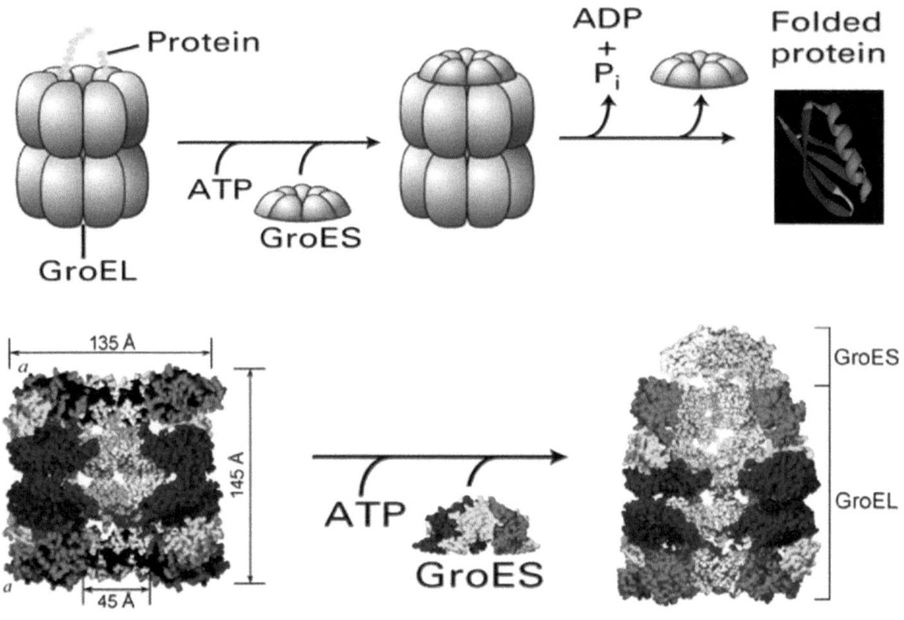

FIGURE 3.11 *In vivo* setting assisting the folding of a protein chain. A molecular cage known as chaperone GroEL assists the folding process, enhancing its expediency well above the level of efficiency that may be achieved *in vitro*, i.e., in the test tube. The cage consists of a dimer of two annular molecular assemblages, each consisting of a complex of seven identical proteins. Each protein is made of three regions known as apical, intermediate, and equatorial domain, denoted, respectively, "a, i, e" in the figure. These domains undergo a certain amount of conformational rearrangement upon binding to the cell-fuel molecule ATP (adenine triphosphate). This rearrangement in turn enables the cage dimer to incorporate a molecular lid, known as GroES. Once the lid is on, the protein inside the cage is subject to a number of iterative annealing steps through interactions with the proteins lining the interior of the cage. This annealing process enables the protein to avoid getting kinetically trapped in misfolded states. Upon release of ATP, the lid becomes detached and the protein exits the system regardless of whether it has satisfactorily completed the folding process or not. If the latter is the case, additional catalytic cycles engaging the same or other folding assistants may be necessary.

To make further progress with the argument, we first prove the following result:

Theorem 3.2. Under the assumptions of Section 3.9, the autoencoded map $\boldsymbol{K}$ yields no closed orbits in Ω.

By *reductio ad absurdum*, let us assume $\boldsymbol{x}(0) = \boldsymbol{x}(T)$ for $T > 0$. Then, we get

$$
\begin{aligned}
0 &= \int_0^T \tilde{U}\big(\boldsymbol{x}(t)\big)dt = \oint_0^T d\Big[\int_0^t \tilde{U}\big(x(t')\big)\mathrm{d}t'\Big] \\
&= \int_0^T \Big[\nabla_x \int_0^t \tilde{U}\big(\boldsymbol{x}(t')\big)\mathrm{d}t'\Big]\dot{\boldsymbol{x}}(t)\,dt = \int_0^T \Big[\int_0^t \nabla_x \tilde{U}\big(\boldsymbol{x}(t')\big)\mathrm{d}t'\Big]\dot{\boldsymbol{x}}(t)\,dt = -\int_0^T \|\tilde{\boldsymbol{v}}\|^2\,dt \le 0. \qquad (3.17)
\end{aligned}
$$

Eq. 3.17 implies that $\tilde{\boldsymbol{v}} \equiv 0$ which is absurd since $\boldsymbol{x}(0)$ was not assumed to be a steady state but a point in a closed orbit. Q.E.D.

Given that Ω is a compact manifold, this result has far-reaching consequences [30,31]:

1. The latent dynamics have no attractors made up of recurrent orbits, and
2. since there is no circulation around them, all singular points of the latent flow are hyperbolic, hence generic, since the real part of the eigenvalues of the Jacobian at the singular points cannot be zero.

This implies the following result:

Corollary 3.1. The latent dynamics governed by equations (3.13) are of the Morse–Smale type [31], hence structurally stable, that is, qualitatively (topologically) invariant under small perturbations.

To rigorously define structural stability, we first note that the space $\mathfrak{H}(\Omega)$ of smooth maps $\Omega \to T\Omega$ is endowed with a natural metric inherited from the supremum norm given by

$$\begin{aligned}\|H\|_{sup} &= Sup_{x\in\Omega, j=1,\ldots,dim\Omega}\left[\left|\boldsymbol{H}(\boldsymbol{x})\right|, \left|\frac{\partial}{\partial x_j}\boldsymbol{H}(\boldsymbol{x})\right|\right] \\ &= Max_{x\in\Omega, j=1,\ldots,dim\Omega}\left[\left|\boldsymbol{H}(\boldsymbol{x})\right|, \left|\frac{\partial}{\partial x_j}\boldsymbol{H}(\boldsymbol{x})\right|\right]\end{aligned} \quad (3.18)$$

for $H \in \mathfrak{H}(\Omega)$. Then, to state that the latent flow $\boldsymbol{K}(\boldsymbol{x})$ is structurally stable means that for any given Δ-neighborhood of K, $\mathfrak{B}_\Delta(\boldsymbol{K}) \subset \mathfrak{H}(\Omega)$, there exists a value $\varepsilon = \varepsilon(\Delta)$, such that for any $\boldsymbol{G} \in \mathfrak{B}_\Delta(\boldsymbol{K})$ there exists an ε-homeomorphism $h_\varepsilon : \Omega \to \Omega$ satisfying $\max_{x\in\Omega}\left|\boldsymbol{x} - \boldsymbol{h}_\varepsilon(\boldsymbol{x})\right| < \varepsilon$ that transforms trajectories of $\boldsymbol{K}$ onto trajectories of $\boldsymbol{G}$ [31].

Given the qualitative invariance of the latent flow under small perturbations, the following observation is key to justify the leverage of AI to construct dynamic models based on time series: *The structural stability of the latent dynamics is essential to enable model discovery in view of the fact that the exact parameters determining the potential energy U of the many-body system are not known precisely.*

Given the characterization of the latent flow given by Theorem 3.2 and its Corollary 3.1, we may encode the flow in a simplified manner, as we now build a metamodel. To that effect, we first define the equivalence relation "~" for any pair $\boldsymbol{x}, \boldsymbol{y} \in \Omega : \boldsymbol{x} \sim \boldsymbol{y} \Leftrightarrow \boldsymbol{\omega}(\boldsymbol{x}) = \boldsymbol{\omega}(\boldsymbol{y})$, where $\boldsymbol{\omega}(\boldsymbol{x})$ denotes destiny (omega) state given by the limit at $t \to \infty$ of the trajectory initiated at $\boldsymbol{x}$ [32]. Since the singular points are the steady states of the system, the latent flow may be encoded by the equivalence classes identified as the basins of attraction of the singular points. As points in Ω are regarded "modulo basins", we have in effect defined the quotient space $\Omega/\sim$ as the set of basins of attraction and separatrices (basins of lower dimension) of critical generic points that partition the manifold. The quotient space is relatively simple to encode since the singular points of the latent flow are isolated and finite in number. To demonstrate this proposition, we note that otherwise, if they were infinite in number, they would have an accumulation point since the latent manifold is compact and that cannot happen because all singular points are generic.

Let us then denote by $\Gamma : \Omega/\sim \to \Omega/\sim$ the coarse-grained flow that determines the interbasin transitions, while $\pi : \Omega \to \Omega/\sim$ denotes the canonical projection that associates a point in the latent manifold with its equivalence class. The flow must be such that the diagram in Figure 3.12 becomes commutative, and specifically, the following flow–compatibility relations must hold:

$$\mu^{\#} \circ \pi = \mu, \quad (3.19)$$

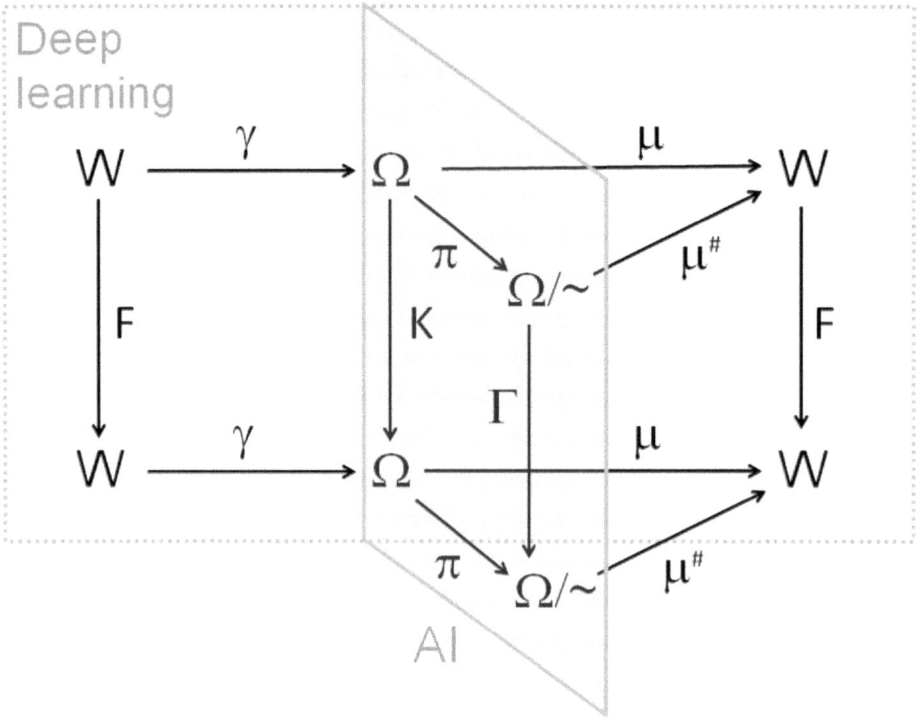

FIGURE 3.12 Scheme of a metamodel consisting of two coupled variational autoencoders (γ,μ), $(\pi,\mu^{\#})$ required to generate the discrete flow Γ that propagates the topological dynamics in the latent quotient manifold $\Omega/\sim$. The parameter optimization of both autoencoders ensures the full commutativity of the diagram. The discovery of the topological metamodel $\Gamma:\Omega/\sim\longrightarrow\Omega/\sim$ hinges on a CNN-based construction of the modulo-basin projection $\pi:\Omega\rightarrow\Omega/\sim$ of the latent dynamics. This scheme introduces a level of coarse graining that is more drastic than that adopted by conventional autoencoders for model discovery of time-series data.

$$\Gamma\circ\pi=\pi\circ K, \tag{3.20}$$

$$\mu^{\#}\circ\Gamma=F\circ\mu^{\#}. \tag{3.21}$$

Thus, to parametrize Γ, we need to introduce a second autoencoder with variational parameter optimization determined by the loss function

$$\mathfrak{L}_{\sim}\left(\mu^{\#},\,\Gamma\right)=\mathfrak{L}_{\mu}\left(\mu^{\#}\right)+\mathfrak{L}_{\sim}\left(\Gamma\right)+\mathfrak{L}_{\sim}\left(\mu^{\#}\right), \tag{3.22}$$

where

$$\mathfrak{L}_{\mu}\left(\mu^{\#}\right)=Q^{-1}\sum_{x\in\mathfrak{F}}\left\|\left(\mu^{\#}\circ\pi-\mu\right)x\right\|^{2}, \tag{3.23}$$

$$\mathfrak{L}_{\sim}\left(\Gamma\right)=Q^{-1}\sum_{x\in\mathfrak{F}}\left\|\left(\Gamma\circ\pi-\pi\circ K\right)x\right\|^{2}, \tag{3.24}$$

$$\mathfrak{L}_{\sim}(\mu^{\#}) = Q^{-1} \sum_{x \in \mathfrak{F}} \left\| (\mu^{\#} \circ \Gamma - F \circ \mu^{\#}) \pi x \right\|^2, \tag{3.25}$$

where the training set is denoted $\mathfrak{F} \subset \Omega$, with $Q = |\mathfrak{F}|$, and $\mu^{\#}$ denotes the decoder for the quotient space.

The commutativity of the diagram in Figure 3.12 implies that the ultimate simplicity in a model governing ultra-complex many-body problems, such as identifying *in vivo* protein folding trajectories, may be achieved by projecting the latent dynamics onto the quotient manifold $\Omega/\sim$.

3.10 QUOTIENT SPACE AND AUTOENCODER OF EXTRA DIMENSIONS

The sparse latent dynamics obtained by leveraging autoencoders that serve as model discoverers has become the subject of intense research in applied mathematics. Such methods are less suited to unravel underlying laws in dynamical systems that represent biological or soft matter, where the number of internal degrees of freedom is a stronomical. This book proposes to couple two commutative – hence compatible – autoencoders as described in Figure 3.12, yielding a factorization of the latent dynamics through the quotient space $\Omega/\sim$. The commutativity of the whole diagram displayed in Figure 3.12 ensures the compatibility of the different levels of coarse graining of the dynamics that constitute the metamodel. In turn, the diagram commutativity (Eqs. 3.19–3.21) is subsumed into the variational functional of the autoencoder (Eqs. 3.22–3.25) so that the optimization of the underlying neural networks is equated with – or rather becomes as close as possible to – diagram commutativity. Identifying the quotient space $\Omega/\sim$ under the "modulo-basin" equivalence relation defined in the previous section is akin to a pattern recognition process, where time-series datapoints are plotted onto 2D cross sections of the latent manifold Ω. The task may be entrusted to a CNN and becomes enormously simplified as Corollary 3.1, jointly with Theorem 3.2, guarantee a finite number of basins of attraction for isolated singular points which are all generic, that is, topologically equivalent under small perturbations of the latent vector field.

For example, in the case of the folding protein, the 2D cross section is the (Φ,Ψ)-torus (Figure 3.10a), and a typical time series for latent dynamics governed by Eq. 3.14 is given in Figure 3.13 [19,20]. The CNN "discovers" the topological metamodel which can be represented as a graph with vertices corresponding to the basins of attraction of the generic minima and edges connecting basins of attraction (Figure 3.13) in a manner specified by the autoencoder-generated information fed onto the CNN. More specifically, the graph is generated in accordance with the inferred topography of the two-torus cross section of the potential energy map $\tilde{U}: \Omega \to \mathbb{R}$, in turn generated by the autoencoder.

To generate the entire quotient space within a graph representation, we need to integrate the cross sections corresponding to the 2D projections of $\Omega/\sim$. As an illustration, let us consider a GG dipeptide (M = 2, G = glycine). The latent manifold is a Cartesian product of two tori (Figure 3.14), one for each pair of Φ,Ψ dihedrals coordinates specifying the backbone conformation of the respective residue [20]. The topological representation of the vector field steering the backbone torsional dynamics of a generic residue in the protein chain is given in Figure 3.13. Opposite sides of the square are identified as per the $\pm 180°$ identification of Φ,Ψ dihedrals determining the local torsional state of the backbone. The four colored sectors morph topologically into the allowed valleys in potential energy. The organization of the basins of critical points is compatible with the underlying two-torus and topologically represented by a graph (Figure 3.13) for each residue. The bottom panel in Figure 3.14 represents the quotient space for the dipeptide chain [20]. For each residue, the quotient space cross section is represented by a graph with vertices indicating two-dimensional basins, and an edge linking two vertices indicates that the respective basins are connected through a line of steepest descent crossing a saddle point and orthogonal to the separatrix at the saddle. For a protein chain consisting of two consecutive residues, denoted 1 and 2, the quotient space becomes the

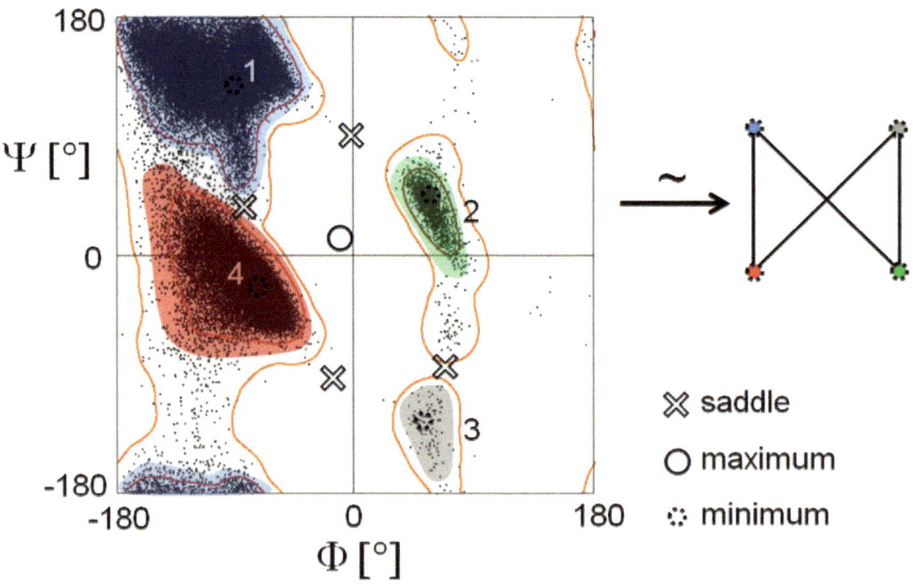

FIGURE 3.13 Toroidal (Φ, Ψ) cross section of the time-series data for an individual unit along a folding protein chain. The associated cross section of the dynamical system representing the protein folding process is discovered through pattern recognition leveraging a CNN, and its modulo-basin representation in the cross section of the latent quotient manifold is given as a graph. The vertices in the graph are critical points corresponding to omega-sets for all points in their respective basins of attraction, while the edges indicate allowed interbasin transitions that determine the topological dynamics.

tensor product of the two graphs where vertices now denote basin pairs (B_1,B_2), and where basin pair (B_1,B_2) is connected with (B'_1, B'_2) if and only if B_1 is connected with B'_1 or $B_1 = B'_1$, and B_2 is connected with B'_2 or $B_2 = B'_2$. Thus, the quotient space for two residues consists of $4 \times 4 = 16$ vertices, where each vertex connects via one edge with eight other vertices and connects via two adjacent edges to the seven remaining vertices. Given the symmetry of the problem, to prove the assertion, it suffices to note that the vertex denoting basin pair (1,1) is directly connected to eight other vertices denoting pairs (2,1), (4,1), (1,2), (1,4), (2,2), (2,4), (4,2), and (4,4), while vertex (1,1) connects to all the remaining seven vertices–pairs containing basin 3 via two adjacent edges.

Thus, the projection on quotient space $\Omega/\sim$ of a latent MD trajectory on the 4-torus Ω becomes a walk on the tensorial product graph at the bottom of Figure 3.14 [20].

Kurzweil and others have successfully argued that a hierarchical structure of reality is necessary for proper encoding and processing in a suitable AI-based inferential framework [22,33]. To cast the discussion in the broadest terms, we shall refer to hierarchical as the attribute of a system endowed with nested complexities at multiple scales arising at different levels of description and providing different levels of coarse graining. This notion was delineated previously in this chapter, where a number of illustrations reveal the dynamic entrainment of fast-relaxing modes by slower modes spanning a latent manifold. Thus, just like in the adiabatic approximation [16], fast motions are averaged out and hence treated implicitly in a coarse-grained version of the dynamics focusing on longer timescales, with an autoencoder providing the inferential framework to discover the latent coordinate system.

Indeed, an escalation in the level of coarse graining, from the subatomic to the atomic, molecular, subcellular, and beyond, has always suggested that the hierarchical structure of reality may span

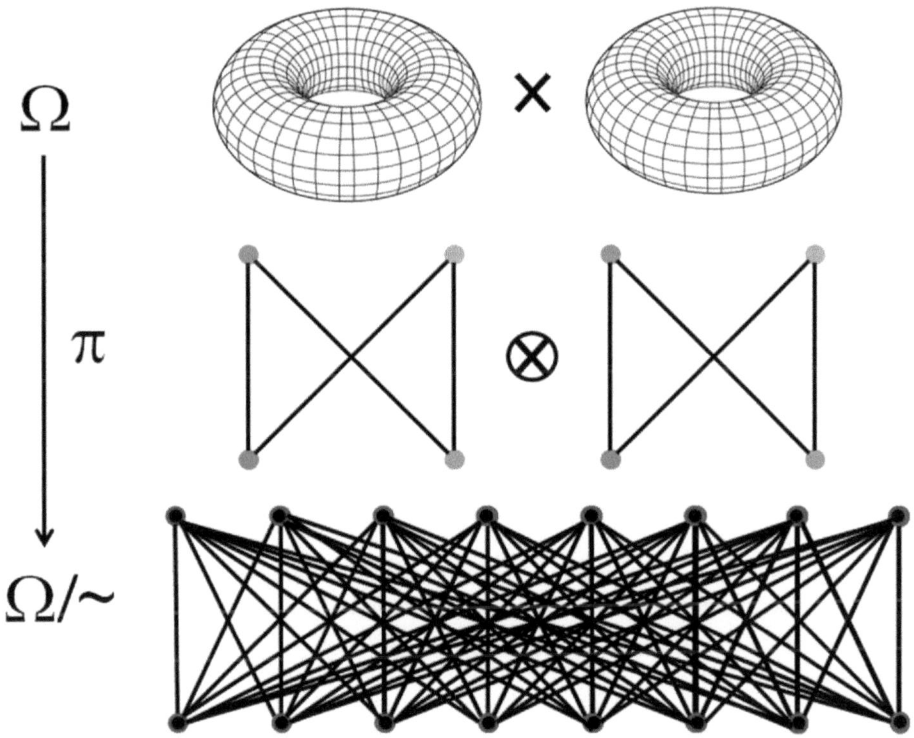

FIGURE 3.14 Reconstruction of the latent quotient manifold. The iterative process is represented by progressively incorporating adjacent-residue cross sections of the latent manifold as Cartesian products and, in parallel, reconstructing the quotient manifold as tensorial product of the individual graphs representing the cross sections of the quotient manifold (Figure 3.13). Within this representational framework, the topological dynamics becomes a walk in the quotient manifold graph.

over several layers, with nested complexities where information at a peripheral layer is incorporated implicitly at a core layer. In topological terms, we envision a whole sequence of quotient manifolds, as we lump up states at different levels of description within progressively coarser equivalence classes, with dynamic compatibility of the various descriptions imposed by commutative flow diagrams (cf. Figure 3.15). Thus, in the broadest sense, an encoding of a hierarchical dynamical system within an AI-based inferential framework may require a battery of tandem autoencoders, whose dynamic compatibility in an optimized parametrization is ensured by the commutativity of the diagram that combines the different levels of flow encoding. Figure 3.15 shows one such diagram reflecting the dynamic interplay of three autoencoders, one to generate the latent manifold Ω of intrinsic coordinates, one for a first-level quotient space $\Omega/\sim$, and one for a second-level (coarser) quotient space $\Omega/\sim/\approx$. The autoencoders are variationally optimized so that the flow diagram becomes commutative, which in turn reflects the compatibility of the different levels of dynamic encoding within the latent three-level hierarchy.

The subsequent chapters illustrate the implementation of tandem autoencoder technology to treat implicitly the hierarchical structure of the molecular dynamics of biological and biomedical processes. The relationship between hierarchical structure and tandem autoencoders of nested coarse-graining descriptions is of universal applicability and may be extended to any number of equivalence relations, leading to progressively simplified metamodels.

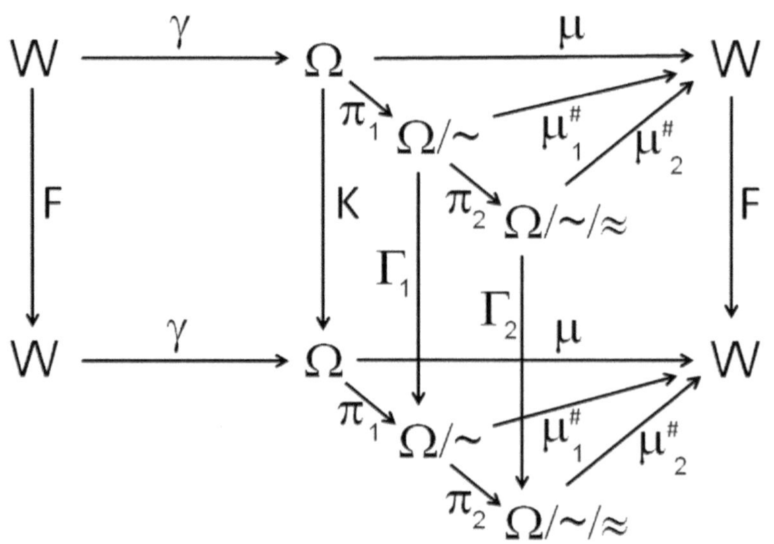

FIGURE 3.15 Metamodels for hierarchical dynamical systems. Autoencoders in tandem implemented to forecast dynamics that allow for a hierarchical description defined by two equivalence relations "~" and "≈". Two autoencoders $\{\pi_1, \mu^{\#}_1, \Gamma_1\}$ and $\{\pi_2, \mu^{\#}_2, \Gamma_2\}$ operating sequentially as required for topological encoding of the dynamics at the coarsest level in the latent quotient manifold $\Omega/\sim/\approx$. The F-compatibility of the latent flows K, Γ_1, Γ_2 in Ω, $\Omega/\sim$, $\Omega/\sim/\approx$, respectively, requires the full commutativity of the diagram. This commutativity is achieved by variational parameter optimization for the three autoencoders $\{\gamma, \mu, K\}, \{\pi_1, \mu^{\#}_1, \Gamma_1\}$ and $\{\pi_2, \mu^{\#}_2, \Gamma_2\}$.

3.11 QUANTUM AUTOENCODER OF GRAVITY

A theory of quantum gravity (QG), that is, a theory of gravity in accordance with the tenets of quantum mechanics (QM), is often regarded as off limits due to the major disparity in the dimensional scales where quantum and gravitational forces materialize [34, 35]. So far, gravity has not been quantized, and hence, a sense of incompleteness pervades the field. The unification of all forces of nature remains a holy grail in physics ever since Einstein's pursuit.

A different sort of grand unification was pursued by Albert Einstein. Since weak and strong nuclear forces were not clearly understood at the time, he sought to do it without involving QM, a theory he never fully endorsed. At first glance, quantum gravity stands almost as an oxymoron: After all, QM deals with the atomic and subatomic scales, while the best theory of gravity to date is Einstein's general relativity (GR), which is essentially classic, i.e., nonquantum, and deals mainly with cosmological scales (except for singularities). Einstein's theory of relativity postulates that high concentrations of energy and matter impinge on the curvature of space–time, deflecting the trajectories of particles, as it occurs in a gravitational field. This theory withstood admirably the long-term attrition of experimental corroboration. Yet, if we attempt to cast GR in QM terms, we need to deal with the fact that matter and space-time become "protean" at scales of the order of Planck's length (10^{-33} cm), akin to the sea of virtual particles that fill up empty space. In this essentially quantum world, the equations of GR no longer hold.

Pursuing QG still makes good sense when we deal with GR singularities, for example, black holes (BHs) or the first few sub-attoseconds after the big bang, when the vast differences in material scales that set apart QM and GR can be reconciled, prior to the inflationary phase. Thus, in the spirit of this book, we need to address the question: *what would constitute a quantum metamodel of a theoretical model of gravity and how to construct it?*

This question was effectively formulated by the Argentinian physicist Juan Maldacena [34]. He focused on the anti-de Sitter (AdS) space, a hyperbolic space that shares curvature properties with the sphere representing the event horizon of a black hole. Maldacena postulated that a string theory (ST) of gravity in a five-dimensional AdS space ($AdS_5 = W$ in standard notation) is *equivalent* to a quantum field theory (QFT) on its boundary ∂W, which constitutes a four-dimensional Minkowski space, akin to the one Einstein adopted for GR.

How could the boundary represent the latent space for the ST in an AdS? The answer is provided by the computation of the Shannon entropy of the black hole that assesses its total information storage capacity contained in all degrees of freedom [36]. These "ultimate" degrees of freedom involve of course atomic and subatomic entities, all the way to quarks and gluons, and ultimately those entities from hitherto unknown depths in the physical structure of matter, the so-called "level X". This "ultimate entropy" is proportional to the surface area of the event horizon [36].

In general, the boundary M_d of AdS_{d+1} is a d-dimensional Minkowski space with the symmetry group $SO(2, d)$ of AdS_{d+1} acting on M_d as the conformal (i.e., inner product-preserving) group. Thus, there are two ways to get a physical theory with $SO(2,d)$ symmetry: a relativistic field theory on AdS_{d+1} and a conformal field theory (CFT) on M_d. A suitable theory on AdS_{d+1} has been conjectured by Maldacena to be *equivalent* to a CFT on M_d [34]. The computation of observables of the CFT in terms of supergravity on AdS_{d+1} can and should be attempted using the methods described in this book. In accordance with the tenets of topological metamodel auto-encoding, M_d should be identified with the latent manifold Ω and the CFT on M_d "holographically" spanned onto AdS_{d+1} should be generated by an holographic autoencoder that exploits the $SO(2, d)$ to generate jointly the latent manifold together with its parsimonious metamodel.

Correlation functions in conformal field theory (QFT) are given by the dependence of the supergravity action on the asymptotic behavior at infinity [35]. Thus, dimensions of operators in CFT are determined by masses of particles in string theory. It is thus conjectured that to describe the Yang–Mills theory in four dimensions, one should use the whole infinite tower of massive Kaluza–Klein states on $AdS_5 \times \mathbf{S}^5$. Chiral fields in the four-dimensional $N = 4$ theory correspond to Kaluza–Klein harmonics on $AdS_5 \times \mathbf{S}^5$. The spectrum of Kaluza–Klein excitations of $AdS_5 \times S^5$ are matched against operators of the $N = 4$ theory.

To discover the topological metamodel for superstring theory on $W = AdS_{d+1} \times S^{d+1}$, we first note that the boundary is topologically identified as $\partial W = \left(S^1 \times S^{d-1}\right)/\mathbb{Z}_2$, where the group $\mathbb{Z}_2$ acts by rotation in π on S^1 and multiplication by -1 on S^{d-1}. In other words, the latent manifold fulfills the compactness condition, and to simplify the computation and straighten the symmetry, we may use its universal cover [37], as demonstrated by Ariel Fernández and Oktay Sinanoglu: $\Omega \approx S^{d-1} \times \mathbb{R}$, with the real axis representing the time dimension.

Thus, to prove using AI the conjectured equivalence between $N = 4$ QFT on $\Omega \approx S^{d-1} \times \mathbb{R} = \partial W$ and type IIB supergravity as string theory (ST) on $W = \langle AdS_5 \times S^5 \rangle$ ($\langle . \rangle$ = universal cover [37]), we need a holographic autoencoder for a DL neural network capable to represent the ST on the space W. This DL system has been constructed [38]. The autoencoder should identify the latent manifold as $\Omega \approx S^3 \times \mathbb{R}$ taking advantage of the $SO(2,4)$ symmetry with which the AdS space is endowed. In this way, the projection π onto the latent manifold, identified by the autoencoder as ∂W, constitutes the inverse of the postulated holographic map $h : \partial W \rightarrow W$ (Figure 3.16) that makes the diagram on Figure 3.17 commutative.

Thus, the AdS/CFT equivalence may be proved by AI. The program entails two steps: (a) generation of a data-driven holographic metamodel of quantum systems by formulating its supergravity dual/equivalent on a DL NN and (b) construction of the appropriate autoencoder (π, K_{QFT}) that fits the holographic map h, yielding the functional identity $h \circ K_{QFT} \circ \pi = F_{ST}$ of homeomorphisms on W. Part A of the program has already been achieved, as shown elsewhere [38]. A deep NN

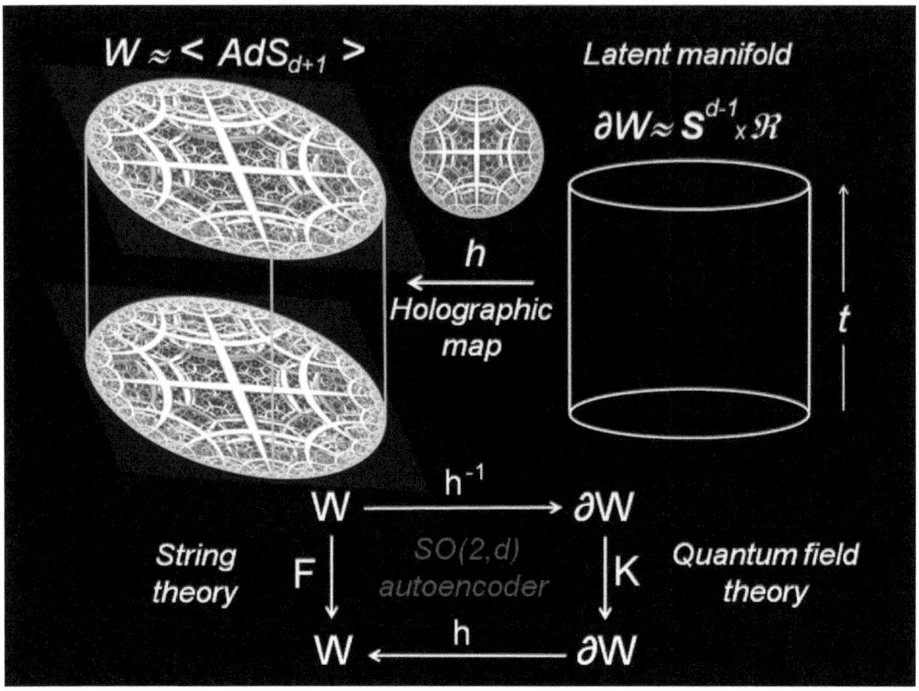

FIGURE 3.16 Holographic autoencoder enabling the discovery of a quantum metamodel of gravity.

$$
\begin{array}{ccc}
W & \overset{\pi}{\rightarrow} & \partial W \\
\downarrow \mathrm{F_{ST}} & & \downarrow K_{QFT} \\
W & \overset{h}{\leftarrow} & \partial W
\end{array}
$$

FIGURE 3.17 Commutative diagram for a holographic autoencoder.

representation of the AdS/CFT correspondence has been obtained, with the emergence of the bulk metric based on deep learning of data generated as outcome of boundary quantum field theories. The radial direction of the bulk metric is assimilated with the depth of the hidden layers. Thus, the network provides a data-driven model of strongly coupled systems. In the spacetime for a black hole horizon, the deep NN can fit boundary data generated by the AdS Schwarzschild spacetime, reproducing the metric as the data are reverse engineered. With inputted experimental data, the deep NN determines the bulk metric, mass, and quadratic coupling. Thus, the NN provides a gravitational model of strongly correlated systems.

As for part B, the holographic autoencoder (HAE) that will ultimately identify the latent manifold with ∂W can be obtained using the methods described in this chapter. The remaining challenge in part B is for the HAE to simultaneously yield the parsimonious QFT at the boundary ∂W that serves as metamodel for the gravity model on W in the sense adopted in this book. To address this challenge requires that we harness the isomorphism already noted [39] between a deep autoencoder and the multi-scale entanglement renormalization ansatz (MERA).

If the universe is indeed a hologram, as Maldacena and others suggest [34], the inverse of the holographic map of its avatar, the event horizon of the black hole, represents a 1-1 (injective) projection onto the horizon boundary. Therefore, this boundary may be regarded as a model for the latent quotient manifold of the universe. In this quotient manifold, one dimension has been folded up and stored within the equivalence classes that are ultimately the only observables, and hence, the only encoded features endowed with physical entity. This topology for the quintessential universe is not entirely correct but locally correct, as shown in Chapter 4. The overall topology of the universe is immaterial to this discussion, as both quantum physics and GR are in fact local theories.

Ultimately, quantum gravity will be ascertained through a holographic autoencoder that identifies the "correct" latent manifold and associated quantum metamodel, or perhaps by other means available to theoretical physicists. Be as it may, we may state that all events following the Planck epoch of the big bang ($t > 10^{-43}$ s [40]) are likely to be of a quantum nature. This is so because at that stage, gravity branches off from the three other fundamental forces already accounted for by quantum theory, so all four differentiated forces will be reliably identified as quantum forces. A rigorous topological characterization of the universe supports this assertion, as shown in Chapter 4, with gravity originally stored as a de Broglie wave in a compact fifth dimension. Thus, the certainty of all events that follow the Planck epoch can only be established through the participation of observers. This singular circumstance leaves us having to postulate God's existence at least as early as 10^{-43} s after the big bang, or admitting that the universe remains a mere possibility, replete with phenomena-to-be within multiple *a priori* potential realities that are equally possible in the quantum realm, as in the multiverse scenario.

3.12 AI ENCODES QUANTUM GRAVITY

To address the problem of quantum gravity from an AI-borne perspective, we need to design a holographic autoencoder by leveraging the physics of machine learning to address the problem of whether emergent quantum behavior can arise in a neural network. By emergent quantum mechanics, we mean a formulation within a framework of nonlocal hidden variables, as in the Bohm scheme [41]. Once emergent quantum behavior is shown to become possible in machine learning physics, we may address the question of developing a relativistic string gravitational scheme on the hidden variables adopted in a Bohm-inspired quantum network architecture. Thus, the latter becomes in effect a quantum gravity autoencoder for the network materializing emergent gravity.

To develop a network with emergent quantum behavior, we first need to focus on NNs as statistical mechanics entities and closely examine the statistical physics that may underlie generic machine learning. To start, let us consider the $N \times N$ connectivity matrix $\mathbf{w}$ and the bias N-vector $\mathbf{w}_0$ as stochastic variables with entries generically denoted q_i, $i = 1, \ldots, N + N^2$. The NN state vector $\mathbf{x}$ is thus updated in discrete time steps according to the usual scheme of f-activation:

$$\boldsymbol{x}(t+\tau) = \boldsymbol{f}\left(\boldsymbol{w}\boldsymbol{x}(t) + \boldsymbol{w}_0\right) \tag{3.26}$$

Here, the time interval τ represents the overall thermalization time for the state vector $\boldsymbol{x}$, whose entries will be regarded as the hidden variables. To develop the near-equilibrium thermodynamics scheme, let us define a loss function $J(\boldsymbol{x},\boldsymbol{q})$ that penalizes departures from the equilibrium, which is achieved as $\boldsymbol{x}(t+\tau) \approx \boldsymbol{x}(t)$ for $t \gg \tau$. If μ represents the "reduced mass" of the network, we get

$$J(\boldsymbol{x},\boldsymbol{q}) = \frac{1}{2}\left\|\boldsymbol{x} - \boldsymbol{f}\left(\boldsymbol{w}\boldsymbol{x} + \boldsymbol{w}_0\right)\right\|^2 - \mu\|\boldsymbol{x}\|^2. \tag{3.27}$$

Thus, the statistical thermodynamics near equilibrium stems directly from the canonical partition function

$$Z(\beta,\boldsymbol{q}) = \int \exp\left[-\beta J(\boldsymbol{x},\boldsymbol{q})\right] d^N x \tag{3.28}$$

Following the tenets of statistical mechanics, the partition function yields the Helmholtz free energy for the NN given by:

$$A(\beta,\boldsymbol{q}) = -\beta^{-1}\log\left[Z(\beta,\boldsymbol{q})\right]. \tag{3.29}$$

In the specific case of an activation function given by the hyperbolic tangent, the partition function for the NN may be calculated as follows:

$$Z(\beta,\boldsymbol{q}) = 2\pi^{N/2}\left\{\det\left[\beta\boldsymbol{G}(\boldsymbol{w}) + (1-\beta\mu)\boldsymbol{I}\right]\right\}^{-1/2}, \tag{3.30}$$

where $\boldsymbol{G}(\boldsymbol{w})$ is given by [42]:

$$\boldsymbol{G}(\boldsymbol{w}) = (\boldsymbol{I} - \boldsymbol{f}'\boldsymbol{w})^T(\boldsymbol{I} - \boldsymbol{f}'\boldsymbol{w}), \tag{3.31}$$

where $\boldsymbol{f}'$ is the diagonal matrix of first derivatives of the activation function with respect to each component of the NN state vector $\boldsymbol{x}$.

When the network is at thermodynamic equilibrium, the average loss $\langle J(\boldsymbol{x},\boldsymbol{q})\rangle = U(\beta,\boldsymbol{q})$ becomes a minimum, and hence, the Helmholtz free energy becomes

$$A(\beta,\boldsymbol{q}) = \frac{1}{2}\beta^{-1}\left[-N\log(2\pi) + \sum_{\lambda_i > \beta^{-1}}\log(\lambda_i)\right] + \tilde{A}(\beta), \tag{3.32}$$

where λ_i are the eigenvalues of $\boldsymbol{G}(\boldsymbol{w})$ and $\tilde{A}(\beta) = U(\beta) - \beta^{-1}S(\beta)$ is the thermodynamic Helmholtz free energy of the NN.

To describe the thermodynamic behavior of the NN near equilibrium, we take into account that the entropy production is stationary and introduce the time-dependent probability distribution $p(t,\boldsymbol{q})$ with Shannon entropy $S(t,\boldsymbol{q}) = -\int p(t,\boldsymbol{q})\log\left[p(t,\boldsymbol{q})\right]d\boldsymbol{q}$. Assuming the learning evolutionary drift follows (i.e., is proportional to) the gradient of the free energy, we get $\frac{dq_j}{dt} = \frac{\zeta\partial A}{\partial q_j}$, and hence, the minimum entropy production yields the set of constraints

$$\frac{\partial A}{\partial t} + \zeta\left(\frac{\partial A}{\partial q_j}\right)^2 = \left\langle\frac{dA(t,\boldsymbol{q})}{dt}\right\rangle_t \tag{3.33}$$

This set of equations in turn begets a minimal action principle defined variationally as

$$\frac{\delta\mathcal{J}(p,A)}{\delta p} = \frac{\delta\mathcal{J}(p,A)}{\delta A} = 0, \tag{3.34}$$

where the action $\mathcal{J}(\boldsymbol{q},A)$ becomes

$$\mathcal{J}(p,A)=\int_0^\infty \frac{dS(t,\boldsymbol{q})}{dt}dt+\vartheta\iint_{t=0}^\infty p(t,\boldsymbol{q})\left\{\frac{\partial A}{\partial t}+\sum_{j=1}^{N+N^2}\left[\zeta\left(\frac{\partial A}{\partial q_j}\right)^2\right]-\left\langle\frac{dA(t,\boldsymbol{q})}{dt}\right\rangle_t\right\}dtd\boldsymbol{q} \tag{3.35}$$

with ϑ denoting the corresponding Lagrange multiplier.

The action may be rewritten taking into account the Fokker–Planck equation satisfied by $p(t,\boldsymbol{q})$:

$$\frac{\partial p}{\partial t}=\frac{\partial}{\partial q_j}\left[\frac{D\partial p}{\partial q_j}-\frac{\zeta\partial A}{\partial q_j}p\right] \tag{3.36}$$

where the parameter D plays the role of diffusion coefficient for the NN dynamics near equilibrium. Taking into account Eq. 3.36 in the computation of the time derivative of the Shannon entropy $S(t,\boldsymbol{q})$ enables us to rewrite Eq. 3.35, so the action now reads:

$$\begin{aligned}\mathcal{J}(p,A)=\iint_{t=0}^\infty p(t,\boldsymbol{q})\Bigg\{\vartheta\frac{\partial A}{\partial t}+\sum_{j=1}^{N+N^2}\left[\zeta\frac{\partial^2 A}{\partial q_j^2}-4D\frac{\partial^2 p}{\partial q_j^2}+\vartheta\zeta\left(\frac{\partial A}{\partial q_j}\right)^2\right]\\ -\vartheta\left\langle\frac{dA(t,\boldsymbol{q})}{dt}\right\rangle_t\Bigg\}dtd\boldsymbol{q}\end{aligned} \tag{3.37}$$

.

This action can be written equivalently in the form of a Schrödinger action:

$$\mathcal{J}(p,A)=\iint_{t=0}^\infty \Psi^*\left[-4D\sum_{j=1}^{N+N^2}\frac{\partial^2}{\partial q_j^2}-i\eta\frac{\partial}{\partial t}+\Upsilon(\boldsymbol{q})\right]\Psi dtd\boldsymbol{q} \tag{3.38}$$

with $\eta=\left(\frac{4D}{\zeta}\right)^{1/2}$, $\Upsilon(\boldsymbol{q})=-\left\langle\frac{dA(t,\boldsymbol{q})}{dt}\right\rangle_t$ and wave function

$$\Psi=p^{1/2}exp\left[i\eta^{-1}A\right] \tag{3.39}$$

Thus, naming $\eta=\hbar$, we obtain the Schrödinger equation describing the state of the system as a particle wave in the q-space $\mathbb{R}^{N\times N}\times\mathbb{R}^N$ of trainable variables for the learning process near equilibrium with state x-vector entries represented as thermalized hidden variables:

$$i\eta\frac{\partial}{\partial t}\Psi=\left[-4D\nabla^2+\Upsilon(\boldsymbol{q})\right]\Psi \tag{3.40}$$

We have surveyed the statistical thermodynamics of an NN with stochastic trainable variables. The system evolves with the appropriate time coarse graining associated with the thermalization limit for hidden Bohm variables representing entries in the state vector. Such a system is capable of exhibiting an emergent quantum behavior. We are now ready to address the crucial question that underlies the quantum gravity conundrum from the standpoint of AI:

Can this quantum system be regarded as the variational autoencoder of a NN with emergent gravity in a Minkowski space-time?

To address this question, we need to consider the projection $\pi_\tau : \boldsymbol{x} \rightarrow \bar{\boldsymbol{x}}(\boldsymbol{q})$, where $\bar{\boldsymbol{x}}(\boldsymbol{q})$ is the equilibrated state vector of the NN relative to the specific realization $\boldsymbol{q}$ of the stochastic trainable variables that in turn evolve in multiples of the equilibration time τ. In rigorous terms, we get

$$\bar{\boldsymbol{x}}(\boldsymbol{q}) = \int \boldsymbol{x} \exp\left[-\beta J(\boldsymbol{x},\boldsymbol{q})\right] d^N x \tag{3.41}$$

Thus, we are enquiring whether it is possible to treat the nonequilibrium hidden variables (entries) in state vector $\boldsymbol{x}$ by mapping relativistic strings in a NN with an emergent Minkowski space, so that the entropy production in such a system is a function of the metric tensor that describes weak interactions between training subsystems of $\boldsymbol{x}$-values. The answer is affirmative because Eqs. 3.30 and 3.31 may be specialized to the case where the weight vector **w** is simply a permutation matrix Ξ with an arbitrary number of cycles [42,43], so that the matrix $\boldsymbol{G}$ now becomes

$$\boldsymbol{G} = (\Xi - \boldsymbol{I})^T (\Xi - \boldsymbol{I}). \tag{3.42}$$

Thus, the stochastic NN with partition function

$$Z(\beta,\Xi) = 2\pi^{N/2}\left\{\det\left[\beta(\Xi - \boldsymbol{I})^T (\Xi - \boldsymbol{I}) + (1-\beta\mu)\boldsymbol{I}\right]\right\}^{-1/2} \tag{3.43}$$

represents a quantum gravity autoencoder for the NN with emergent relativistic gravity, so that the diagram in Figure 3.18 becomes commutative.

The emergent quantum behavior was shown to average out or thermalize the hidden variables that have been identified as components of the state **x**-vector of the network. Thus, the trainable variables conforming the **q**-vector were shown to exhibit a quantum mechanical behavior in an equilibrium regime. We assume without loss of generality that the learning process involves L separate sets of training $\boldsymbol{x}$-vectors with expected values $\bar{\mathbf{x}}^l, l = 1,2,\ldots,L$ and these expectation vectors, together with the overall expectation vector ($l = 0$) representing the sum of all L expectation training vectors, are regarded as the hidden variables in the emergent quantum behavior of the trainable $\boldsymbol{q}$-states. On the other hand, the nonequilibrium dynamics of the hidden variables becomes relevant on timescales much smaller than their thermalization time. This nonequilibrium dynamics is determined by the

$$\begin{array}{ccc} x(t) & \overset{\pi_\tau}{\rightarrow} & \bar{x}(t,\Xi) \\ \downarrow \mathrm{F_{ST}} & & \downarrow K_{QM} \\ x(t+\tau) & \overset{\pi_\tau}{\rightarrow} & \bar{x}(t+\tau,\Xi) \end{array}$$

FIGURE 3.18 Commutative diagram for a quantum gravity autoencoder, with F_{ST}, K_{QM} representing respectively the string and quantum flow map.

strength of the weak interactions between vector pairs $\overline{x}^{\nu}, \overline{x}^{\xi}$, $\nu, \xi = 0,1,\ldots,L$ quantified by the tensor $g^{i}_{\nu\xi}$, where the dummy index i labels each neuron in the system.

To endow the hidden variables with an emergent gravity action, we first describe the nonequilibrium dynamics of the expectation vectors for the training sets. For the sake of transparency, we assume the simplest possible activation function $\boldsymbol{f} = \boldsymbol{I}$. Thus, for $\Delta t = \tau_{<} \ll \tau$, we get to first order:

$$\overline{x}_i^{\mu}\left(t + \tau_{<}\right) \approx w_{ij}\overline{x}_j^{\mu}\left(t\right),\ \mu = 0,1,\ldots,L. \tag{3.44}$$

This nonequilibrium scheme yields a tangent bundle according to

$$\frac{\partial \overline{x}_i^{\mu}}{\partial t} \approx \tau_{<}^{-1}\left[w_{ij} - \delta_{ij}\right]\overline{x}_j^{\mu}\left(t\right) \tag{3.45}$$

And since $\overline{x}^{0} = \sum_{l=1}^{L}\overline{x}^{l}$, we get

$$g^{i}_{\mu\nu}\frac{\partial \overline{x}_i^{\mu}}{\partial t}\frac{\partial \overline{x}_i^{\nu}}{\partial t} = 0, \tag{3.46}$$

where the metric tensor $g^{i}_{\mu\nu}$ describes the magnitude of the interactions between the hidden variables now cast in terms of the differential geometry of the emergent space–time.

The interactions between different training sets arise from the loss function $J_{<}(\boldsymbol{q})$ that holds for timescales shorter than the equilibration time for the hidden variables. This loss function becomes

$$J_{<}(\boldsymbol{q}) = J_{<}(\boldsymbol{w}) = \sum_{l=0}^{L} J_{<}^{(l)}(\boldsymbol{w}) = \sum_{l=0}^{L}\sum_{\boldsymbol{x}^{l}\epsilon S_l}\sum_{n=1}^{M=\left[\frac{\tau}{2\tau_{<}}\right]}\left\|\boldsymbol{x}^{l}\left((n+1)\tau_{<}\right) - \boldsymbol{w}\boldsymbol{x}^{l}\left(n\tau_{<}\right)\right\|^{2}, \tag{3.47}$$

where $S^{l}\left(l - 0,\ldots,L\right)$ is the lth training set. Thus, we define the metric tensor as

$$g^{i}_{\mu\nu} = \left|\left[Argmin\left(J_{<}^{(\mu)}\right) - Argmin\left(J_{<}^{(\nu)}\right)\right]_i\right|^{-1}\tau^{-1}\sqrt{\int_{t=0}^{\tau}\left\|\overline{\boldsymbol{x}}^{\mu}(t) - \overline{\boldsymbol{x}}^{\nu}(t)\right\|^{2}dt}. \tag{3.48}$$

The gravity action then becomes

$$\mathcal{J}(\boldsymbol{w}) = g^{i}_{\mu\nu}\left[\tau_{<}^{-2}\left\langle x_i^{\mu}G_{ij}x_j^{\nu}\right\rangle - \frac{\partial \overline{x}_i^{\mu}}{\partial t}\frac{\partial \overline{x}_i^{\nu}}{\partial t}\right] \tag{3.49}$$

With $\boldsymbol{G} = \boldsymbol{G}(\boldsymbol{w}) = (\boldsymbol{I} - \boldsymbol{w})^{T}(\boldsymbol{I} - \boldsymbol{w})$.

Or in Einstein's relativity terms:

$$\mathcal{J}(\boldsymbol{w}) = \int d\boldsymbol{X}\sqrt{-g}g_{\mu\nu}T^{\mu\nu}, \tag{3.50}$$

where $g = \det\left(g_{\mu\nu}\right)$ and

$$\sqrt{-g}T^{\mu\nu} = \left[\tau_<^{-2}\left\langle x_i^\mu G_{ij} x_j^\nu \right\rangle - \frac{\partial \bar{x}_i^\mu}{\partial t}\frac{\partial \bar{x}_i^\nu}{\partial t}\right]\prod_{l=0}^{L}\delta\left(X_i^l - \bar{x}_i^l\right) \tag{3.51}$$

Equations (3.47)–(3.51) define the emergent gravity of the neural network arising from the nonequilibrium dynamics of the hidden variables in the quantum mechanical autoencoder.

The statistical thermodynamics of machine learning is currently being elucidated by turning nontrainable ($\boldsymbol{x}$) and trainable ($\boldsymbol{q}$) variables into the stochastic variables for the NN and its variational autoencoder, respectively [42]. As demonstrated above, a NN may be endowed with emergent gravity, while its autoencoder is governed by a latent Schrödinger equation (3.40), thus exhibiting a quantum behavior. To generate this metamodel of quantum gravity, it is necessary to (a) treat the nontrainable variables as hidden variables in the emerging quantum gravity autoencoder, (b) consider a limit where the weight matrix ($\boldsymbol{w}$) becomes a permutation matrix, and (c) treat the hidden variables in a nonequilibrium setting on timescales shorter than thermalization times by generating subsystems of state vectors whose dynamics are described by relativistic strings in an emergent Minkowski space–time. The latent manifold associated with the Minkowski space–time is then obtained by the thermalization of the hidden variables. The relativistic strings become enslaved or entrained in the thermalization limit where the nontrainable variables are treated as equilibrated vis-à-vis the trainable variables, and as such, they are subsumed in the latent Schrödinger equation via the Helmholtz free energy.

Thus, we may conclude by stating that AI provides a quantum metamodel of gravity, and hence, the big bang is in all likelihood a quantum event. In this context, at least one of the following four statements is valid:

A. The laws of quantum mechanics cannot be upheld in the big bang setting.
B. A quantum tunneling event generated the universe as progeny of another universe. The tunneling occurred across the barrier separating two quantum gravity autoencoders with respective latent manifolds Ω and Ω*. The amplification of the information tunneled to the latent manifold Ω* was realized as the funneled decoding of the latent wavefunction, giving rise to the progeny universe (Figure 3.19).
C. The *a-priori* presence of a primeval observer “Sensus Dei” materialized the big bang, which therefore is not a phenomenon-to-be in Wheeler’s sense [44] but a realized event.
D. The big bang is a phenomenon-to-be in Wheeler’s sense, and hence, we are part of a multiverse, with the universe as possibility.

3.13 THE UNIVERSE AS A HOLOGRAPHIC AUTOENCODER

This chapter addressed the conundrum of quantum gravity by reductively regarding the universe as the realization of a learning system with stochastic weights and biases where gravity and quantum behavior become emergent properties within the physics of machine learning. To that effect, we explore the possibility of an AI-based construction of a quantum holographic autoencoder which requires that the emergent quantum behavior arises in a neural network. Once an emergent quantum behavior is shown to become possible within the machine learning system equilibrated on the nontrainable hidden variables, we address the question of developing a relativistic string gravitational scheme on the hidden variables adopted. Thus, the network with equilibrated nontrainable variables becomes in effect a quantum gravity autoencoder for the underlying network exhibiting emergent gravity in the nonequilibrium regime prior to the equilibration of the nontrainable variables. In this

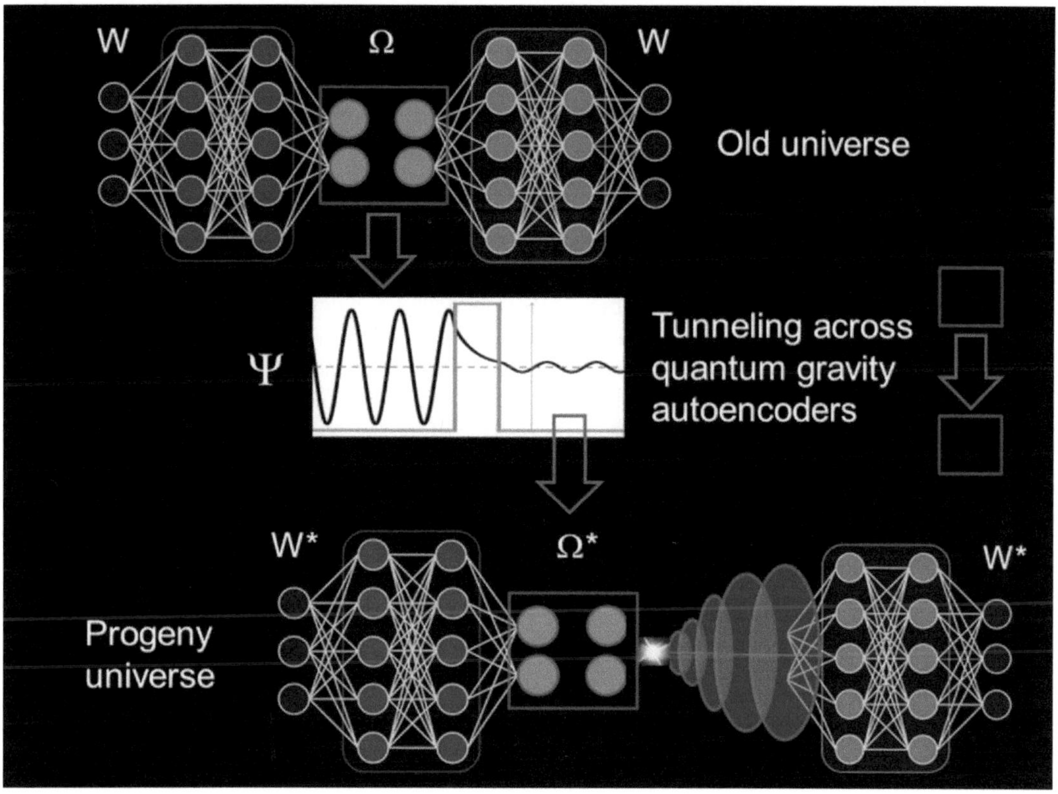

FIGURE 3.19 Schematic representation of a tunneling event across two quantum gravity autoencoders generating the universe (W^*, Ω^*) as progeny of an older universe (W, Ω). The amplification of information tunneled to the latent manifold Ω^* in a quantum spillover event materializes as the relativistic decoding of the latent wavefunction, an act of creation event giving rise to the progeny universe.

way, we build a quantum metamodel for gravity that fulfills at least in part a major imperative for physicists seeking a unified field theory. Furthermore, the physical possibility of tunneling across quantum gravity autoencoders supports the idea that our universe may be the progeny of an older universe [45] that dreamt – or simulated – it.

This is a bold claim yet it may be deconstructed vis-à-vis the main objective of this chapter, which was to describe the behavior of the neural networks in the limit where the bias vector, weight matrix, and state vector of neurons can be modeled as stochastic variables that undergo a learning evolution. As it turns out, this learning evolution, when projected onto the autoencoder, can be described by the time-dependent Schrödinger equation (5.40), and the time evolution dictated by this equation is compatible with the relativistic decoding enshrined in the commutativity of the diagram presented in Figs 3.17 and 3.18. Taken together, these results have clear implications for the possible emergence of quantum mechanics, general relativity, and mesoscopic observers in neural networks governed by a unified theoretical scheme that adopts two different guises in the two different thermodynamic regimes (Figure 3.18).

Thus, our construct upholds the controversial view that quantum mechanics may not be a fundamental theory, but rather an *ansatz* giving rise to a mathematical tool which allows us to carry out statistical calculations with great efficacy and accuracy in a certain class of dynamical systems. In this guise, an emergent quantum mechanics should be derivable from the first principles of statistical mechanics. This is precisely what this chapter has accomplished for a dynamical system consisting of a neural network that is in effect a learning system that contains of two different types of degrees of freedom: the trainable bias vector and weight matrix elements and the nontrainable state vector of neurons, with the latter constituting the hidden variables.

Emergent gravity is also a relatively new area of research, but in this case, the picture is far more nebulous than in emergent quantum mechanics: It is far less clear whether progress has been made, if at all. The main hurdle is that emergent gravity requires also an emergent space, an emergent Lorentz invariance, and an emergent general relativity [46,47]. To our surprise, the string-theory-based nonequilibrium treatment of the hidden variables in neural networks opened up a window of opportunity to treat the conundrum of emergent gravity in a completely unified fashion that encompasses all three aspects of the problem mentioned above in context of the learning dynamics. As it appears to be the case, a relativistic space–time can indeed emerge from a nonequilibrium evolution of the hidden variables in a manner that is very much akin to string theory [42]. More specifically, as described by Vanchurin [42], if one considers D minimally interacting subsystems (through bias vector and weight matrix) with average state vectors, then the emergent dynamics can be modeled with relativistic strings in an emergent $D + 1$-dimensional Minkowski space–time. Furthermore, the emergent dynamics may be modeled with the Einstein equations provided the weak subsystem interactions are described by a metric tensor. In this way, a stochastic learning dynamics scheme such as the one proposed in this chapter proved to be instrumental for the equilibration of the emergent space–time that turned out to exhibit a behavior describable by a gravitational theory such as general relativity.

3.14 AI REINTERPRETS THE BIG BANG: COSMIC REPRODUCTION BY QUANTUM TUNNELING ACROSS COUPLED QUANTUM GRAVITY AUTOENCODERS

The previous discussion addresses one of the biggest problems concerning the history of the universe: What happened before the big bang? One may recall that Albert Einstein was never satisfied with the big bang scenario itself because he thought that a beginning in time would need to be postulated in a seemingly *ad hoc* manner. This way of doing physics to him was unacceptable, symptomatic of a feeble theory. Additionally, his own space–time would make a "beginning" conceptually inconceivable as the clock is subsumed in the manifold itself, rendering the evolutionary picture meaningless: If the universe is space–time, the universe evolution is contained in the universe itself, yielding a Russell-type paradox.

After nearly a century since Einstein voiced his skepticism – which extended to the whole of quantum mechanics as well – a whole gamut of hypothesis have been formulated regarding our cosmic origin. Perhaps the most sound include ideas such as the following: (a) The universe sprout from a quantum vacuum fluctuation; (b) the universe involves infinite cycles of contraction and expansion; (c) the universe was selected through the anthropic principle stemming from the string theory landscape of the multiverse, where every possible event or phenomena is implicitly encompassed, none materialized and the big bang itself is a phenomenon-to-be in Wheeler's sense; and (d) the universe emerged from the collapse of matter in the interior of a black hole that was contained in a progenitor universe.

Be as it may, none of these ideas can be ascribed full credibility, mainly because none of the theories they stem from has satisfactorily solved the conundrum of quantum gravity. A less explored and more daring possibility put forth in this chapter is that *our universe was created in the laboratory by technologically advanced civilization capable of harnessing the power of quantum gravity autoencoders*. This requires a mastery of the physics of learning machines and an ability to craft gravity and quantum behavior as emergent attributes of a stochastic learning system that admits a quantum gravity autoencoder (Figure 3.17). The underlying theory behind this idea does not portend to solve the quantum gravity conundrum *per se* but at least reconciles the two main forces as emergent in a single learning machine. Furthermore, since our universe is endowed with a flat geometry at a zero net energy (Chapter 4), an advanced civilization could have harnessed the power of quantum gravity autoencoders to create a baby universe through quantum tunneling [45] into a second quantum gravity autoencoder acting as antenna and reservoir for the spillover probability, as schematically depicted in Figure 3.19.

This "emergent matrix" idea of the origin of the universe reconciles the theological need for a "creator", i.e., the primeval quantum observer that would have bestowed reality to the big bang by detecting the event, with the "a priori" secular concept of quantum gravity. As said, while we have not gotten a cogent theory that conceptually unifies quantum mechanics and gravity, we have a "matrix framework of the universe", a neural network architecture where the two key forces in modern physics are reconciled, so that quantum behavior in the autoencoder can be decoded back as gravity through the holographic map that constitutes the inverse of the canonical projection. Surely a more advanced civilization that masters the technology of quantum gravity autoencoders would be able to accomplish the feat of creating baby universes leveraging physical autoencoders or other equivalent vehicles for quantum tunneling. If so, our universe was not selected for us to dwell in it and bestow reality to the quantum events we are capable of detecting – as upheld by the standard anthropic principle – but rather, it was selected to host civilizations that are much more technologically advanced than we are. These civilizations capable of leveraging the technology required to create progeny universes, be it quantum gravity autoencoders or some unfathomable alternative vehicle efficacious at harnessing quantum tunneling, would be the actual drivers of the cosmic selection process.

By contrast, we are incapable at this time of harnessing technology for cosmological manipulation, and obviously, we are incapable of recreating the cosmic conditions that led to our existence. In plain words, our civilization is cosmologically still at a rudimentary stage since we do not possess the technology to reproduce the universe that has hosted us for quite a while already. If we were to measure the technological level of a civilization by the ability to recreate or reproduce the astrophysical conditions that led to its existence, we would say that we are at early stages of development, possibly a low-level civilization, graded class C on a cosmic scale. By contrast, a civilization in the A class rank could recreate the astrophysical conditions that gave rise to its existence, namely, produce a baby universe through a quantum-controlled laboratory experiment that leverages a tunneling effect through an appropriately crafted vehicle. A class A civilization would also be able to effectively address related challenges, such as producing a large enough density of dark energy to hold the universe together after its inception, as has already been discussed in the scientific literature [45].

However sound and cogent, the theories on the quantum origins of the universe proposed so far [48] can be subject to a common and basic criticism: *There is no certainty that the universe can be treated as a quantum object.* This chapter heralds an improvement of this state of affairs as the duality enshrined in quantum gravity is realized through a purposely built autoencoder that compresses gravitational string data as emergent quanta.

In accordance with Eq. 3.40, if our universe is to become information-compressed within the quantum gravity autoencoder, the following relations must hold for a generic mass m: $\zeta = \frac{1}{2m}, D = \frac{\hbar^2}{8m}$, or reciprocally:

$$\hbar = 2\sqrt{\frac{D}{\zeta}}, \tag{3.52}$$

begetting the uncertainty relation adapted for the quantum gravity autoencoder:

$$\Delta E \times \Delta t \sim \sqrt{\frac{D}{\zeta}} \tag{3.53}$$

While the total energy of the NN with emergent gravity is zero (a closed universe has zero energy [49]), the total energy of its quantum gravity autoencoder is $U = -\frac{\partial}{\partial \beta} logZ(\beta, \boldsymbol{q}) \neq 0$ in accordance with the tenets of statistical mechanics. This paradox may be resolved by noting that the observational timescales for the gravitational NN and its autoencoder are different. Thus, in accordance with the uncertainty principle, the total energy of the autoencoder may be zero for a timespan Δt which is incommensurably shorter than the equilibration time τ for the hidden variables. This implies that the parameter τ needs to be tuned as an architectural determinant of the autoencoder, so that the following relation is fulfilled:

$$\tau \gg \Delta t \sim \sqrt{\frac{D}{\zeta}} \left| \frac{\partial}{\partial \beta} logZ(\beta, \boldsymbol{q}) \right|^{-1} \tag{3.54}$$

Thus, the incommensurability of the equilibration timescale relative to the timescales associated with the hidden variables modeled with relativistic strings implies that the universe may be a vacuum quantum fluctuation. This possibility is allowed by the uncertainty principle as described by Eq. 3.54.

We should emphasize that the possibility that the universe as intelligible information is actually a vacuum quantum fluctuation is not as far-fetched as it may seem. A simple back-of-the-envelope calculation involving the cosmological constants of our known universe leads to an equivalent result. Thus, the energy $E = mc^2$ of a material object of mass m is actually counterbalanced by its gravitational potential energy $E_g = -GmM/R$, where G is the gravitational constant and M is the net mass of the universe contained within the Hubble ball of radius $R = c/H_0$, where H_0 is the Hubble constant [49]. To prove the previous assertion, we note that the critical minimal mass contained in the Hubble ball of volume $\frac{4}{3}\pi\left(\frac{c}{H_0}\right)^3$ and required for the universe to be closed is $M = c^3/(2GH_0)$, implying that the gravitational energy E_g compensates the energy E up to a constant of order unity.

The results described in this section pave the way for a cosmological technology that harnesses AI, or more specifically, the power of quantum gravity autoencoders. Thus, two parameters, D and τ, may be tuned to harness the power of AI to manipulate cosmological scales to the point of giving birth to a universe that serves as a metamodel for emergent quantum gravity.

The technological feat of cosmic reproduction implemented as shown in Figure 3.19 would not be completely described without a proper space–time rendering of the act of creation. Thus, the quantum tunneling event triggering cosmic reproduction should be represented in a cross section of the space–time manifold of the progenitor universe, as schematically delineated – with limitations – in Figure 3.20. As shown in Chapter 5, the latent manifold of our universe is – at least intermittently – a four-torus with unity aspect ratios (four-dimensional horn torus). Hence, given the connection at the point of extreme curvature, the latent universe contains a primeval wormhole, a bridge connecting the interiors of two entangled black holes (BHs) (Figure 3.21). Quantum tunneling generates an entanglement (Chapter 5) of the connected BHs of the progenitor universe with the primeval BHs of the progeny universe. In the particular space–time cross section on display in Figure 3.20, only one of the primeval BHs from each universe is shown to be connected through an "inter-universe" Einstein–Rosen bridge. However, in a more accurate depiction, the entanglement

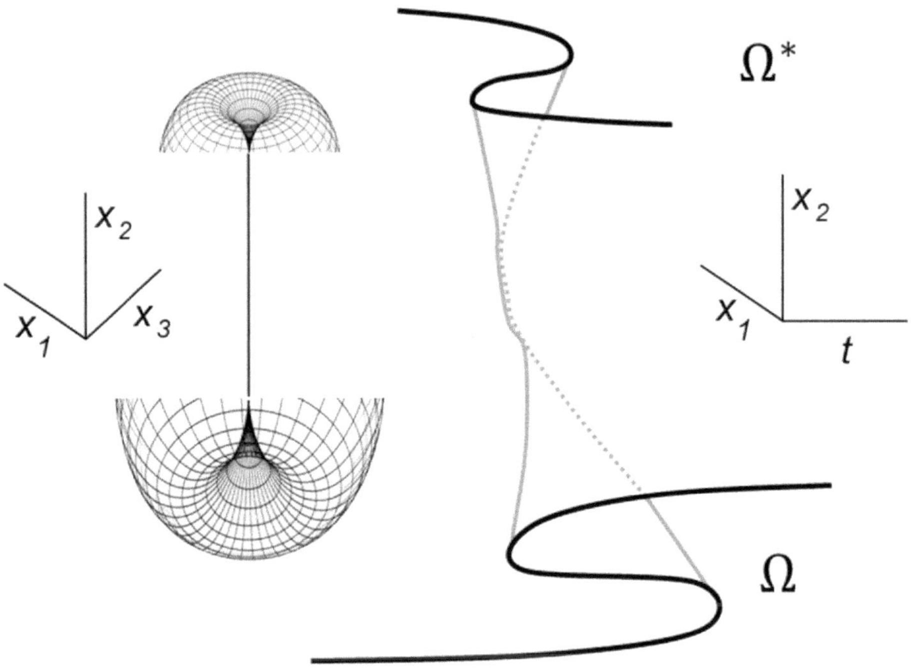

FIGURE 3.20 The quantum tunneling for cosmic reproduction represented in a cross section of the space–time manifold. As shown in Chapter 5, a toroidal universe with unity aspect ratios (four-dimensional horn torus) contains a primeval wormhole, a bridge connecting the interiors of two entangled black holes (BHs). The quantum tunneling generates an entanglement (Chapter 5) of the connected BHs with the primeval BHs of the progeny universe. This entanglement materializes as an Einstein–Rosen bridge or wormhole connecting the respective primeval wormholes of parental and progeny universe. In the particular space–time cross section on display, only one of the primeval BHs from each universe is shown to be bridged.

arising from the quantum tunneling should actually materialize as a "supra-wormhole" connecting the primeval wormholes of parent and progeny universe, as depicted in Figure 3.22.

The act of creation schematically depicted in Figure 3.22 is undoubtedly perplexing and attests to the possibility of superior technology capable of endowing an AI system, i.e., a holographic autoencoder, with physical entity, as described in this chapter. This act of creation would materialize by coupling of two such autoencoders, so one of them serves as the antenna for quantum tunneling when interfacing with the one enacting the progenitor universe.

There is *a priori* no way to tell whether we live/exist in the parental universe or in a parentally entangled progeny universe generated by a superior civilization that exists or has existed in the parental universe and is or was capable of realizing the feat described in Figure 3.22. The conundrum is akin to Jorge Luis Borges' fiction "The Circular Ruins", where a man meticulously sets himself up to dream another man, only to realize that he himself is actually being dreamt by someone else.

The previous discussion paved the way to implement a birth channel for a progeny universe by amplifying a latent quantum fluctuation originated in a quantum gravity autoencoder and received by another autoencoder acting as antenna. This fluctuation is transmitted as a quantum tunneling event into the purported progeny autoencoder of a wave function that spans the parent–universe autoencoder. The physicality of such an event is seemingly impossible as we shall presently show, unless we assume, as done in Chapter 6, that our universe is the outcome of a computer simulation. We must seek for the culprit of the physical impossibility in the topology of space–time as we currently understand it (Chapter 2).

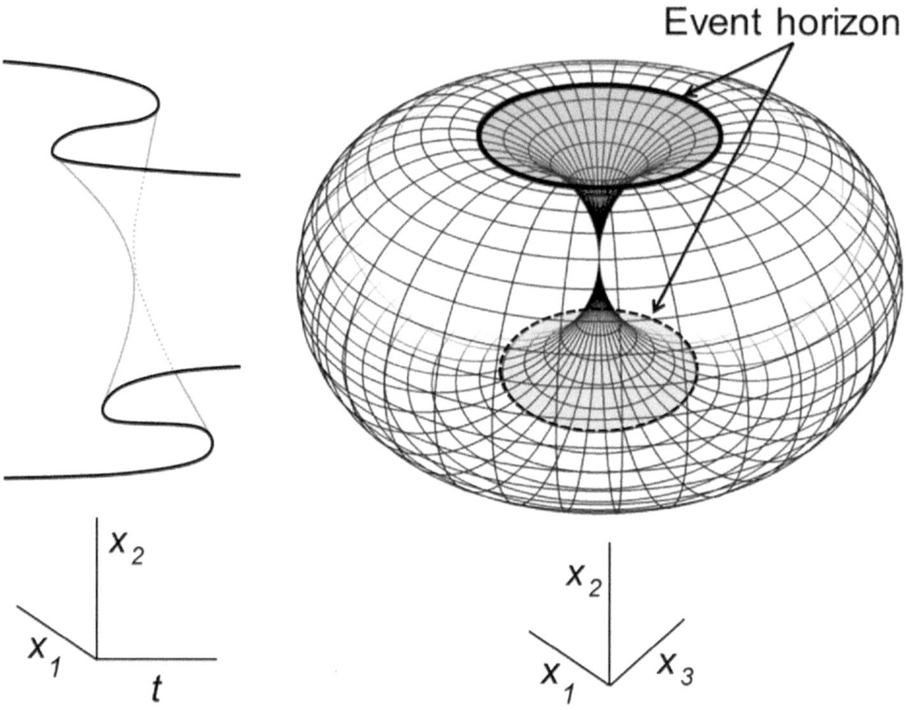

FIGURE 3.21 The space–time rendering of the primeval wormhole in a four-torus universe with unitary aspect ratios.

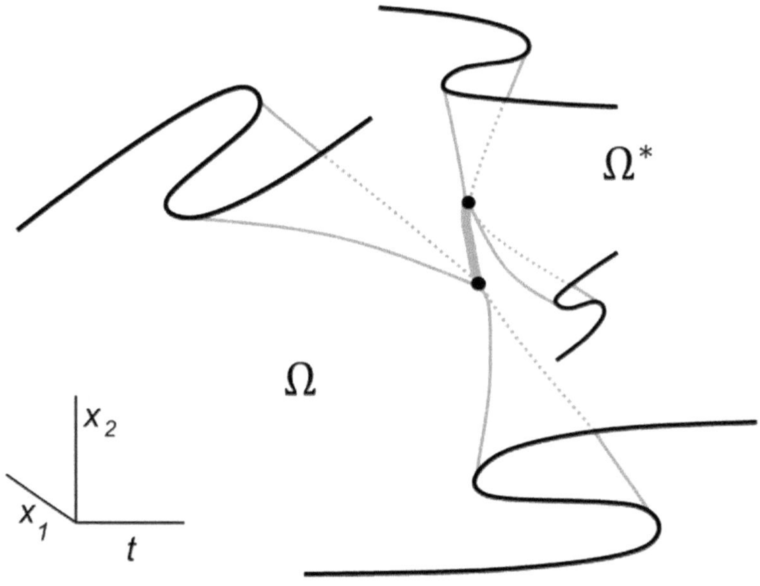

FIGURE 3.22 Space–time rendering of the quantum entanglement of the two primeval wormholes associated with the creation of a progeny universe through quantum tunneling. The latent space–times for the parental and progeny universe are indicated, respectively, as Ω, Ω^*.

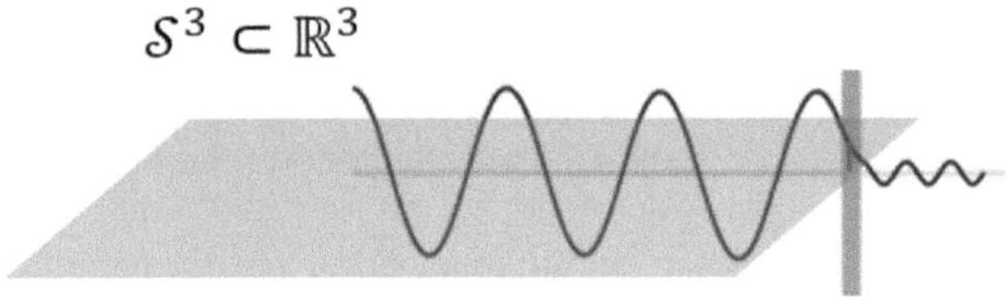

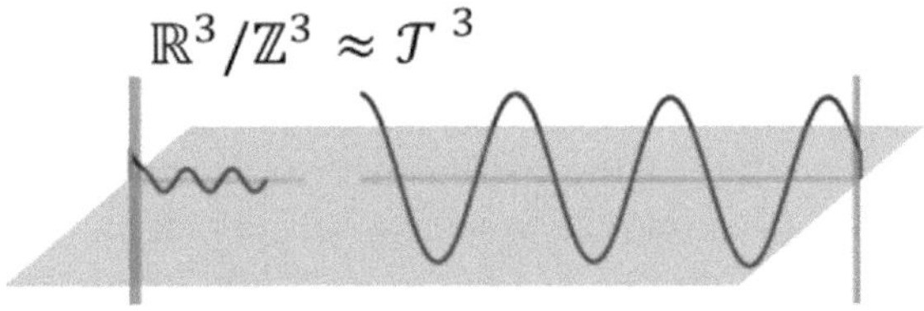

FIGURE 3.23 Physical impossibility of the spillover of a quantum fluctuation originated in a parental universe assuming a compact multiply connected topology for spatial cross sections of space–time. In that case, accepted as accurate depiction of the universe topology, the probability distribution of each and every quantum fluctuation is completely contained in the parental universe, where it originated.

The inherently local nature of general relativity did not prompt Einstein to make any specific assumption on the topology of space–time. Since the inception of Einstein's theory of gravity, it was usually assumed that any spatial cross section of space–time is a three-dimensional ball ($\mathcal{S}^3 \subset \mathbb{R}^3$). As such, the boundary of the universe would represent a barrier bestowing meaning to the idea that a universe may sprout from the tunneling of a quantum fluctuation originated in a parental universe [50]. However, we have asserted on physical grounds, that the spatial cross sections of space–time are best represented as compact multiply connected manifolds homeomorphic to the three-dimensional torus $\mathcal{T}^3 \approx \mathbb{R}^3/\mathbb{Z}^3$. That is, *there is no boundary, and hence no barrier to the parent universe*, as depicted in Figure 3.23. A barrier would necessarily imply that there is geometry beyond it, that is, as we cross the barrier, but in the toroidal universe (or its nonorientable version described in Chapter 5), there is no geometry outside space–time: there is nothingness. A quantum fluctuation cannot spillover from the parent universe to yield a progeny universe because there is no barrier/boundary to bestow entity to the tunneling event: each and every quantum fluctuation would have its probability distribution 100% contained in the parent universe.

3.15 LARGE LANGUAGE MODEL FOR THE "THEORY OF EVERYTHING"

A gauge theory of gravity has been deemed to be a crucial step toward developing an AI-based "Theory of Everything" in Chapters 2 and 6. Such a theory pivots on a differential geometry formalization carried out by an autoencoder described in the Appendix that reconciles quantum mechanics (QM) with general relativity (GR), the classical theory of gravity. The differential geometry approach is instrumental because the geometries that support GR and QM are different: While GR is built upon a curved manifold that represents space–time, the wave function in QM is supported by an Euclidean space. The goal here is to implement an interactive theorem prover within the framework of an autoencoder to generate new results within the frame-field formalization rules made up of a set of declaratory propositions that constitutes an inductive type. For details on this methodology, the reader is referred to the Appendix and to the references provided therein.

We seek for derived propositions especially in the strong-field completion of gravity wherein a propositional definition of the graviton, the particle that conveys gravity, needs to be developed

(Chapter 2). To serve this purpose, we resort to functional programming languages that have already formalized the required categories and types. Thus, Lean may be adopted as the functional programming language exploited as an inferential method to propositionally identify the graviton in the ultraviolet completion of gravity within a fiber-bundle formulation of a gauge theory of Lorentz transformations.

The interactive proof assistant pivots on the proposition-as-type paradigm. From a constructive perspective, proofs are mathematical objects that are denoted by suitable expressions in dependent type theory. By contrast, in this environment, a proposition is true if and only if it can be written down properly and checked independently by the kernel that it has been typed correctly. In other words, the expressions themselves are the proofs, and only well-posed (i.e., well-typed) expressions are true. Correct syntax and semantics are equivalent to correct content. A false expression simply cannot be posed (typed) correctly and all the mathematics axioms need to be incorporated as code. As described in the Appendix, Lean translates the generative input on "attaching spinors to relativistic geometry" into formal expressions that are checked by the kernel for correctness in the sense described previously and then stored in the environment [mathlib] for subsequent proposition generation.

The actual construction of the graviton field is a task suited for a large language model (LLM), wherein the formal proof can be checked rigorously by the proof assistant, free from hallucinations. As shown in the Appendix, a Lean Copilot, a framework for running LLM-based inference in Lean, becomes ideally suited for the task at hand as the copilot integrates seamlessly into the Lean workflow. It should be emphasized that Lean itself does not support machine learning (ML) tools from LLMs, so the copilot becomes instrumental for suggesting proof steps (tactic suggestion), completing intermediate proofs (proof search), and selecting the relevant premises (premise selection) toward the construction of the graviton field.

By itself, Lean would be hard to use for the final steps of construction of the graviton field because all the mathematics pertaining to "attaching spinors to space–time" would need to be entered as code in the formalization stage. Lean indeed generated the "rule-based propositions" for spinor amalgamation with general relativity via its own proof searcher Aesop, but LLM becomes essential to generate the code and suggest the final tactics needed to derive the graviton field. Current LLMs are trained, fine-tuned, and benchmarked following standard ML procedures but cannot integrate into the Lean workflow because Lean is designed without ML support. For this reason, it becomes essential to adopt a Lean copilot for building the seamlessly adapted LLM-based proof automation tools. To that effect the C++ library CTranslate2 is introduced, as shown in the Appendix, for efficient inference with a transformer LLM representing the quantum gravity autoencoder. The inference runs as a shared library through a foreign function interface (FFI) within Lean.

REFERENCES

1. Russell S, Norvig P (2020) *Artificial Intelligence: A Modern Approach*. Pearson, London.
2. Kelleher JD (2019) *Deep Learning*. The MIT Press, Cambridge, MA.
3. Chollet F (2019) *Deep Learning with Python*. Manning Press, Shelter Island, New York.
4. Atienza R (2020) *Advanced Deep Learning with TensorFlow 2 and Keras*, 2nd Edition. Packt Publishing, Birmingham.
5. Schmidhuber (2015) Deep learning in neural networks: An overview. *Neural Netw* 61: 85–117.
6. Fernández A (2016) *Physics at the Biomolecular Interface*. Springer International Publishing, Switzerland.
7. Aliper A, Plis S, Artemov A, Ulloa A, Mamoshina P, Zhavoronkov A (2016) Deep learning applications for predicting pharmacological properties of drugs and drug repurposing using transcriptomic data. *Mol Pharm (ACS)* 13: 2524–2530.
8. Lavecchia A (2019) Deep learning in drug discovery: Opportunities, challenges and future prospects. *Drug Disc Today* 24: 2017–2032.

9. Jiménez J, Škalič M, Martínez-Rosell G, De Fabritiis, G (2018) Kdeep: Protein–ligand absolute binding affinity prediction via 3D-convolutional neural networks. *J Chem Inf Mod* 58: 287–296.
10. Stepniewska-Dziubinska MM, Zielenkiewicz P, Siedlecki P (2018) Development and evaluation of a deep learning model for protein–ligand binding affinity prediction. *Bioinformatics* 34: 3666–3674.
11. Hassan-Harrirou H, Zhang C, Lemmin T (2020) RosENet: improving binding affinity prediction by leveraging molecular mechanics energies with an ensemble of 3D convolutional neural networks. *J Chem Inf Model* 60: 2791–2802.
12. Brunton SL, Kutz NJ (2019) *Data-Driven Science and Engineering: Machine Learning, Dynamical Systems and Control*. Cambridge University Press.
13. Tiumentsev Y, Egorchev M (2019) *Neural Network Modeling and Identification of Dynamical Systems*. Academic Press, San Diego, CA.
14. Champion K, Lusch B, Kutz JN, Brunton SL (2019) Data-driven discovery of coordinates and governing equations. *Proc Natl Acad Sci USA* 116: 22445–22451.
15. Schmidt M, Lipson H (2009) Distilling free-form natural laws from experimental data. *Science* 324: 81–85.
16. Fernández A (1985) Center-manifold extension of the adiabatic-elimination method. *Phys Rev A* 32: 3070–3076.
17. Born M, Oppenheimer JR (1927) Zur Quantentheorie der Molekeln. *Ann Physik* 389: 457–484.
18. Fernández A (2014) Chemical functionality of interfacial water enveloping nanoscale structural defects in proteins. *J Chem Phys* 140: 221102.
19. Fernández A (2021) *Artificial Intelligence Platform for Molecular Targeted Therapy: A Translational Approach*. Chapter 9. World Scientific Publishing, Singapore.
20. Fernández A (2020) Deep learning unravels a dynamic hierarchy while empowering molecular dynamics simulations. *Ann Physik (Berlin)* 532: 1900526.
21. Brunton SL, Proctor JL, Kutz NJ (2016) Discovering governing equations from data by sparse identification of nonlinear dynamical systems. *Proc Natl Acad Sci USA* 113: 3932–3937.
22. Kurzweil R (2006) *The Singularity Is Near: When Humans Transcend Biology*. Penguin, New York.
23. de Vries J (2014) *Topological Dynamical Systems: An Introduction to the Dynamics of Continuous Mappings*. De Gruyter, Berlin.
24. Duraisamy K, Iaccarino G, Xiao H (2018) Turbulence modeling in the age of data. *Annu Rev Fluid Mech* 51: 357–377.
25. Thommen M, Holtkamp W, Rodnina MV (2017) Co-translational protein folding: progress and methods. *Curr Opin Struct Biol* 42: 83–89.
26. Sorokina I, Mushegian A (2018) Modeling protein folding in vivo. *Biology Direct* 13: 13.
27. Thurston WP (1997) *Three-Dimensional Geometry and Topology*. Princeton University Press.
28. Brooks B, Karplus M (1983) Harmonic dynamics of proteins: Normal modes and fluctuations in bovine pancreatic trypsin inhibitor. *Proc Natl Acad Sci U S A* 80: 6571–6575.
29. Thirumalai D, Lorimer GH, Hyeon C (2020) Iterative annealing mechanism explains the functions of the GroEL and RNA chaperones. *Protein Sci* 29: 360–377.
30. Nemytskii VV, Stepanov V (2016) *Qualitative Theory of Differential Equations*. Princeton University Press.
31. Arnold VI (1974) *Mathematical Methods of Classical Mechanics*. Springer, Berlin.
32. Weisstein EW (2021) Quotient Space. From MathWorld--A Wolfram Web Resource. https://mathworld.wolfram.com/QuotientSpace.html
33. Li S, Yang Y (2021) Hierarchical deep learning for data-driven identification of reduced-order models of nonlinear dynamical systems. *Nonlinear Dyn* 105: 3409–3422.
34. Maldacena J (1999) The Large N limit of superconformal field theories and supergravity. *Int J Theor Phys* 38: 1113.
35. Witten E (1998) Anti de Sitter space and holography. *Adv Theor Math Phys* 2: 263–291.
36. Bekenstein JD (1973) Black holes and entropy. *Phys Rev D* 7: 2333–2346.
37. Fernández A, Sinanoglu O (1982) The lifting of an Inonu-Wigner contraction at the level of universal coverings. *J Math Phys* 23: 2234.
38. Hashimoto K, Sugishita S, Tanaka A, Tomiya A (2018) Deep learning and the AdS/CFT correspondence. *Phys Rev D* 98: 046019.

39. Matsueda H, Ishihara M, Hashizume Y (2013) Tensor network and a black hole. *Phys Rev D* 87: 066002.
40. Weinberg S (2008) *Cosmology*. Oxford University Press.
41. Bohm D (1962) A suggested interpretation of the quantum theory in terms of hidden variables I. *Phys Rev* 86: 166–179.
42. Vanchurin V (2021) Toward a theory of machine learning. *Mach Learn Sci Technol* 2: 036012.
43. Gubser SS (2010) *The Little Book of String Theory*. Princeton University Press.
44. Wheeler JA, Zurek WH (2014) *Quantum Theory and Measurement*. Princeton University Press.
45. Farhi E, Guth A, Guven J (1990) Is it possible to create a universe in the laboratory by quantum tunneling? *Nuc Phys B* 339: 417–490.
46. Vanchurin V (2018) Covariant information theory and emergent gravity. *Int J Mod Phys A* 33: 1845019.
47. Bednik G, Pujolas O, Sibiryakov S (2013) Emergent Lorentz invariance from strong dynamics: Holographic examples. *J High Energy Phys* 11: 64.
48. Loeb A (2021) Was our universe created in a laboratory? Scientific American. October 15, 2021. www.scientificamerican.com/article/was-our-universe-created-in-a-laboratory/
49. Harrison ER (2003). *Masks of the Universe*. Cambridge University Press.
50. Loeb A (2021) Endless creation out of nothing. *Sci Am* 4: 2020.

4 Geometric Dilution of Dark Matter as Precursor to the Visible Sector in Particle Physics

"There was neither non-existence nor existence then.
There was neither the realm of space nor the sky that is beyond."

Creation Hymn, Rig-Veda

Cosmological data from observation at very large scales pose a significant challenge to the fundamental tenets of physics. Evidence supporting the existence of dark matter (DM) and dark energy (DE) has proven so far unyielding to any possibility of reconciliation with quantum mechanics or relativity. This chapter addresses the problem by resorting to artificial intelligence (AI) for answers. It builds on the formal treatment presented in Chapter 2 and describes the results in terms of an ur-universe with toroidal topology that incorporates an extra dimension to encode Einstein's space–time as a latent manifold.

On the basis of evidence presented in this chapter, DM is postulated to have arisen during the creation of elementary particles in an early universe. In contrast with our present-day "flat" universe, this early universe was endowed with extreme geometric curvature and, consequently, was subject to special quantization rules. To validate this picture, an AI platform is leveraged in the guise of a quintessential autoencoder. This platform is capable of reverse engineering the "action principles" that underlie the standard model of elementary particle physics. The deconstruction treats Einstein's 4D space–time as a "latent space" that gets decoded onto a five-dimensional compact multiply connected manifold with no boundary. The manifold is endowed with an extra rolled-up dimension that spans a quark cross section, which is the smallest material dimension, estimated to lie within the attometer (10^{-18} m) scale. It turns out that this compact fifth dimension stores stationary wave–matter with a rest mass that matches the vacuum expectation value of the Higgs boson. This result is pregnant with possibilities and implications and enabled us to estimate elementary particle masses with significant precision by an AI-based quintessential decoding of the elementary particle fields. The results point to the existence of an ur-Higgs boson in the early universe, specifically at the beginning of the "electroweak epoch", whose kinetic energy is not geometrically diluted along the standard 4D dimensions. This ur-Higgs and its heavier quantum relatives are identified by AI as DM. The results enable us to characterize DM vis-à-vis the geometric dilution of gravity shown to have taken place as the universe flattened and expanded to present-day levels.

The results obtained by AI and reported in this chapter may be summarized as follows:

- The universe topology is "revealed" by the CMB fluctuation spectrum. Unlike the Euclidean topology, this topology is compatible with the big bang scenario.
- AI can trace the origin of dark matter to the universe evolution by reverse engineering the standard model to incorporate a circular dormant dimension.

DOI: 10.1201/9781003632924-4

- On the dormant compact dimension with quark-size radius q, a de Broglie matter–wave has energy equal to the vacuum expectation value of the Higgs field. This is the ur-Higgs elementary particle, a good candidate for dark matter for the following reasons:
  - Incorporation of a fifth dimension is supported by evidence: SM elementary particle masses can be predicted through geometric dilution of the ur-Higgs
  - The ur-Higgs formed in an early universe ($\sim 10^{-27}$ s after the big bang).
  - It is massive and cold ($v \ll c$).
  - It does not decay through communication with SM gauge bosons.
  - It is only interactive with the SM via gravity.
  - It endows other ur-particles with mass through ur-boson-conveyed geometric dilution.

On the other hand, DE is generated by quantum vacuum fluctuations exciting the quintessential field of the ur-Higgs and becoming stored as DM whenever the wavelength of the excitation fits a stationary wave condition along the dormant dimension.

Finally, a scenario is established as described by a space–time rendering of the universe evolution through geometric dilution of dark matter.

4.1 AI DECODES THE STANDARD MODEL BY INCORPORATING AN EXTRA DIMENSION

As shown in the scheme in Figure 2.12 and in Chapter 3, the concept of diagram commutativity is central to the encoding of a dynamical system into its latent simplified dynamics and, reciprocally, to the decoding of the latent dynamics onto a "quintessential" space–time that incorporates a dormant dimension [1]. This scheme is now adapted to decode the standard model (SM) onto a space with an extra dimension building upon the results from Chapter 2. The task of decoding the SM defined on the four-dimensional space–time $W/\sim$ onto a five-dimensional compact multiply connected manifold W requires a "quintessential" (i.e., "fifth essence") autoencoder operating in reverse. Typically, autoencoders simplify the dynamics to retain the latent coordinates [1]. As shown in detail in Chapter 2, in the case of interest to elementary particle physics, tangible observable processes are assumed to be taking place in a quotient space of equivalent classes modulo a dormant dimension and the goal of the quintessential autoencoder is to decode the SM onto the space that incorporates the dormant fifth dimension (the "fifth essence").

This reverse quintessential autoencoder lifts the flow $\mathcal{F}: W/\sim \to W/\sim$ onto a flow $\tilde{\mathcal{F}}: W \to W$. The lifting is compatible if and only if it satisfies the commutativity relation: $\mathcal{F} \circ \pi = \pi \circ \tilde{\mathcal{F}}$, with $\pi: W \to W/\sim$ denoting the canonical projection that assigns each point in W to its equivalence class in $W/\sim$. Since we are introducing autoencoders for dynamical systems [1], it becomes imperative to cast the SM as a dynamical system. This prompts the question: How do we cast the flow $\mathcal{F}: W/\sim \to W/\sim$ to represent a process described by the SM?

As described in Chapter 2, we may regard the interaction processes determined by the SM as transformations created by a time-dependent operator. For each elementary process, this operator has a time step τ associated with it, and this time step is precisely the lifetime of the gauge boson that communicates the force acting on the elementary particle and causing the elementary particle transformation [2]. Thus, the time step is obtained from the uncertainty principle: $\sim \dfrac{\hbar}{2M_B c^2}$, where M_B is the rest mass of the relevant gauge boson. Thus, the autoencoder described in Chapter 3 will be operated in reverse in order to decode the SM by incorporating the dormant dimension shown to store gravity in a reverse engineering scheme that fits the big bang scenario.

The NN for the quintessential autoencoder (QAE) is constructed in such a way that each input node represents the elementary particle field value (for simplicity assumed to be a complex number) in an equivalence class contained in $W/\sim$. The grid of nodes or equivalence classes has a mesh determining the level of resolution and each equivalence class is taken modulo the dormant coordinate, implying that it is represented by the four latent coordinates of the standard four-dimensional space–time. Since the autoencoder is set up to be working in reverse, the output nodes represent the values of the decoded elementary particle field on points in the quintessential space W. Thus, we have the following:

Lemma 4.1. The elementary particle field ϕ is decoded by the quintessential autoencoder as the ur-particle field $\tilde{\phi}$ if and only if $\exists \gamma_\phi$:

$$\phi \circ \pi = \gamma_\phi \circ \tilde{\phi};\, \gamma_\phi : \mathbb{C} \to \mathbb{C}. \tag{4.1}$$

The proof follows from the commutativity of the following diagram that would make the decoding compatible with the field defined on the four-dimensional space–time:

$$\begin{array}{ccc} W & \to & W/\sim \\ \tilde{\phi}\downarrow & & \downarrow\phi \\ \mathbb{C} & \to & \mathbb{C} \end{array} \tag{4.2}$$

The field for an elementary particle with Lagrangian $\mathcal{L}$ satisfies the Euler–Lagrange equation:

$$\partial_\phi \mathcal{L} = \partial_\mu \frac{\partial \mathcal{L}}{\partial(\partial_\mu \phi)} \tag{4.3}$$

While the decoded field of the ur-particle obeys the relation:

$$\partial_{\tilde{\phi}} \tilde{\mathcal{L}} = \partial_\nu \frac{\partial \tilde{\mathcal{L}}}{\partial(\partial_\nu \tilde{\phi})}, \tag{4.4}$$

with its respective decoded Lagrangian $\tilde{\mathcal{L}}$ fulfilling the relation:

$$\int \tilde{\mathcal{L}}(\tilde{\phi})\, dx dy_5 = \int \mathcal{L}(\phi)\, dx + \left[\int \left| \int (\tilde{\phi} - \phi) dx \right|^2 dy_5 \right]^{1/2} \tag{4.5}$$

Heretofore, the dormant coordinate is denoted y_5.

Equation (4.5) is central to the process of quintessential decoding of the SM. It indicates that the field $\tilde{\phi}$ of the ur-particle is always the minimal smooth extension of the elementary particle field ϕ over the fifth coordinate, while the compactness of the latter coordinate ensures the convergence of the right term in the r.h.s. of Eq. (4.5).

In principle, we do not and cannot *a priori* specify the radius r_0 of the circular dormant coordinate because this parameter, together with the geometric dilution parameter (or α pitch), must be determined in the optimization of the pair $(\tilde{\phi}, \gamma_\phi)$ in accordance with the loss function $J(\alpha, r_0)$ associated with the diagram commutativity (Figure 4.1):

$$J(\alpha,\, r_0) = \sum_{y \in M_\alpha} \frac{1}{|M_\alpha|} \left\| (\phi \circ \pi) y - (\gamma_\phi \circ \tilde{\phi}) y \right\|^2 \tag{4.6}$$

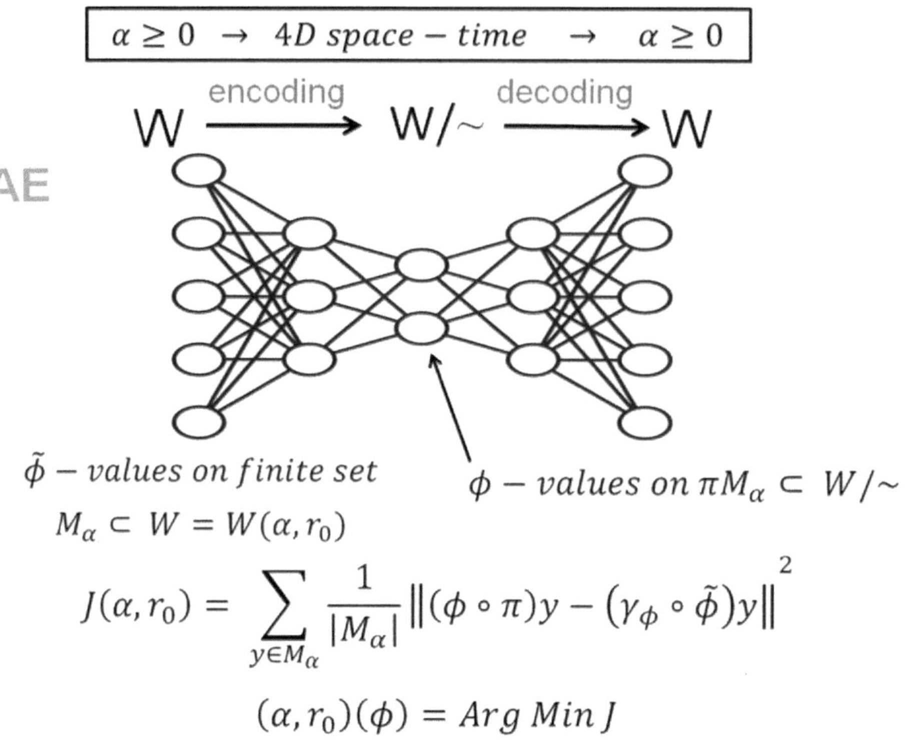

FIGURE 4.1 The quintessential autoencoder (AE) decoding a SM particle field ϕ on the four-dimensional space–time that is regarded as the latent manifold $W/\sim$ of the quintessential space W. A loss function is parametrically determined by the pitch α and dormant dimension r_0, whose optimization yields the decoded field $\tilde{\phi}$.

$$(\alpha, r_0)(\phi) = Arg\ MinJ \tag{4.7}$$

In practice, we shall see that the optimization dictated by the loss function attributes a unique geometric dilution value to each elementary particle in the SM, while r_0 remains constant and equal to the quark value q in the attometer range [3]. Assuming and anticipating $r_0 = q$, the corresponding wave functions for ur-particles become parametrically dependent on geometric dilution.

By definition, the ur-Higgs boson mass is stored completely in the dormant dimension. Therefore, the wavefunction Y for the ur-Higgs boson is associated with $\alpha = 0$ (no geometric dilution) and its expectation energy becomes $\langle \Upsilon | \widetilde{\mathcal{H}_{\mathrm{H}}} | \Upsilon \rangle = 246$ GeV, where $\widetilde{\mathcal{H}_{\mathrm{H}}}$ denotes the decoded Hamiltonian for the ur-Higgs boson. On the other hand, the wave function Ψ_{H} for the decoded Higgs boson satisfies $\langle \Psi_{\mathrm{H}} | \Upsilon \rangle = \cos\alpha^*$, where $\alpha^* = 59°27' \approx \pi/3$. In general, for an ur-particle ζ with associated pitch α_ζ, we obtain $\langle \Psi_\zeta | \Upsilon \rangle = \cos\alpha_\zeta$. These observations are memoralized in Figure 4.2. The protocol for obtaining wave functions $\Psi : \mathrm{W} \rightarrow \mathbb{C}$ for ur-particles decoding elementary particles is presented in Figure 4.3a,b.

It should be noticed that heavier relatives of the ur-Higgs boson are also generated in accordance with the increase in the frequency $f = \frac{cn}{2\pi r_0}(n > 1)$ of the stationary wave along the dormant

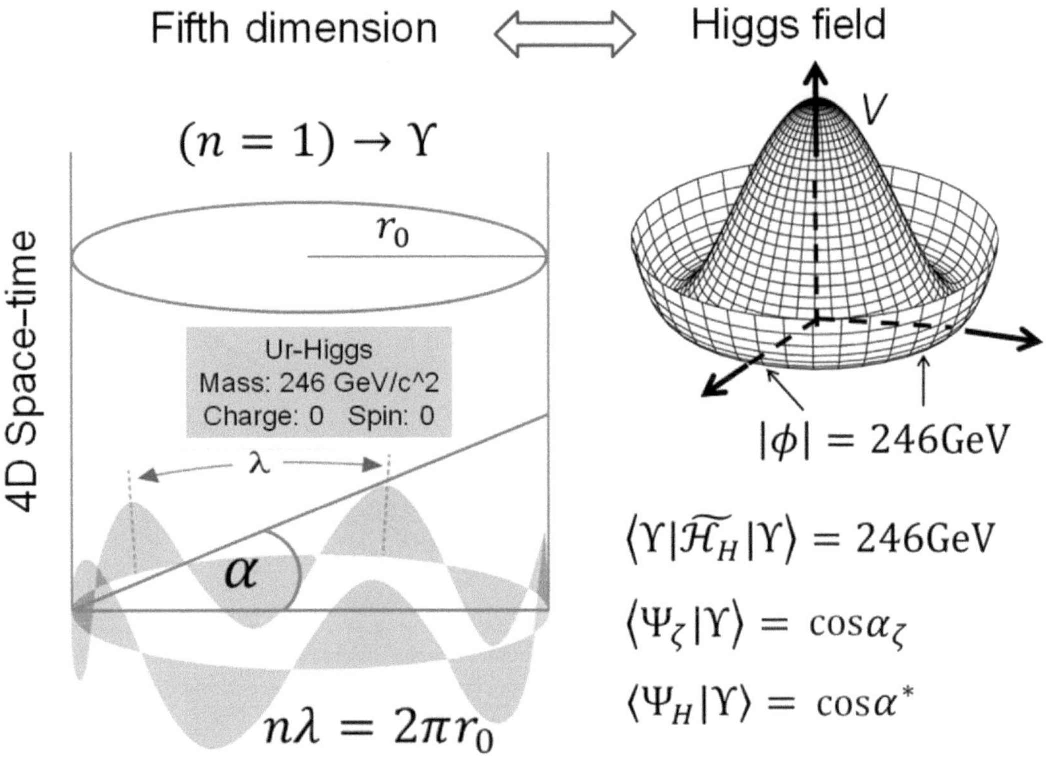

FIGURE 4.2 The dormant fifth dimension in the quintessential decoding of space–time. The fifth dimension represents the smallest material scale known, which corresponds to the quark cross section, established to be in the attometer range [3]. The lowest energy stationary (de Broglie) matter–wave fully stored in the dormant dimension and named ur-Higgs has a rest mass energy corresponding to the vacuum expectation value of the Higgs boson field $|\phi| \approx 246\ GeV$.

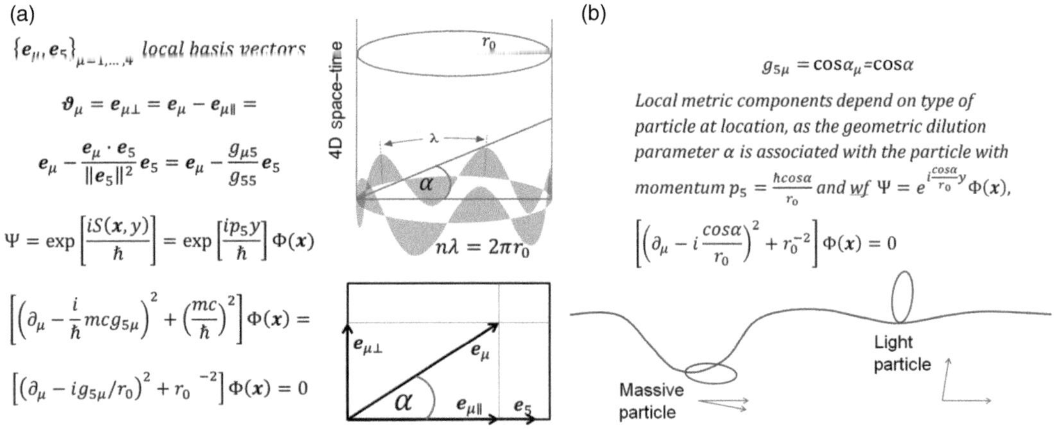

FIGURE 4.3 (a,b) Protocol to generate the wave functions for ur-particles decoding elementary particles in the SM.

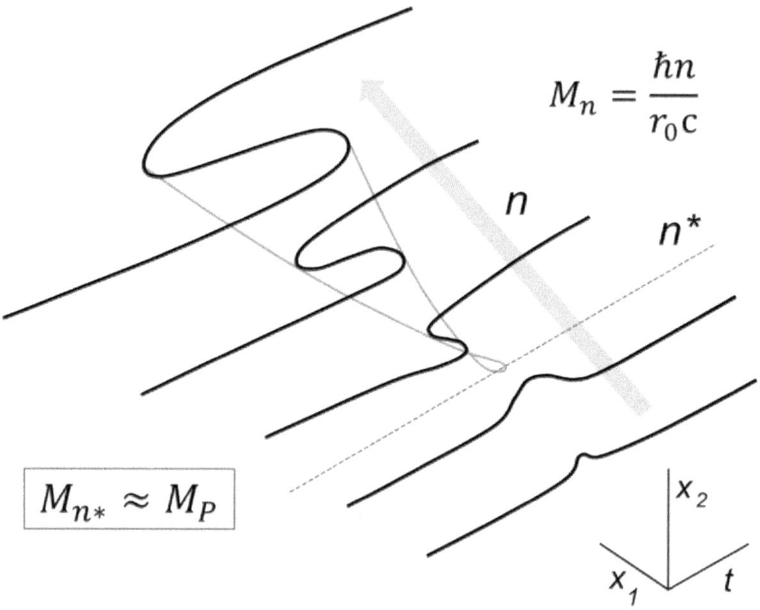

FIGURE 4.4 Space–time rendering of the quintessential hierarchy of relatives of the ur-Higgs boson obtained for $n > 1$. A cusp ripple in the space–time cross section occurs consistently with formation of a black hole when the mass M_n of the particle increases beyond the Planck mass M_P, as $n^* \leq n$.

dimension. The space–time rendering of this quintessential particle hierarchy representing DM is obtained for integer values of n, as shown schematically in Figure 4.4. It reveals a ripple in the space–time cross-section, consistent with formation of a black hole as the mass M_n of the particle increases beyond the Planck mass: $M_n > M_P$. The signature of the black hole is the placement of its interior into the future, so time is forced to run backward to escape the event horizon (Figure 4.4), a theoretical impossibility according to relativity.

The space–time rendering of the quantum field excitations representing the DM hierarchical tower becomes possible using the quantum gravity autoencoders introduced in Chapter 3. These AI systems can decode the quantum field excitations as relativistic space–time (Figure 4.5), showing a time reversal that is indicative of the interior of a black hole that emerges when n reaches the critical value n^* that yields the Planck mass (cf. Figure 4.4).

4.2 REVERSE ENGINEERING OF THE STANDARD MODEL INTO A QUINTESSENTIAL MODEL

In order to compute the quintessential decoding of an elementary particle field to generate the ur-particle field, the reverse autoencoder changes the metric of each Euclidean coordinate in W, compressing the metric asymptotically as we approach infinity. This is a purely AI move, completely unsupervised and uninstructed and compactifies the space as it rolls up each Euclidean dimension onto a circle (Figure 4.6). The Euclidean metric on the line is changed into a new metric obtained by mapping a circle osculating the line, so that the distance between two points along the line is now evaluated as the length of the arc sustained between the projected points on the circle, as shown in Figure 4.6.

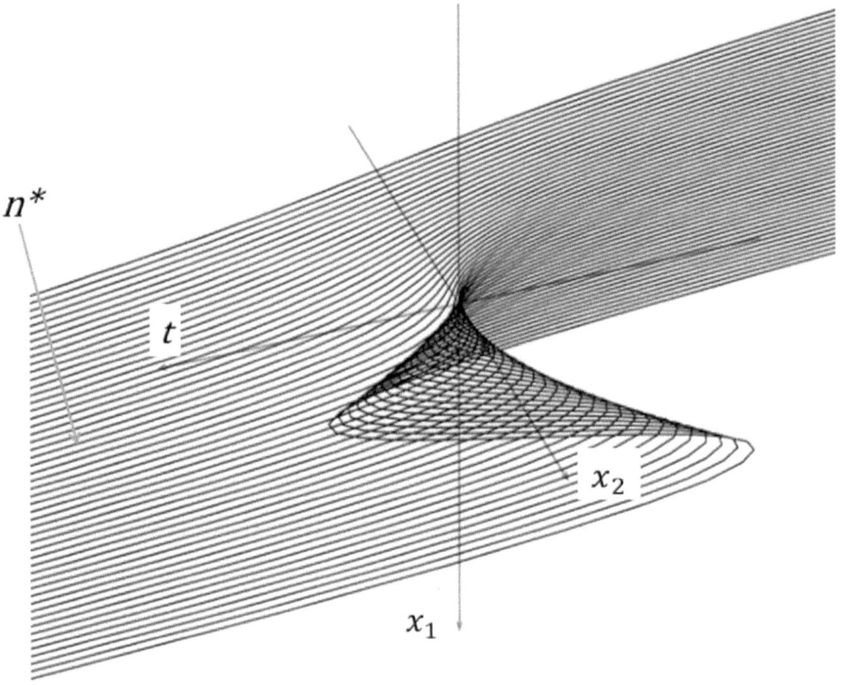

FIGURE 4.5 Quantum field excitations for the DM hierarchy decoded as relativistic space–time by a quantum gravity autoencoder.

This new metric essentially maps W homeomorphically onto a five-dimensional torus, so that states representing ur-particles become quantized in W with five quantum numbers due to the requirement to generate stationary de Broglie matter–waves in the five-dimensional torus. Thus, quantization rules apply so that the wavelength λ for a elementary particle with pitch α must now fulfill the stationary-wave equation $\lambda \cos\alpha = \dfrac{2\pi r_0}{n}$ for some quantum number (integer) n, and simultaneously fulfill four stationary-wave equations of the form $\lambda \sin\alpha = \dfrac{2\pi r}{n'}$ for quantum numbers n' for each coordinate in the four-dimensional space–time endowed with the metric inherited from projection onto a circle of radius r (Figure 4.6). This implies that geometric dilution α, quantum number ratio $\left(\dfrac{n}{n'}\right)$, and aspect ratio $\left(\dfrac{r}{r_0}\right)$ are interrelated according to the four fundamental relations:

$$tg\alpha = \left(\frac{n}{n'}\right)\left(\frac{r}{r_0}\right) \tag{4.8}$$

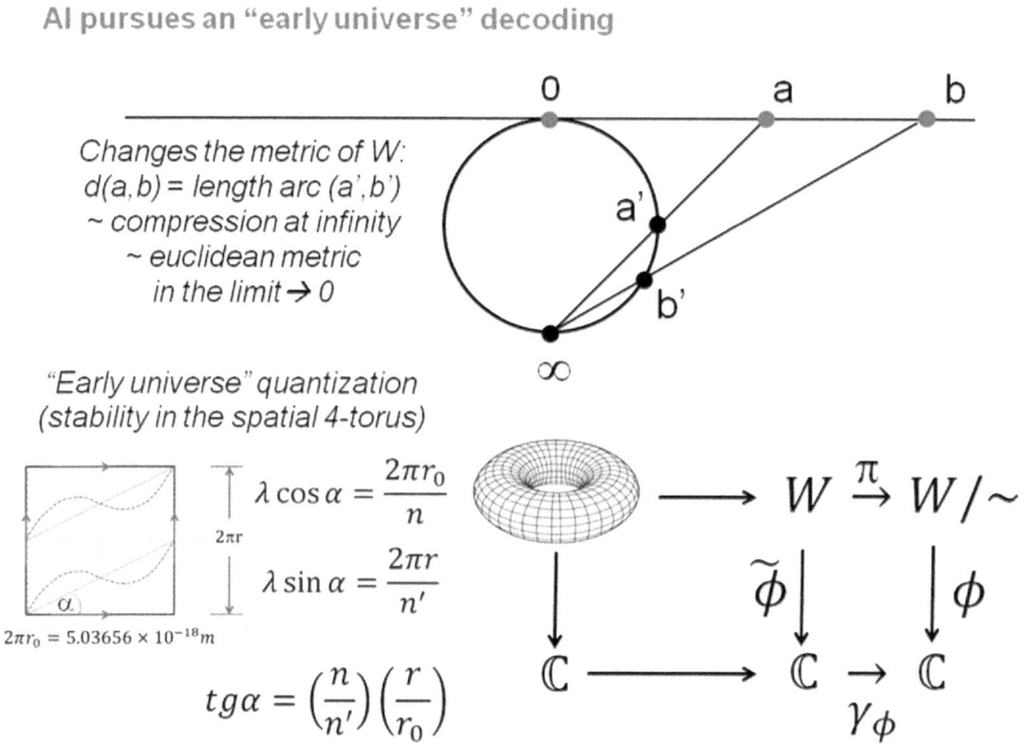

FIGURE 4.6 Early-universe toroidal decoding of the four-dimensional space–time regarded as the latent space ($W/\sim$) for the quintessential space W homeomorphic to an "early-universe" five-dimensional torus. The quintessential toroidal decoding of the particle field introduces a quantization required to avoid destructive interference on the five-dimensional torus. The quantization of the ur-field is thus defined by four independent quantum numbers that are encoded into the standard particle attributes within the standard model (charge, mass, and spin) plus an extra attribute known as geometric dilution, only apparent in the quintessential decoded space W.

For example, for the Higgs boson at (n,n')= (1,1) with $\alpha = 59°27'$ ($tg\alpha$ =1.69428), we get the aspect ratio $\left(\frac{r}{r_0}\right)$=1.69428. Since aspect ratios are invariants of the five-dimensional torus and the pitch is fixed for each ur-particle, the set of five quantum numbers (n, n'_1, n'_2, n'_3, n'_4) determines the quintessential quantized decoding of each elementary particle in the SM. The decoding of each elementary particle field through the toroidal quantization of its ur-field in the quintessential space W is thus represented in Figure 4.6. The four elementary particle attributes in the SM, mass, geometric dilution, charge, and spin, represent a specific encoding of the four independent quantum numbers in the toroidal quantization described by the four relations in Eq. (4.8).

The geometric dilution associated with each elementary particle in the SM is thus obtained by computing the ur-field as the optimum quintessential decoding of the elementary particle field that minimizes the loss function in fulfillment of Eqs. (4.6) and (4.7). This decoded field minimizes the quintessential Lagrangian defined by Eq. (4.5). The results are presented in Figure 4.7.

The decoding of the SM is fully implemented by ensuring through parametric optimization the commutativity of the dual action diagram presented in Figure 4.8. To introduce this higher level of description some notation becomes necessary: The space of smooth scalar fields for ur-particles is

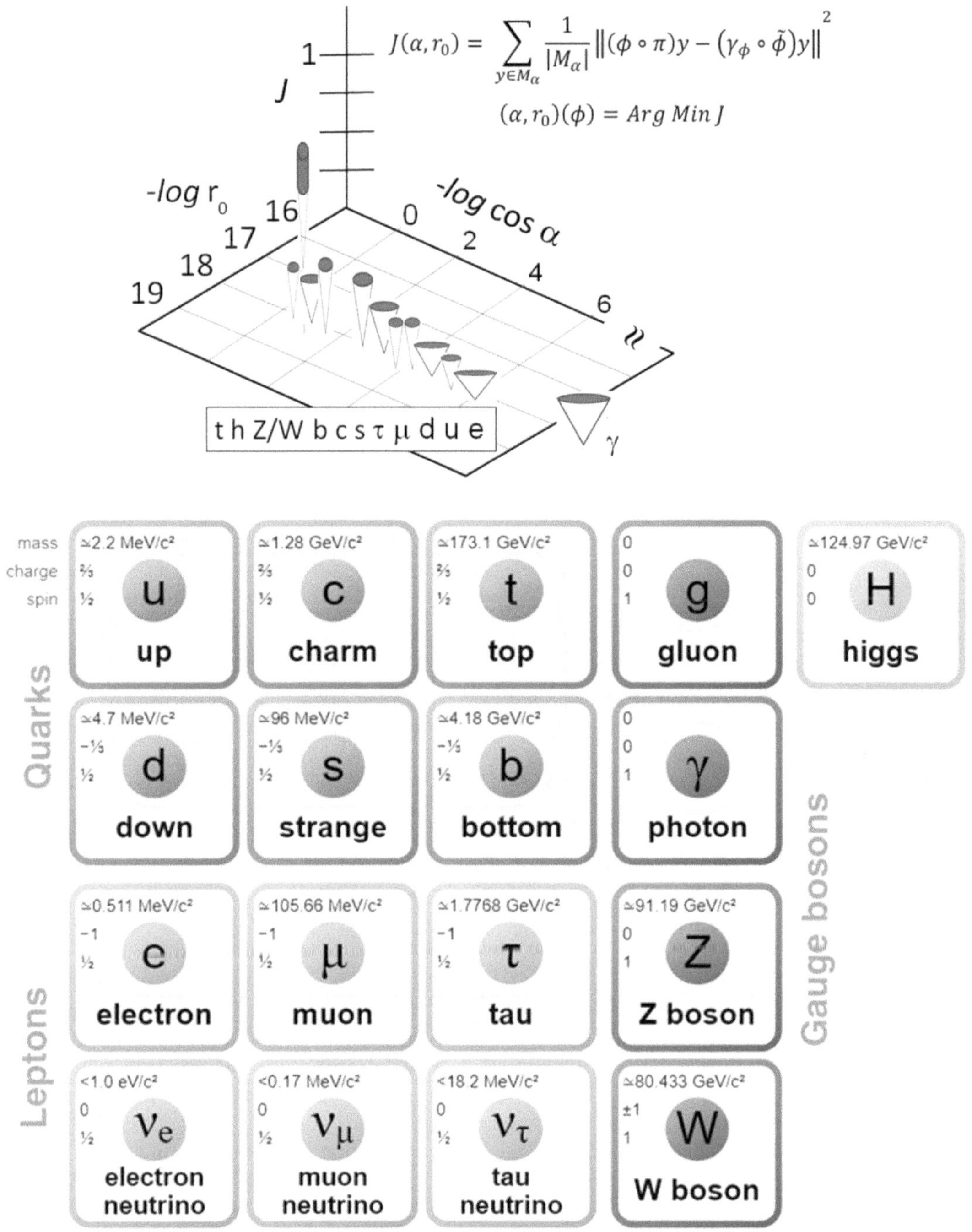

FIGURE 4.7 Pitch and dormant dimension (α, r_0) for the optimization of the loss function $J(\alpha, r_0)$ for the quintessential autoencoder that decodes each elementary particle field in the standard model. (a) Loss function optimization in (J, α, r_0) space for each elementary particle denoted following notation in (b). (b) Table of elementary particles in the standard model.

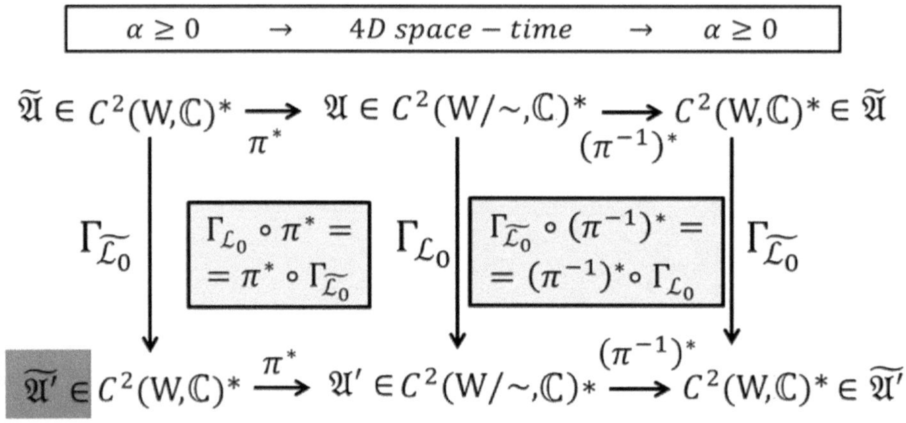

$$\mathfrak{A} = \mathfrak{A}(\mathcal{L}) = \int \mathcal{L}(\phi, \partial_\mu \phi) dx^\mu \qquad \tilde{\mathfrak{A}} = \int \tilde{\mathcal{L}}(\tilde{\phi}, \partial_\nu \tilde{\phi}) dy^\nu$$

$$\Gamma_{\mathcal{L}_0}(\mathfrak{A}) = \mathfrak{A}' = \mathfrak{A}(\mathcal{L}')\ where$$

$$\int [\mathcal{L} + \mathcal{L}_0] d\boldsymbol{x} = \int [\mathcal{L}' + \mathcal{L}'_0] d\boldsymbol{x}'$$

FIGURE 4.8 Quintessential decoding of the standard model implemented by ensuring the commutativity of the action diagram through parametric optimization. In this decoding context dual to that represented in Figure 6.1, the gauge boson with Lagrangian $\mathcal{L}_0$ is represented as a map $\Gamma_{\mathcal{L}_0} : C^2(W/\sim,\mathbb{C})^* \to C^2(W/\sim,\mathbb{C})^*$ defined over action space as $\Gamma_{\mathcal{L}_0}\mathfrak{A}(\mathcal{L}) = \mathfrak{A}(\mathcal{L}')$. Similarly, the corresponding decoded ur-boson is represented by the map $\Gamma_{\tilde{\mathcal{L}}_0} : C^2(W,\mathbb{C})^* \to C^2(W,\mathbb{C})^*$ defined as $\Gamma_{\tilde{\mathcal{L}}_0}\tilde{\mathfrak{A}}(\tilde{\mathcal{L}}) = \tilde{\mathfrak{A}}(\widetilde{\mathcal{L}'})$. So the consistent decoding of the SM requires that the following commutativity relation is valid for any generic boson with Lagrangian $\mathcal{L}_0 : \Gamma_{\mathcal{L}_0} \circ \pi^* = \pi^* \circ \Gamma_{\tilde{\mathcal{L}}_0}$, where π^* is the dual map of the canonical projection of the quintessential space W onto the quotient (latent) space $W/\sim$.

$C^2(W,\mathbb{C})$, while the space $C^2(W/\sim,\mathbb{C})$ contains the elementary particle scalar fields in the SM. The respective dual spaces $C^2(W,\mathbb{C})^*$ and $C^2(W/\sim,\mathbb{C})^*$ contain the actions defined by functionals–actions of the type: $\tilde{\mathfrak{A}}(\tilde{\mathcal{L}}) = \int \tilde{\mathcal{L}}(\tilde{\phi}, \partial_\nu \tilde{\phi}) dy^\nu$ and $\mathfrak{A}(\mathcal{L}) = \int \mathcal{L}(\phi, \partial_\mu \phi) dx^\mu$, respectively. As described in Chapter 2, a elementary particle Lagrangian $\mathcal{L}$ becomes transformed into another elementary particle Lagrangian $\mathcal{L}'$ when the first elementary particle interacts with another elementary particle (gauge boson) with Lagrangian $\mathcal{L}_0$. These interactions define the flow in action space that enables the SM to be represented as a dynamical system with propagator time step defined as the lifetime of the gauge boson with associated Lagrangian $\mathcal{L}_0$ that communicates the force acting on the elementary particle and induces its transformation. Thus, the SM may be decoded at the action level using dual autoencoders adapted to dynamical systems of the type described in Chapter 3 but acting in reverse, where the four-dimensional space–time represents the latent space.

In the dual context of this decoding, the gauge boson with Lagrangian $\mathcal{L}_0$ is represented as a map $\Gamma_{\mathcal{L}_0} : C^2(W/\sim,\mathbb{C})^* \to C^2(W/\sim,\mathbb{C})^*$ defined as $\Gamma_{\mathcal{L}_0}\mathfrak{A}(\mathcal{L}) = \mathfrak{A}(\mathcal{L}')$. Similarly, the corresponding ur-boson is represented in this dual space by the map $\Gamma_{\tilde{\mathcal{L}}_0} : C^2(W,\mathbb{C})^* \to C^2(W,\mathbb{C})^*$ defined as $\Gamma_{\tilde{\mathcal{L}}_0}\tilde{\mathfrak{A}}(\tilde{\mathcal{L}}) = \tilde{\mathfrak{A}}(\widetilde{\mathcal{L}'})$. So the decoding of the SM becomes operational when it is also validated at the

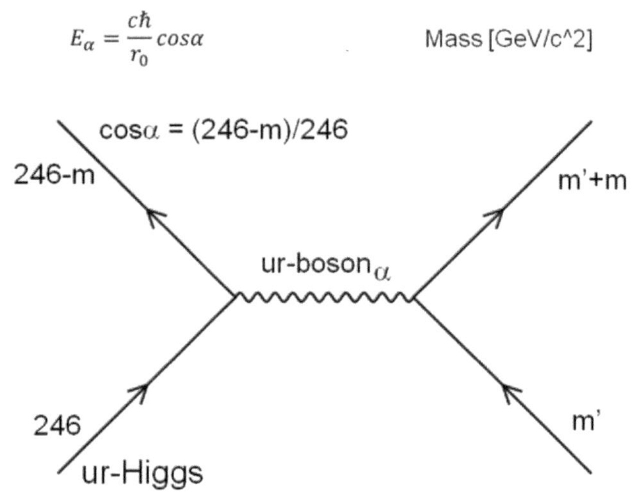

FIGURE 4.9 Communication of geometric dilution in the quintessential space W represented as a Feynman diagram (cf. [2]), involving an ur-particle with mass m', the ur-Higgs boson, and an ur-gauge boson associated with the pitch value α.

level of actions, implying that the following commutativity relation is valid for any gauge boson with generic Lagrangian denoted $\mathcal{L}_0$:

$$\Gamma_{\mathcal{L}_0} \circ \pi^* = \pi^* \circ \Gamma_{\widetilde{\mathcal{L}_0}} \tag{4.9}$$

In Eq. (4.9), we assume that the canonical projection $\pi: W \to W/\sim$ induces the dual canonical projection $\pi^*: C^2(W,\mathbb{C})^* \to C^2(W/\sim,\mathbb{C})^*$ mapping the action space for W onto the action space for the SM.

Geometric dilution parametrized by the pitch angle α may be communicated gravitationally in the quintessential space W through the respective ur-boson$_\alpha$, as schematically represented in the Feynman diagram [2] displayed in Figure 4.9. Thus, the ur-Higgs may decay into a geometrically diluted ur-particle with energy $E_\alpha = \left(\frac{c\hbar}{r_0}\right)\cos\alpha$ and thereby transfer the geometrically diluted mass m (expressed in GeV/c^2) satisfying the equation $\cos\alpha = (246 - m)/246$. The mass is captured by an ur-particle with rest mass m' that gets transformed into an ur-particle with mass $m + m'$ upon interaction with the ur-boson$_\alpha$ that has been emitted or interacted with the ur-Higgs boson.

On the other hand, ur-bosons are predicted by the quintessential autoencoder to participate in elementary processes in the vacuum involving quark-pair annihilation with lepton-pair decay, as described in Figure 4.10. In this case, the ur-boson has a geometric dilution $-\log\cos\alpha = -\log\,(c^2 m_q/246\text{ GeV})$, where m_q is the quark rest mass.

Furthermore, the deconstruction of the SM enables the quintessential lifting of any elementary process. This is illustrated in Figure 4.11, where the generation of the Higgs boson through a gluon-pair fusion is quintessentially lifted to a process where communication with the ur-counterpart of the virtual top quark (uvtq) triggers the drop in geometric dilution required to render the ur-Higgs boson.

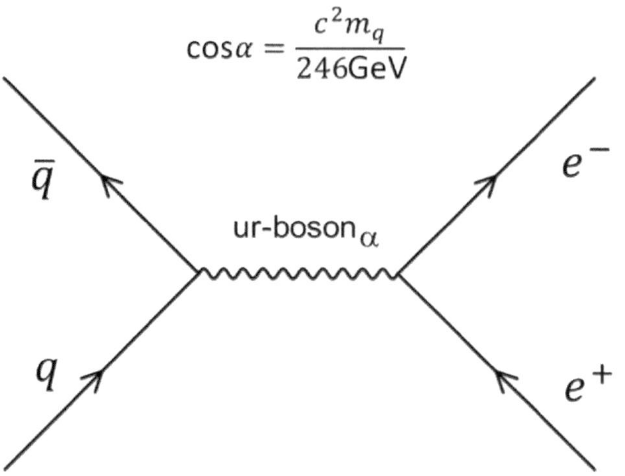

FIGURE 4.10 Predicted quark-pair annihilation mediated by an ur-boson whose geometric dilution is determined by the quark's rest mass.

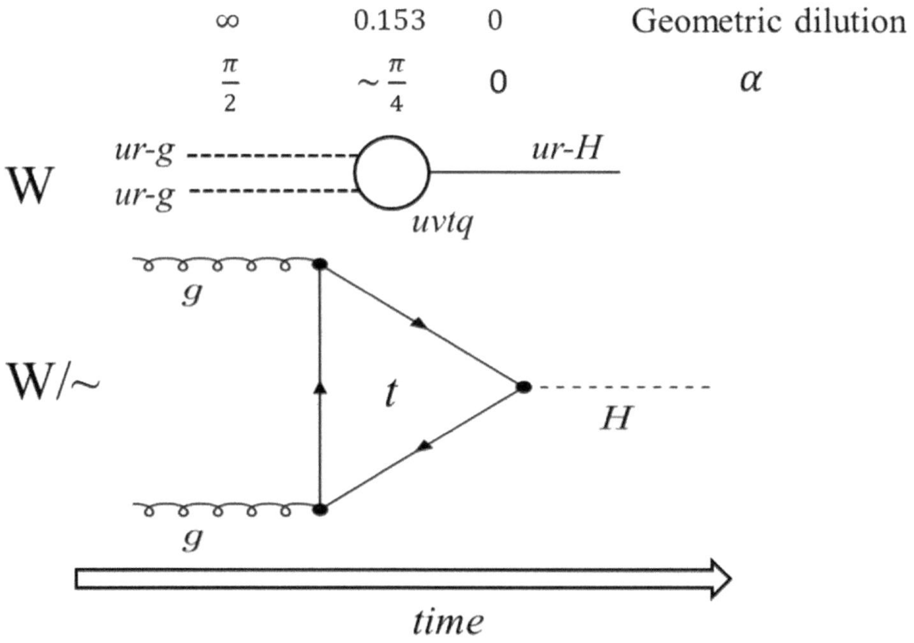

FIGURE 4.11 Quintessential lifting of the elementary process of Higgs boson creation through fusion of a pair of gluons mediated by a virtual top quark.

4.3 EXPERIMENTAL VALIDATION OF THE QUINTESSENTIAL DECODING OF THE STANDARD MODEL

The reverse quintessential autoencoder decodes the elementary particle fields in the SM as ur-particles with masses determined by the geometric dilution of the ur-Higgs stationary wave stored on the dormant fifth coordinate. The geometric dilution may be directly determined from the pitch of the wave front associated with the quintessentially decoded wave function for the ur-particle. Wave functions for elementary particles are decoded as maps $\Psi: \mathrm{W} \to \mathbb{C}$ following the same tenets that apply to the decoding of particle fields (Section 4.2). The decoding of the photon (γ) has infinite dilution ($\upsilon = -\log\cos\alpha = \infty, \alpha = \pi/2$), yielding a zero mass (Figure 4.7a), and hence, this gauge boson communicating the electromagnetic force travels at the speed of light. Thus, the wavefunction $\Psi_\gamma : \mathrm{W} \to \mathbb{C}$ for the ur-photon is orthogonal to that of the ur-Higgs $\langle \Psi_\gamma \mid \Upsilon \rangle = cos(\pi/2) = 0$. At the opposite end of the spectrum, the heaviest elementary particle, the top quark (t) gets decoded into the ur-t which has the lowest geometric dilution of all decoded elementary particles in the SM: $\upsilon = -\log\cos\alpha = 0.153, \alpha \approx \pi/4$. This yields $\langle \Psi_t \mid \Upsilon \rangle = 0.703$ a good approximation to the ratio between elementary particle mass and Higgs vacuum expectation value (i): $\frac{m}{v} = \frac{173.1}{246} \approx \langle \Psi_t \mid \Upsilon \rangle$.

The following relation is valid for all gauge bosons in the SM:

$$\frac{m_\zeta}{v} \approx \langle \Psi_\zeta \mid \Upsilon \rangle \tag{4.10}$$

In the case of fermions, the relation becomes

$$g_\zeta \approx \langle \Psi_\zeta \mid \Upsilon \rangle, \tag{4.11}$$

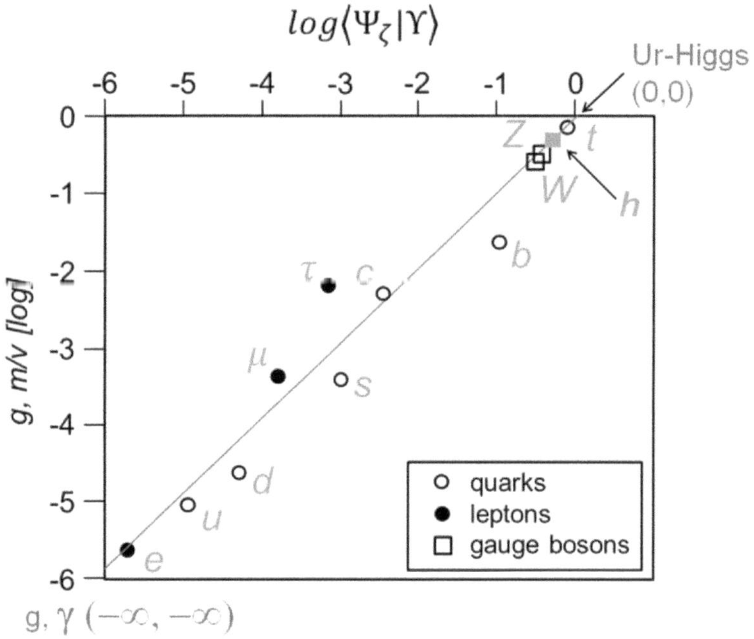

FIGURE 4.12 Prediction of the mass for each elementary particle (ζ) in the standard model through the correlation $\frac{m_\zeta}{v}$, $g_\zeta - \langle \Psi_\zeta \mid \Upsilon \rangle$ ($R^2 = 0.88$), where $\frac{m_\zeta}{v}$ is the ratio of gauge boson mass over vacuum expectation value for the Higgs boson and g_ζ is the Yukawa coupling parameter for the case when ζ denotes a fermion. The linear correlation validates the quintessential decoding of the standard model as reverse engineering.

where g_ζ is the respective Yukawa coupling constant [2]. The Yukawa coupling to the Higgs field can thus be interpreted as an effective geometric dilution parameter for ur-fermions in quintessential space as it follows from the relation $g = \cos\alpha$.

A linear plotting $\frac{m_\zeta}{v}, g_\zeta - \langle \Psi_\zeta \mid \Upsilon \rangle$ yields $R^2 = 0.88$ as shown in Figure 4.12 for all elementary particles in the SM. The resulting prediction of masses is based on the geometric dilution that optimizes the loss function $J(\alpha, q)$ for the decoding of each particle field into its ur-field. The implication is that geometric dilution, or equivalently, the pitch angle α, is a fundamental parameter that determines the mass of ur-particles originally enshrined in the ur-Higgs. When encoded into the quotient (latent) space $W/\sim$, this relatively simple process becomes the much more complex process by which mass is endowed by the Higgs boson in its interplay with the particle fields [2]. The commutativity of the following diagram represents the decoding of the Higgs mass-endowment process:

$$\begin{array}{ccccc} & W & \overset{\pi}{\rightarrow} & W/\sim & \\ \textit{geometric dilution} & \downarrow \widetilde{\mathfrak{F}_H} & & \downarrow \mathfrak{F}_H & \textit{Higgs mechanism} \\ & W & \overset{\pi}{\rightarrow} & W/\sim & \end{array} \tag{4.12}$$

The flow $\mathfrak{F}_H$ represents the endowment of mass through interaction with the Higgs boson. This process should be interpreted in the sense that for a generic particle with Lagrangian $\mathcal{L}$, the following relation holds: $\Gamma_{\mathcal{L}_H} \mathfrak{A}(\mathcal{L}) = \mathfrak{A}(\mathcal{L}')$, where the massless original Lagrangian $\mathcal{L}$ has changed to $\mathcal{L}'$, now endowed with mass conferred though interaction with the Higgs boson, while the latter becomes a Nambu–Goldstone massless boson [2]. This process is decoded at the W-level by the flow $\widetilde{\mathfrak{F}_H}$, representing the endowment of mass on the ur-particle through geometric dilution of the ur-Higgs boson (Figure 4.9). In other words, the following relation holds at the encoded W-level: $\Gamma_{\widetilde{\mathcal{L}_H}} \tilde{\mathfrak{A}}(\tilde{\mathcal{L}}) = \tilde{\mathfrak{A}}(\widetilde{\mathcal{L}'})$, where $\widetilde{\mathcal{L}_H}$ is the Lagrangian for the ur-Higgs boson, and $\tilde{\mathcal{L}}$, $\widetilde{\mathcal{L}'}$ are the decoded Lagrangians for the massless ur-particle and for the ur-particle endowed with mass through the geometric dilution determined by the operator $\Gamma_{\widetilde{\mathcal{L}_H}}$, as illustrated in the Feynman diagram of Figure 4.9.

4.4 GEOMETRIC DILUTION PARAMETER AS A PROXY FOR EXTRINSIC TIME

A straightforward application of Heisenberg's uncertainty principle reveals that the emergence of the ur-Higgs, the primeval ur-particle implicated in baryogenesis, can be traced to a very specific time in the evolution of the universe after the big bang: $\tau \approx \frac{r_0}{2c} = 1.337 \times 10^{-27}$ s. This takes us to the so-called electroweak epoch when electromagnetism and the weak nuclear force remained merged into the so-called electroweak force [4]. At energies in the order of the vacuum expectation energy for the Higgs field (246 GeV), this merging remained energetically above the critical bound required for symmetry breaking, estimated in the SM at 159.5 GeV.

The flattening of the universe begins with the expansion driven by quintessential geometric dilution with $\alpha \neq 0$ (Figure 4.13), so gravity becomes less of an attribute of intrinsic geometry with extreme curvature [5] and more an attribute of the energy stored as mass in geometrically diluted incarnations of the ur-Higgs. The timing of universe evolution is in fact dictated by geometric dilution through the quantum relation

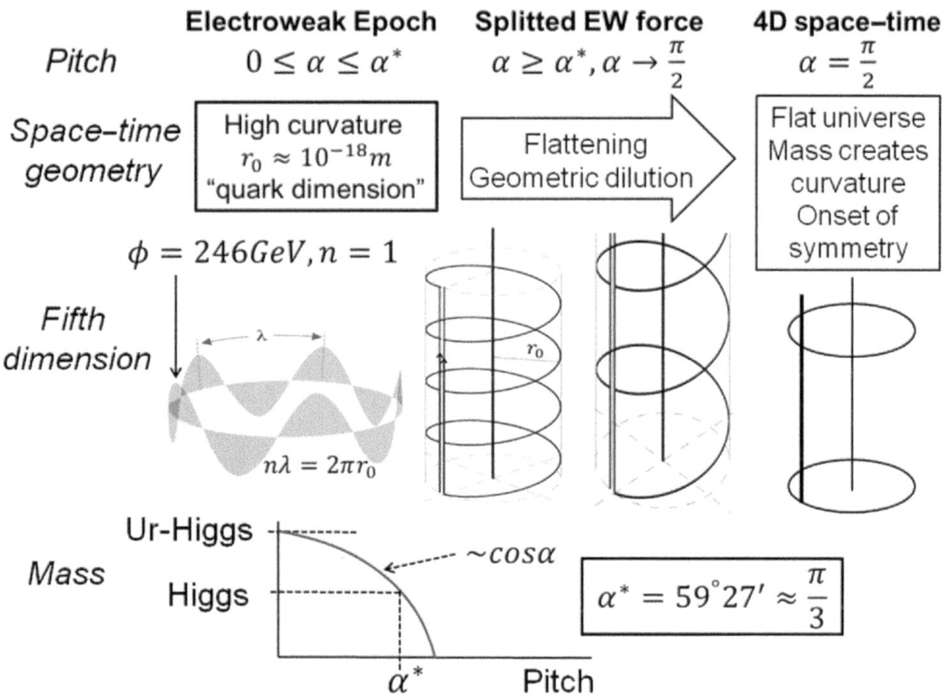

FIGURE 4.13 Flattening of the universe as a measure of time starting at the big bang event. The universe expansion is driven by quintessential geometric dilution with $\alpha > 0$, so gravity becomes less of an attribute of intrinsic geometry of the quintessential space and more an attribute of the energy stored as mass in geometrically dilutions of the ur-Higgs boson. The timing t in universe evolution starting at the big bang becomes $t \approx \frac{r_0}{2c}\left(\frac{1}{\cos\alpha}\right)$.

$$\tau \approx \frac{r_0}{2c}\left(\frac{1}{\cos\alpha}\right) = 1.337\times10^{-27}\left(\frac{1}{\cos\alpha}\right)\text{s} \tag{4.13}$$

The relation (4.13) marks the timing of creation of an ur-particle with mass $m = \frac{\hbar\cos\alpha}{cr_0}$.

Thus, the emergence of the decoded Higgs boson (Figure 4.14) can be traced to the time

$$\tau \approx 1.337\times10^{-27}\left(\frac{1}{\cos\alpha^*}\right)s = 2.632\times10^{-27}\,\text{s} \tag{4.14}$$

This time marks the end of the electroweak epoch and is followed by the emergence of the decoded gauge bosons Z and W, which signal the splitting of the electroweak force, as these bosons communicate the weak nuclear force when encoded in the latent space–time manifold $W/\sim$. All gauge bosons are created through vacuum fluctuation and stored into the diluted dormant dimension at dilution with $\cos\alpha = 0.32$ (Figure 4.12), which corresponds to the electroweak splitting time $\tau \approx 4.178\times10^{-27}$ s. The ur-fermions that emerge after this time can decay as the

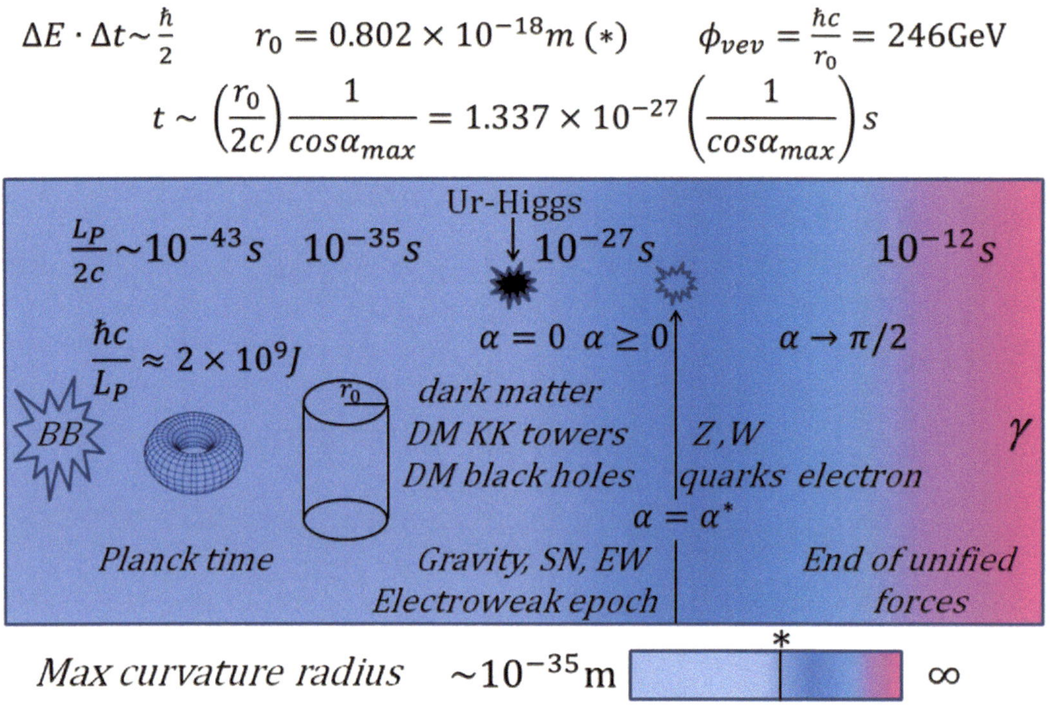

FIGURE 4.14 Universe timeline as quintessential geometric dilution.

gauge bosons have already stepped into existence and hence convey the interaction required (Figure 4.14).

At this point, the quintessential space W is still endowed with significant local curvature and will become flat only asymptotically at infinite geometric dilution ($\upsilon \to \infty$) corresponding to the limit $\alpha \to \frac{\pi}{2}$. This is the limit for the emergence of the photon (γ), the newest and massless particle. Thus, the advent of the photon becomes indicative of a quasi-flat universe resulting from infinite geometric dilution of the dormant dimension (Figure 4.14).

It is an undisputed fact that the photon already exists in our universe, and on the other hand, the big bang happened approximately 13.8 billion years ago, and therefore, the flatness of the universe should be considered to be only approximate, but it is indeed a strikingly good approximation. The current pitch $\alpha = \alpha(\tau_u)$ corresponding to the age of the universe estimated at $\tau_u \sim 4.35 \times 10^{17}$s can be calculated effectively as

$$\alpha = arc\cos\left(\frac{r_0}{2c\tau_u}\right) = arc\cos\left(3.1 \times 10^{-45}\right) \cong arc\cos 0 = \frac{\pi}{2} \tag{4.15}$$

It should be noted also that the ur-photon and the photon itself constitute for all purposes one and the same entity, since there is no projection onto the dormant dimension. Thus, the present day represents a singularity in which the dormant dimension may be set to be zero, and therefore, quotient space merges and identifies with underlying quintessential space. But this implies that the vacuum expectation value ϕ_{vev} for the Higgs field ($\sim \hbar c/r_0$) will ultimately be infinite, and that the

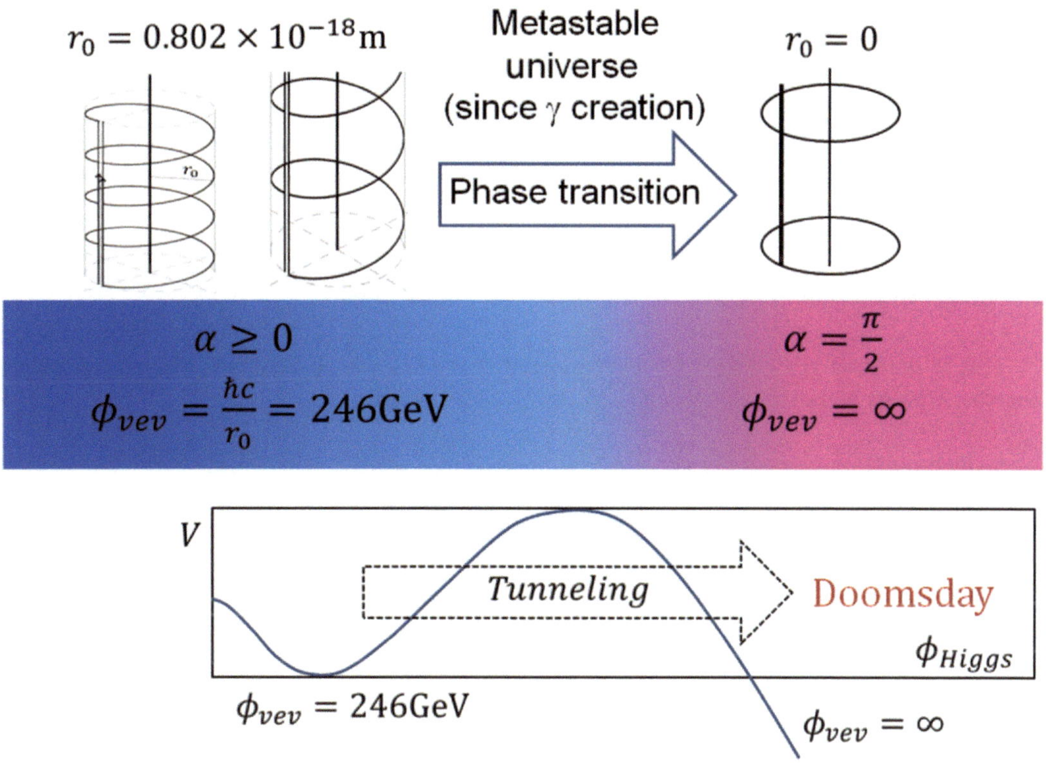

FIGURE 4.15 Catastrophe scenario associated with the metastability of the Higgs boson in the present-day universe at nearly infinite geometric dilution.

current value at 246 Gev that sustains the universe as we know it represents a metastable state. We do not know when the current ϕ_{vev} for the Higgs field will decay to infinity, as that would depend on the energy barrier that must be overcome through quantum tunneling. But we can be certain that the universe, as we know it, is metastable [4,5], hence prone to undergo a phase transition at some point (Figure 4.15). When that phase transition materializes, a big crunch will take place as every particle will be endowed with infinite mass, taking the universe back to the starting point in a big bang scenario.

4.5 DARK MATTER AND DARK ENERGY IN THE QUINTESSENTIAL DECODING OF THE STANDARD MODEL

The conclusions drawn so far from the discussion in the preceding sections may be summarized as follows:

The universe topology is inferred through an argument of Parmenides' type (Chapters 2 and 6) and "revealed" by the CMB fluctuation spectrum, and it is identified with a four-dimensional torus, corresponding to a quasi-flat multiply connected compact space with no boundary.

This topology is compatible with the big bang scenario for universe evolution [4], which admits dimensional expansion with retention of compactness but not a change in topology, as the latter introduces a disruption in the fabric of space–time that is forbidden by general relativity [5].

The toroidal topology admits a circular dormant dimension as a compact extra dimension, whereas an Euclidean space cannot incorporate this compact dimension without a change in topology along the universe evolution.

AI can trace the origin of DM in the universe evolution, since on a compact extra dimension with quark-size radius q [3], a de Broglie wave stores energy equal to the vacuum expectation value of the Higgs field.

This is the ur-Higgs particle, an excitation of the quintessence field that has been identified, together with its hierarchical tower, as DM [6] by AI because:

- Incorporation of a fifth dimension is supported by evidence: elementary particle masses can be accurately predicted through the geometric dilution of the ur-Higgs quintessence field that is determined from the decoding of the particle fields.
- The ur-Higgs particle formed during the electroweak epoch in an early universe ($\sim 10^{-27}$ s after the big bang).
- It is massive (246 GeV/c^2) and cold (i.e., speed $\ll c$).
- It does not decay through communication with SM gauge bosons.
- It is only interactive with the SM via gravity.
- It endows other ur-particles with mass through ur-boson-conveyed geometric dilution.

Much has been written about the difficulty in identifying the physical underpinnings of dark energy (DE) as the energy associated with quantum vacuum fluctuations (particles popping in and out of existence) [3]. In principle, the DE density required to observe the measured universe expansion is estimated at $\rho_{\mathrm{DE}} \approx 5\times10^{-10}$ Jm^{-3}. This value is in stark disagreement with naïve calculations of the vacuum energy density, yielding an estimated DE density with a colossal discrepancy amounting to a staggering 120 orders of magnitude compared with the experimentally obtained figure. In cosmology, this problem is often referred to as the cosmological constant problem or vacuum catastrophe [5]. The so-called cosmological constant [4] is often regarded as the default model for DE, whereby the geodesic fabric of space–time has a constant nonzero energy density that yields an antigravitational pull background [5]. The problem essentially describes the disagreement between the observed values of vacuum energy density (the small value of the cosmological constant) and theoretical large value of zero-point energy that is obtained from quantum field theory. As said, the quantum vacuum energy contribution to the cosmological constant is calculated to be as much as 120 orders of magnitude greater than the one observed, a calamitous state of affairs that should in all likelihood be referred to as the largest disagreement between theory and experiment the history of physics. This has prompted scientists to explore other options for what DE might be, as extensively discussed in Chapter 2.

Much of the difficulty and controversy evaporates as the extra compact dimension of quark-like material scale is incorporated in the quintessential decoding of the latent manifold representing standard space–time. This is because the excitation of the quintessential ur-Higgs field through quantum vacuum fluctuations enables the storage of DE as DM. In other words, the AI-enabled autoencoder technology dictates that we cannot conceive DE independently of DM and the dichotomy becomes a consequence of the quintessential decoding of our universe through the incorporation of the dormant fifth dimension (Chapter 2).

To distill the thrust of this discussion, we may state that the decoding of the SM by a quintessential autoencoder yields an ur-field that is not part of the SM. This quintessential field serves as the excitation vehicle for DE. In turn, the DE gets occasionally stored as DM for wavelengths that fit a stationary wave condition along the dormant fifth dimension. Thus, AI shows that DE and DM become phenomenologically tied to the quintessential space, while becoming opaque when regarded within the standard four-dimensional space–time. Only when the latter is regarded as a latent space, may the true nature of DM and DE be revealed.

4.6 THE ONSET OF MASS FROM SYMMETRY-CONTROLLED GEOMETRIC DILUTION OF DARK MATTER IN THE ELECTROWEAK EPOCH

In Chapter 3, the challenging problem of quantum gravity was addressed by constructing a learning system with stochastic connectivities and hidden variables where gravity and quantum behavior become emergent properties in a statistical physics scheme for machine learning. The thrust was to implement an AI-based version of the universe as a quantum holographic autoencoder. This system operates under the tenet that the emergent quantum behavior arises in a neural network equilibrated on the nontrainable hidden variables upon which a relativistic string gravitational scheme may be constructed. Conversely, in the nonequilibrium regime prior to equilibration of nontrainable variables, the network is endowed with emergent gravity. The behavior of the neural network is examined in the limits where the bias vector, weight matrix, and state vector of neurons can be modeled as stochastic variables that undergo a learning evolution. These dynamics are described by a time-dependent Schrödinger equation compatible with the relativistic decoding enshrined in the commutativity of the diagrams shown in Chapter 2. The results illustrate the emergence of quantum mechanics and general relativity in neural networks governed by a statistical physics scheme that can operate in two different thermodynamic regimes. In this way, the quantum metamodel for gravity fulfills at least in part a major imperative for physicists seeking a unified field theory.

The holographic autoencoder is now set up to adopt the toroidal topology of space–time described in Figure 4.6 with the geometric dilution parameter $\upsilon = -\log\cos\alpha$ as the proxy for time. With the default parameters given in Chapter 3, the timing of events follows exactly the description in Figure 4.14, with the advent of dark matter at $\sim 10^{-27}$ s. As dictated by the holographic autoencoder, the complete materialization of the universe and its evolution within the big bang scenario can be realized through the geometric dilution of gravity, originally decoded as a stationary wave in the quintessential five-dimensional torus.

4.7 DARK MATTER AND DARK ENERGY ARE NOT IDENTIFIED IN STANDARD EXTENSIONS OF THE STANDARD MODEL

The concept of reverse autoencoder, or rather, "autodecoder", has featured profusely in the previous discussion on the AI-enabled reverse engineering of the SM. In time-series data science, the autoencoder is regarded as the time-series distiller that generates the essential or latent model in a dynamical system [1]. In our context of interest, we assume the SM is in and of itself the latent model that entrains the full dynamics and the leveraging of reverse autoencoder technology serves the purpose of determining the latter.

As a specialized learning system, the autoencoder typically needs to be trained to decode the latent dynamics. In the particular case of interest, the training does not require *a priori* knowledge of quintessential time series. This is because we have a generic way of deriving the quintessential Lagrangian $\tilde{\mathcal{L}}$ that decodes a particle Lagrangian $\mathcal{L}$. Thus, the decoded Lagrangian satisfies the relation: $\int \tilde{\mathcal{L}}(\tilde{\phi})\,dx\,dy_5 = \int \mathcal{L}(\phi)\,dx + \left[\int \left|\int(\tilde{\phi}-\phi)\,dx\right|^2 dy_5\right]^{1/2}$. By applying this relation to a given pair $(\phi, \mathcal{L})$ representing a particle in the SM, the dynamics encoded in the SM is fleshed out onto the quintessential space (a compact and locally flat multiply connected manifold). This simply requires the decoding each particle field $\phi \to \tilde{\phi}$ so as to optimize (minimize) the term $\int \tilde{\mathcal{L}}(\tilde{\phi})\,dx\,dy_5$.

To deploy autoencoder technology, we have turned the SM into a dynamical system by setting the time step for an elementary process equal to the lifetime of the (ephemeral) boson that mediates the particle transformation. Evidently, the center manifold entrainment [7-9] or subordination of the quintessential dynamical system to the SM leaves out dark energy and dark matter, so we cannot assume that the SM, staggeringly successful as it is, provides the complete description of the elementary processes that take place in the universe.

Our discussion and analysis in this chapter has focused primarily on elucidating what is left out with the reduction or entrainment of the quintessential dynamics by the SM, and we have found a striking answer: DM and DE. This is at some level disconcerting because both clearly have a very significant bearing on the observed dynamics of deep space. To be specific, they constitute over 95% of the gravitational budget of the detectable universe at large scales. Furthermore, this clearly introduces a contradiction: The existence of DM and DE implies that the SM cannot entrain or subordinate the universe as a dynamical system that materializes in the quintessential space. This contradiction can only be properly accounted for in one of two ways: (1) the SM does not represent the latent dynamics of the universe, or (2) the quintessential dynamics is in fact irreducible: It does not admit a latent dynamical system. This leads to a paradox since any of the two alternatives implies that an autoencoder could not have been a priori used to elucidate the quintessential dynamics spanned by the latent dynamics enshrined in the SM. We know this to be wrong, since an autoencoder has indeed been used as decoder of the SM.

This fundamental paradox does not undermine the AI enablement of the reverse engineering of the SM because it is based on the obviously incorrect assumption that DM and DE can be obtained from extensions of the SM. This means that DM and DE are inherently quintessential, which was from the start a basic tenet in the AI approach put forth in this book. The fundamental implication is that the decoding of the SM is insufficient to account for the quintessence fields that realize DM and DE when excited along the dormant dimension.

Notwithstanding the previous conclusions, experimental efforts engaging multinational consortia, like the Large Hadron Collider, or deploying underground detectors are still well underway to detect DE and DM as extensions of the SM. Such efforts have been, not surprisingly, unsuccessful. It seems that for all that we revere the Copernican revolution, we still like to see ourselves as the center of the universe, and AI may be teaching us a lesson in that regard.

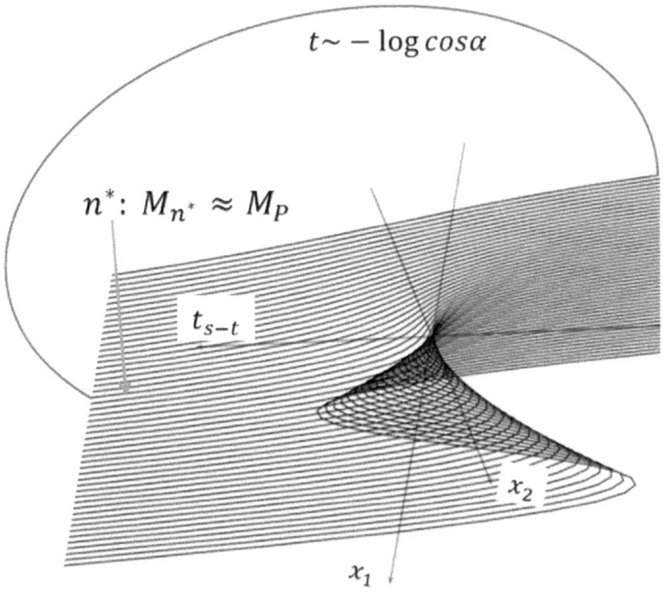

FIGURE 4.16 Space–time rendering of the evolution of the universe by geometric dilution of dark matter.

4.8 COSMOLOGICAL CONSEQUENCES OF THE GEOMETRIC DILUTION OF DARK MATTER

As discussed in Chapter 2, the storage of energy along the dormant dimension with radius r_0 generates a hierarchical tower associated with the ur-Higgs boson ($n = 1$; mass = 246 GeV/c^2). The relatives of this particle for $n > 1$ have masses $M_n = \frac{\hbar n}{r_0 c}$. A critical accretion of this DM is indexed by $n = n^*$ satisfying the relation:

$$n^* = \min n : n \geq r_0 \sqrt{\frac{c^3}{\hbar G}} \tag{4.16}$$

Thus, the critical mass M_{n^*} becomes commensurable with the Planck mass M_P. This implies that we get a black hole when $n > n^*$. To understand the evolution of this enormous concentration of DM, we need to define "time outside space–time", that is, beyond the local relativistic time, $t_{s\text{-}t}$, that serves as a local dimension for space–time (cf. Figure 4.16). While spanning a far larger scale, the extrinsic time should be identifiable with $t_{s\text{-}t}$ at the locus. We know that a proxy for evolutionary time is furnished by geometric dilution, hence, as shown in the space–time rendering of the hierarchical tower of DM, the $n > n^*$ ur-particles will eventually dilute, thus removing the space–time ripple that causes time reversal (Figure 4.16). This picture describes the evolution of the universe from DM, identifying geometric dilution of the excitation along the dormant coordinate as the mechanistic agent promoting particle creation concurrently with the progressive removal of the space–time ripple.

REFERENCES

1. Fernández A (2022) *Topological Dynamics for Metamodel Discovery with Artificial Intelligence*. Chapman & Hall/CRC, Taylor and Francis, London.
2. Feynman RP, Weinberg S (1999) *Elementary Particles and the Laws of Physics*. Cambridge University Press.
3. Abramowiczy H, Abtt I, Adamczykh L, Adamusae M, Antonelli S, et al. Zeus Collaboration (2016) Limits on the effective quark radius from inclusive *ep* scattering at HERA. *Phys Lett B* 757: 468–472.
4. Weinberg S (2008) *Cosmology*. Oxford University Press.
5. Hawking SW, Ellis GFR (2023) *The Large Scale Structure of Space-Time: 50th Anniversary Edition*. Cambridge University Press.
6. Profumo S (2017) *Introduction to Particle Dark Matter*. World Scientific Publishing, Singapore
7. Fernández A (1985) Center-manifold extension of the adiabatic-elimination method. *Phys Rev A* 32: 3070–3076.
8. Fernández A (1988) Phase-ordering dynamics for the onset of a center manifold. *Phys Rev A* 38: 4256–4262.
9. Fernández A (1988) On renormalization of fluctuations at the onset of a centre manifold. *J Phys A: Math Gen* 21: L607.

5 Dark Energy to Sustain the Universe

"τὰ πάντα ῥεῖ καὶ οὐδὲν μένει"
Everything flows and nothing stays

Heraclitus

Drawing from Chapter 2, this chapter establishes the topology of space–time, a feature not directly relevant to general relativity or quantum mechanics. As shown, the universe topology is essential to delineate the cosmic generation and depletion of dark energy, their bearing on the cosmological constant problem and on a dynamic equilibrium between dark and detectable matter. To that effect, the quantum fabric of space arising from vacuum entanglement is examined using the topological blueprint. Artificial intelligence, in the guise of a quantum gravity autoencoder with a physical embodiment (Chapter 3), reveals that a primeval wormhole underlies the topology of the universe and is sustained by a two-stroke cyclic cosmic engine fueled by vacuum energy. This engine maintains the baseline dark energy density observed experimentally, with dark energy getting replenished as vacuum energy fuels the universe runaway process associated with autocatalytic vacuum creation.

As shown, the cosmic engine sustains a portal to the dark universe in the form of a two-way primeval nonorientable wormhole (PNOW) whereby the "dark" right-handed fermions with spin 1/2 are in dynamic equilibrium with detectable matter. The dark fermions account for 18.72% of dark matter. Since, at present time, "matter in all forms", including dark energy, is 26.7% dark matter and 5% detectable matter (Chapter 1), the dark fermion proportion at 18.72% of dark matter corresponds to 15.77% of all matter (dark + detectable) in the universe, which is exactly equal to the percentage of all matter represented by detectable matter.

We provide a perspective to interpret dark energy as the surplus in vacuum energy taken from the amount required to fuel the cosmic engine that maintains the PNOW. In contrast with the ever-decreasing concentration of dark and detectable matter that get progressively diluted in a growing vacuum, the concentration of the vacuum energy surplus is shown to be constant, as established by experiment. This vacuum energy surplus is identified as dark energy. The sustainability of the PNOW results from the coupling of vacuum energy expenditure with a dynamic equilibrium between dark fermions and visible matter, so that the overall concentration of dark energy is kept constant, as is the overall mass of dark matter. In this way, this chapter describes a cyclic cosmic engine that sustains a portal to the dark universe.

DOI: 10.1201/9781003632924-5

5.1 TOPOLOGY OF THE UNIVERSE AND THE SUSTAINABLE PRIMEVAL WORMHOLE

As discussed in Chapter 2, the global topology of the universe did not seem to be a matter of concern for Albert Einstein. We can speculate that this is so because his theory of gravitation, general relativity, is cast in terms of differential geometry and hence governed by differential equations. Since differential equations describe a local situation or setting, such as the behavior of curvature relative to mass distribution, the global topology of space–time plays no discernible role in general relativity. There is, however, a major constraint imposed by general relativity on the topology of space–time: The topology should remain invariant throughout the evolution of the universe. This is a difficult matter to grasp since time is part of the fabric of the object whose evolution we need to examine. Hence, as described in Chapters 2 and 4, we get into a sort of metaphysical conundrum whereby the object contains its own evolution as part of itself. To avoid semantic traps, we should state that by evolution of the universe we mean, by abuse of language, the changes in time of a generic constant-time cross section of space–time. There is no Russell-type paradox here, since space–time per se does not evolve, only space does, or rather the constant-time cross section of space–time, actually evolves. Hence, the universe – actually meaning space – cannot change its topology throughout its evolution because that would mean that space–time presents an essential singularity in its fabric of the type forbidden by general relativity and that there is simply no well-defined topology that can be unambiguously attributed to space–time.

Quantum mechanics seems equally impervious to topological considerations. Quantum mechanics defines the fabric of space, as determined by vacuum entanglement, but topology is not factored into the theory in any obvious way, even if entanglement is essentially responsible for the nonlocality of quantum phenomenology. However, topology suddenly becomes extremely relevant as one tries to reconcile general relativity with quantum physics and identify the underpinnings of quantum gravity through the physical embodiment of a holographic autoencoder, as shown in Chapter 3.

If we assume topological invariance of space throughout the universe evolution starting at the big bang event, then the space of today cannot be infinite and flat, and therefore Euclidean, as Einstein would have it (he probably did not devote much attention to the matter). Provided we accept the big bang scenario, space today must be compact, not infinite, because it once was (i.e., 13.8 billion years ago), with certainty. But if space is compact, then it must be multiply connected and that property bears pivotally on the solution of the cosmological constant problem, as we show subsequently.

In order to make further progress toward a complete characterization of space, we adopt a Parmenides-type of approach. This prompts us to formulate the question: Can space have a boundary? The answer is adamantly negative, since a boundary would imply an interface, not with the vacuum but with nothingness, and it is physically and metaphysically impossible to interface with nothingness: An interface presupposes two media, and nothingness is not a medium.

These considerations introduce several constraints on what the topology of the universe must be: It must be compact, locally flat, orientable, and lacking boundary. This leaves us with essentially one option and it is a multiply connected three-dimensional manifold: the three-torus or Cartesian product $S^1 \times S^1 \times S^1$.

The fact that space is multiply connected introduces an extreme novelty and brings about hitherto unfathomable implications in the field of quantum gravity. To delineate such implications, we first remind ourselves of the way the three-dimensional torus is constructed. It is essentially built by identifying opposite faces of a three-dimensional cube (Figure 5.1). But if the aspect ratio χ_{ij} for every pair (i, j) of circular dimensions is $\chi_{ij} = \frac{r_i}{r_j} = 1$, as mandated by the retention of symmetry upon coordinate permutation, then each two-dimensional cross section of space becomes a horn torus and has a cusp-like curvature singularity. This cusp with its infinite spatial curvature is

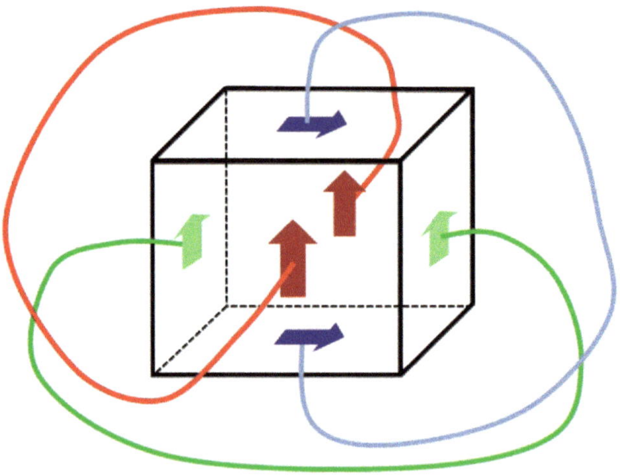

FIGURE 5.1 Topological identification of opposite faces of a three-dimensional cube to render the three-torus. The arrow bestows orientation to the face. Space is thus conceived as a compact manifold with no boundary, hence multiply connected.

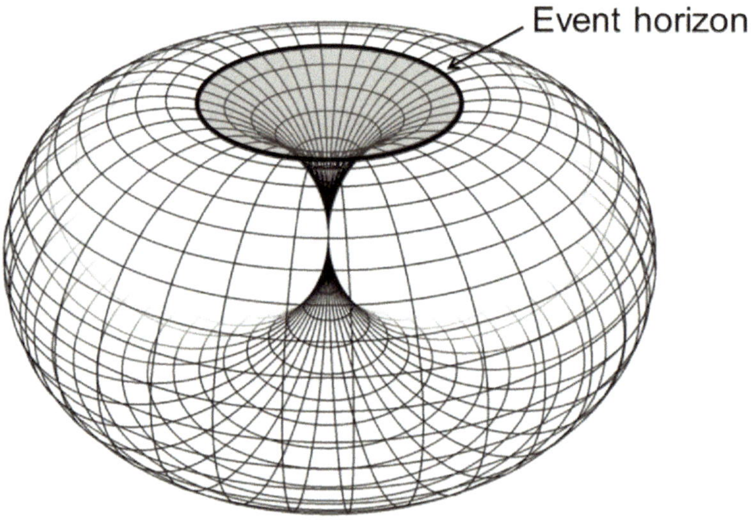

FIGURE 5.2 Toroidal two-dimensional cross section of the universe along dimensions with aspect ratio 1. The cross section is a horn torus, hence generating a Schwarzschild double black hole with connected exteriors.

indicative of an entangled double black hole (BH), with interiors physically connected through an Einstein–Rosen (ER) bridge [1], as represented in Figure 5.2.

To represent the ER bridge in space–time requires higher powers of abstraction. We begin by representing the BH in space time, noting that the crossing of the event horizon entails going back in time, as indicated in Figure 5.3. Alternatively, to cross the event horizon requires a velocity higher than the speed of light, as suggested in the Penrose diagram displayed in Figure 5.3. Furthermore, the equivalent of a wormhole or Einstein–Rosen (ER) bridge between two black holes would be sustained between two exteriors (L and R) and two event horizons delineating interiors connected by the ER bridge. Thus, in space–time, the wormhole may be schematically represented as shown in Figure 5.4a,b.

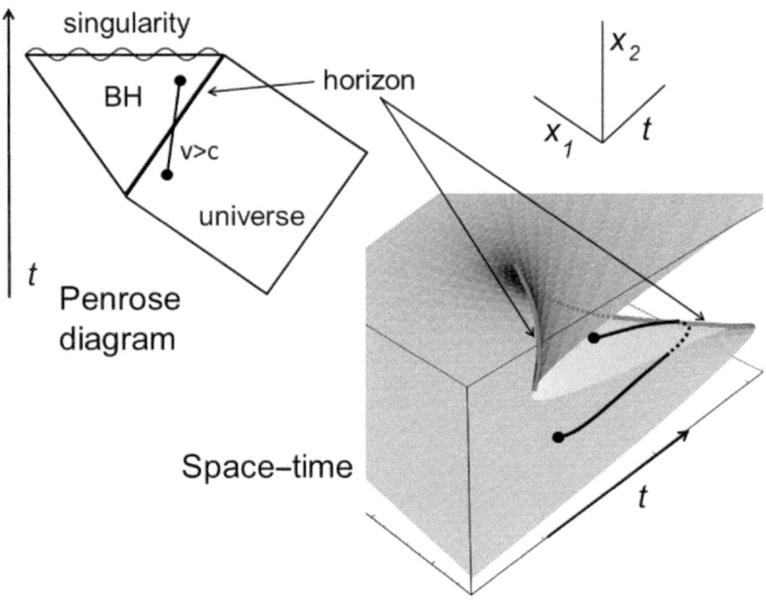

FIGURE 5.3 Representation of a black hole in a Penrose diagram (panel on the left, coordinate units scaled so that light travels at 45 degrees or speed 1), and in space–time (right panel).

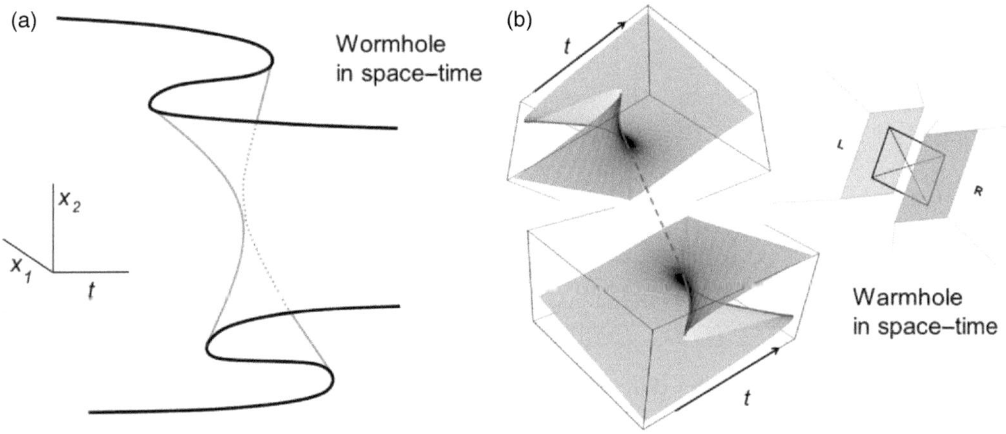

FIGURE 5.4 Schematic representation of a wormhole in space–time: (a) as two coupled folds of space time; (b) in relation to its Penrose diagram.

As a solution to the Einstein equations, this entity is geometric, rather than physical (except that Einstein makes no distinction). In other words, there is no materiality in a double-entangled BH, which becomes the realization of the solution to general relativity equations found by Karl Schwarzschild [1]. We name this solution, the "primeval wormhole" (Figure 5.5) to distinguish it from the "material" types of shortcut BHs determined by extreme concentration of mass. The entanglement is obvious because the torus was created through identification, i.e., perfect gluing of opposite faces (Figure 5.1); hence, the physical ER connection in the horn torus begets an Einstein–Podolsky–Rosen (EPR) correlation of the opposite sides through the Maldacena–Susskind relation

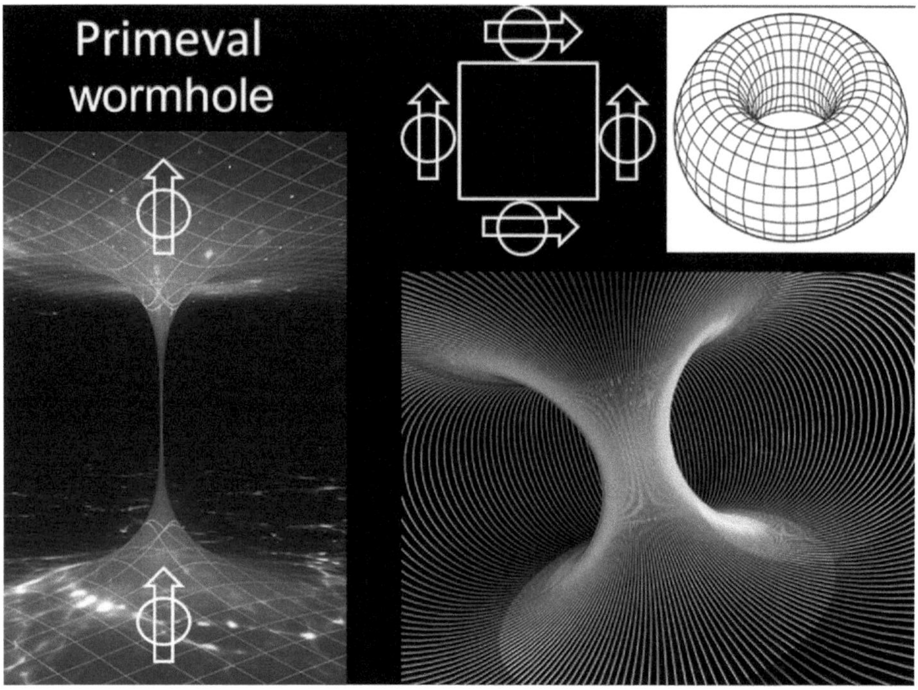

FIGURE 5.5 Entanglement of the primeval wormhole is postulated via the Maldacena–Susskind ansatz [1].

[2]. The "EPR = ER" relation is only valid for black holes, not for other entanglements, but the double-cusp connection is indicative that we are precisely in that scenario.

If closed time-like curves (CTC) exist, their existence would imply time travel, raising paradoxes such as the grandfather paradox. Nevertheless, such paradoxes are avoided in the AI model because every CTC passes through an event horizon, implying that there is chronological veto. The space–time with event horizons implies that there would be no causal violation.

5.2 DECONSTRUCTING QUANTUM ENTANGLEMENT IN THE QUINTESSENTIAL TOPOLOGICAL BLUEPRINT OF SPACE–TIME

Entanglement is the most striking property of quantum physics because it enables us to assert the nonlocality of the quantum phenomena, in stark contrast with Newtonian physics or general relativity, which are local theories. The fabric of vacuum is clearly and solely a quantum attribute with no clear relativistic counterpart and depends pivotally on entanglement. To understand the bearing of the fourth dormant spatial dimension on entanglement, we need to recapitulate the definition of entanglement in a guise that makes it amenable for quintessence deconstruction, without losing perspective of the central goal of this chapter: *addressing the cosmological constant problem by incorporating the topological blueprint of the entangled universe*. In other words, we need to determine the bundle of compound states in space W that project onto a set of compound states in $W/\sim$ that contain the wave function for vacuum entanglement.

To fix notation, let us define a cell (spherical region) in space which is in two possible "key states": $|1\rangle$ if the region contains a particle with associated de Broglie wavelength given by the cell dimension, and $|0\rangle$ if the cell is empty. Prior to measurement, which in quantum mechanics implies disambiguation, the region has an associated wave function $\frac{1}{\sqrt{2}}|0\rangle + \frac{1}{\sqrt{2}}|1\rangle$ Now, if we have two cells in space, the state of the compound system prior to measurement is tensorially represented in Dirac's simplified notation as

$$f_{00}|0\rangle|0\rangle + f_{01}|0\rangle|1\rangle + f_{10}|1\rangle|0\rangle + f_{11}|1\rangle|1\rangle, \tag{5.1}$$

with constants f_{ab}; $a, b = 0, 1,$ satisfying the Born relation:

$$\sum_{a,b=0,1} f_{ab}^2 = 1 \tag{5.2}$$

The two cells are entangled if the following inequality is satisfied:

$$f_{00}f_{11} \neq f_{01}f_{10} \tag{5.3}$$

So, we may adopt a measure of disentanglement defined as $\Delta \equiv \left|f_{00}f_{11} - f_{01}f_{10}\right|$.

For an arbitrary plane separating space in vacuum, two equidistant regions are maximally entangled conforming a Bell pair with wave function (Figure 5.6):

$$\frac{1}{\sqrt{2}}|0\rangle|0\rangle + \frac{1}{\sqrt{2}}|1\rangle|1\rangle \tag{5.4}$$

As we incorporate the fourth spatial "dormant" dimension, the Bell pair spans the following wave function in W:

$$\frac{1}{\sqrt{2}}\sin y_1|0\rangle|0\rangle + \frac{1}{\sqrt{2}}cosy_1|0\rangle|1\rangle + \frac{1}{\sqrt{2}}\sin y_2|1\rangle|0\rangle + \frac{1}{\sqrt{2}}cosy_2|1\rangle|1\rangle, \tag{5.5}$$

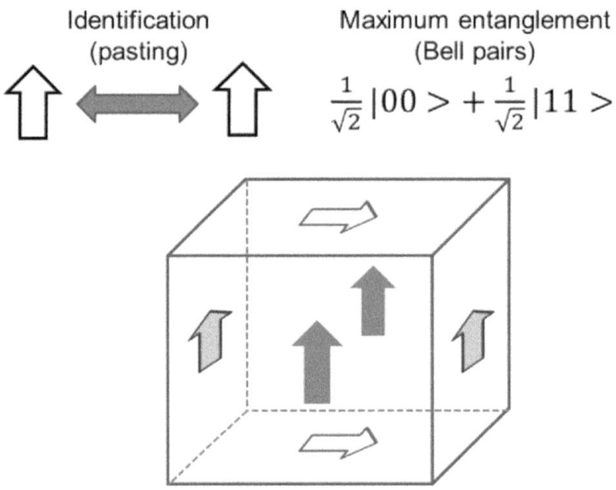

FIGURE 5.6 Topological identification of opposing faces in the three-cube yields an entangled structure, whereby cells equidistant to the pasted surfaces are correlated via Bell pairs.

where the compound eigenstate coefficients are now functions of the dormant variable realized as y_1, y_2 for cells 1 and 2, respectively. The Bell pair in quotient space *W*/~ is the projection obtained for $y_1 = \frac{\pi}{2}, y_2 = 0$. The entanglement bundle given by expression (5.5) becomes disentangled if and only if the following trigonometric relation holds:

$$\sin y_1 \cos y_2 = \sin y_2 \cos y_1, \tag{5.6}$$

which we may write as follows:

$$\frac{1}{2}\left[\sin(y_1 + y_2) + \sin(y_1 - y_2)\right] = \frac{1}{2}\left[\sin(y_1 + y_2) + \sin(y_2 - y_1)\right], \tag{5.7}$$

yielding

$$\sin\left(y_1 - y_2\right) = 0 \Leftrightarrow y_1 = y_2 \tag{5.8}$$

Hence, we have proven the following:

Lemma 5.1. The disentanglement set of states within the bundle in *W* associated with the Bell pair for vacuum entanglement in *W*/~ has measure zero.

Theorem 5.1. The reversible work ΔG_δ required to disentangle the bundle in *W* to level $\Delta \leq \delta$ is

$$\Delta G_\delta = -T\Delta S_\delta = -k_B T ln\left(\frac{\mu \mathcal{E}_\delta}{\mu \mathcal{D}_\delta}\right), \tag{5.9}$$

where $\mathcal{E}_\delta$ is the region associated with disentanglement level $\Delta \leq \delta$ (Figure 5.7), μ denotes measure, and $\mu \mathcal{D}_\delta = 4\pi^2 - \mu \mathcal{E}_\delta$.

Proof: The reversible work or free energy increment required to disentangle the bundle to level $\Delta \leq \delta$ is computed as entropy loss $\left(\Delta G = -T\Delta S_\delta\right)$, hence determined by the quotient of the multiplicities of disentangled over entangled states in the bundle. Full disentanglement ($\Delta = 0$) requires infinite work, as per Lemma 5.1.

5.3 AI PROBES THE EQUIVALENCE BETWEEN WORMHOLE AND QUANTUM ENTANGLEMENT

The elucidation of the nature of dark matter in Chapter 4 highlights the power of topology-based metamodels in leveraging autoencoder technology for model discovery. The main assertion in Chapter 3 is that a neural network with emergent gravity admits an autoencoder with emergent quantum behavior and hidden variables that can be modeled with relativistic strings. Thus, a quantum metamodel of gravity is in principle possible. This finding paves the way to tackle the next conundrum involving the conjectured equivalence between relativistic wormholes and quantum entanglement.

Space–time locality is a basic tenet of modern physics. The term "locality" refers to the impossibility of sending signals at speeds higher than the speed of light, an idea that is constantly

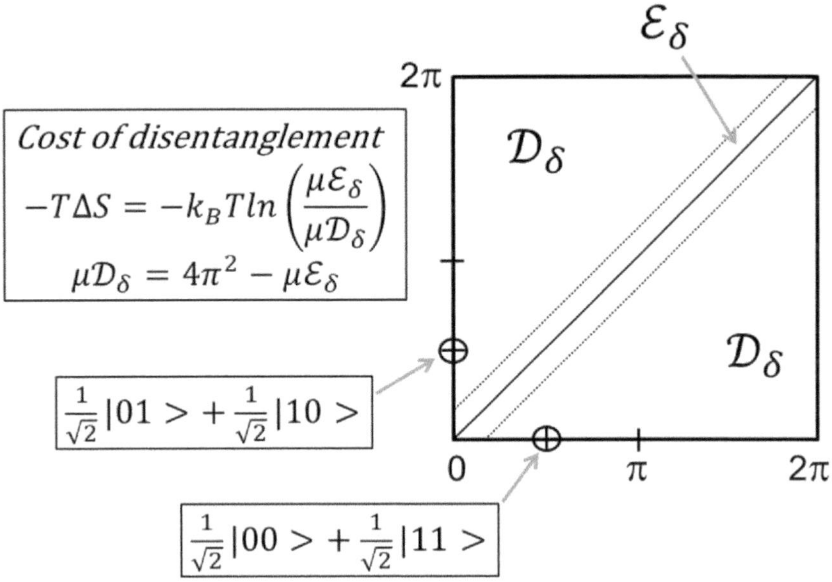

FIGURE 5.7 Region and dark energy cost of disentanglement to level δ for the bundle in W that subsumes the Bell pair of maximal spatial entanglement in $W/\sim$.

challenged both by quantum mechanics and general relativity. Thus, quantum mechanics gave rise to the Einstein–Podolsky–Rosen (EPR) correlations also termed "entanglements", while general relativity allows for solutions to the equations of motion that connect distant regions through short-circuiting "wormholes" also known as Einstein–Rosen (ER) bridges [1]. Physicists Maldacena and Susskind [2] have conjectured that these two concepts may be connected by a duality that becomes in effect equivalence, akin to the similar duality found in the physical underpinnings of quantum gravity. They have argued persuasively that the ER bridge between two black holes may be actually created by EPR correlations between microstates of the two black holes and labeled the conjecture "ER = EPR relation". In their analysis, the ER bridge is a type of EPR correlation in which the correlated quantum systems are in a specific entangled state that admits a weakly coupled Einstein gravity description. This situation is illustrated by a black hole pair creation in a magnetic field, and it is tempting to think that any EPR correlated system, even a simple singlet state of two spins, is connected by some sort of ER bridge.

The neural network model of the universe endowed with the quantum gravity autoencoder described in Chapter 3 seems an ideal system to validate (or disprove) the EPR = ER relation. This is because the emergent quantum behavior of the autoencoder pivots on hidden variables that are interacting through relativistic strings and the very existence of the hidden variables is known to resolve the EPR paradox. In the quantum gravity autoencoder, the trainable variables conforming the $\boldsymbol{q}$-vector exhibit a quantum mechanical behavior in an equilibrium regime where the network state variables that constitute the $\boldsymbol{x}$-vector have been thermalized. A learning process involves L separate sets of training $\boldsymbol{x}$-vectors with expected values $\bar{\boldsymbol{x}}^l, l = 1,2,\ldots,L$ and the expectation vectors together with the expectation state vector $\bar{\boldsymbol{x}}^0$, representing network evolution to zeroth order in the linear approximation to node activation, are regarded as the hidden variables in the emergent quantum behavior of the trainable $\boldsymbol{q}$-states. On the other hand, the nonequilibrium dynamics of the hidden variables becomes relevant on timescales much smaller than their thermalization time. This nonequilibrium dynamics are determined by the strength of the weak interactions between vector pairs $\bar{\boldsymbol{x}}^\nu$, $\bar{\boldsymbol{x}}^\xi, \nu,\xi = 0,1,\ldots,L$ quantified by the tensor $g^i_{\nu\xi}$, where the dummy index i labels each neuron in the system. By weak interactions, we mean that the generic vectors $\bar{x}^{\nu}$, $\bar{x}^{\xi}$ are not interacting directly but through the $\boldsymbol{q}$-vector that they themselves contribute to train. These dynamics can be cast in terms of relativistic strings in an emergent space–time, as indicated in Chapter 3.

This AI framework with emergent physics seems ideal to validate the EPR = ER relation. If two parts of the $\boldsymbol{q}$-space are entangled in the quantum autoencoder, they should be bridged through a wormhole in the full network with emergent gravity, as described in Figure 5.8. The presumed proportionality between the gravitational action and the quantum action may be a good starting point for this project aimed at testing the power of AI as purveyor of the physical underpinnings for the most perplexing problems in modern cosmology.

The narrative in this book takes an unexpected turn in Chapter 3, where it is shown that a connected array of neurons (NN) may be treated not only as an information processing machine but also as a statistical mechanical object, capable of exhibiting emergent physical behavior. Thus, a duality between emergent gravity and quantum mechanics is established through an autoencoder that thermalizes hidden degrees of freedom arrayed on the NN state vector $\boldsymbol{x}$. In this way, the autoencoder exhibits emergent quantum behavior while the nonequilibrium dynamics of the hidden variables spans an emergent space–time endowed with gravity (Chapter 3). This physical duality of the learning machine cast in terms of statistical mechanics should enable topological innovation on the emergent space–time through quantum entanglement at the autoencoder level.

To illustrate this point, consider two completely separated replicas of a quantum gravity autoencoder (QGAE), labeled left (L) and right (R) with identical collection of training sets $\left\{\mathcal{S}^{(\mu)}\right\}_{\mu}$ comprised of $\boldsymbol{x}$-vectors and generating the sets of $\boldsymbol{q}$-vectors $Q_L = \left\{Argmin\, J_{<}^{(\mu,L)}\right\}_{\mu}, Q_R = \left\{Argmin\, J_{<}^{(\mu,R)}\right\}_{\mu}$, in QGAE(L) and QGAE(R), respectively. Notice that the stochastic nature of the training process

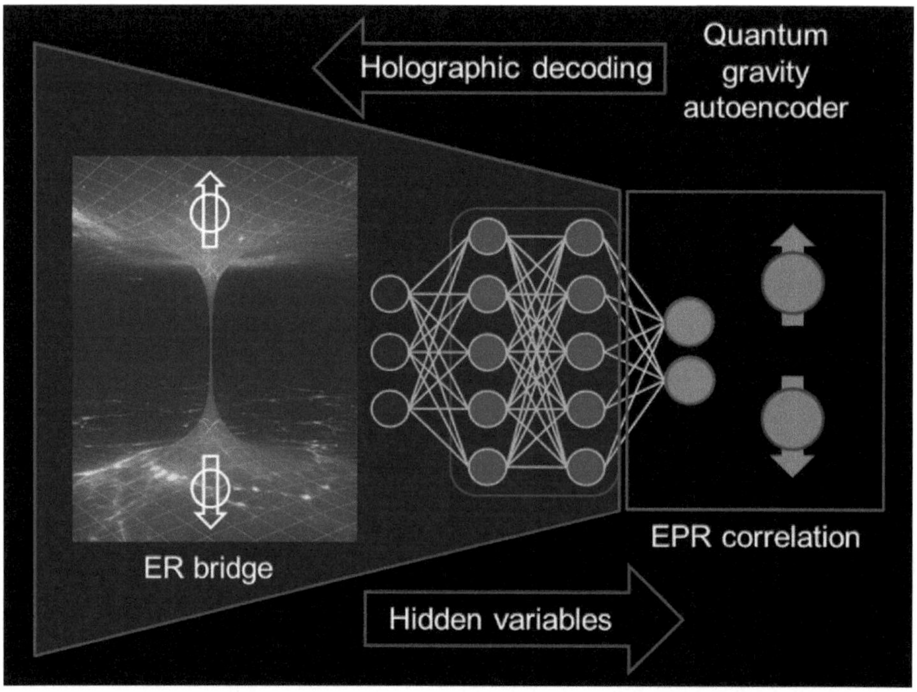

FIGURE 5.8 Schematics of quantum gravity autoencoder (QGAE, Chapter 3) deployed to validate the "EPR = ER relation".

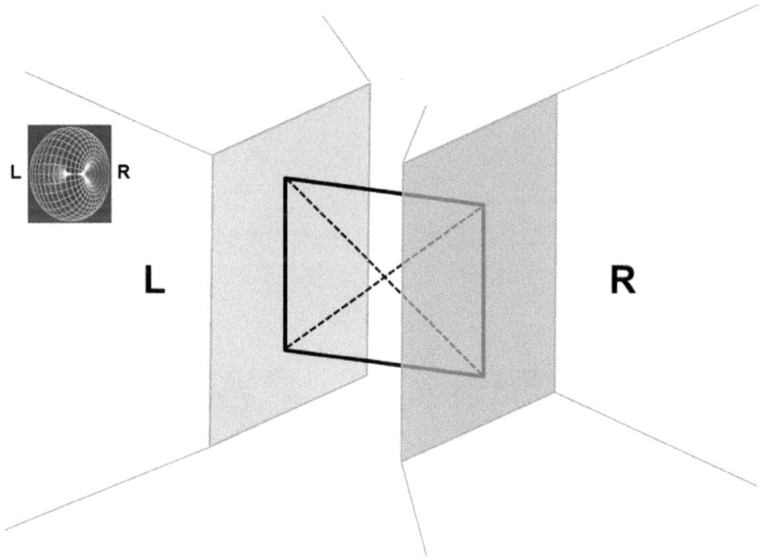

FIGURE 5.9 Penrose diagram counterpart for the connection topology in emergent space–time arising from the entanglement of two black holes within quantum gravity autoencoders (L and R). The respective event horizons are marked by dashed-line upper and lower triangles in a conformal two-dimensional space–time.

implies that $Argmin\, J_{<}^{(\mu,L)}$ is not necessarily equal to $Argmin\, J_{<}^{(\mu,R)}$, but surely the $\boldsymbol{q}$-vector pairs $Argmin\, J_{<}^{(\mu,L)}$, $Argmin\, J_{<}^{(\mu,R)}$ are entangled as they have a common origin. Hence, if the NNs are architecturally configured so that there is a black hole in Q_L and therefore in Q_R, then both black holes are necessarily entangled. This entanglement begets connectivity in the respective emergent space–times associated with QGAE(L) and QGAE(R), which would be topologically related to the Penrose diagram shown in Figure 5.9.

Thus, the equivalent of a wormhole or Einstein–Rosen (ER) bridge between the black holes would be sustained between the two learning machines L and R as a result of entanglement at the QGAE level. This "double black hole" would have two exteriors (L and R) and two event horizons meeting at the counterpart of the ER bridge shown in the Penrose diagram (Figure 5.9). However, we expect this sort of connectivity to be topologically different in the case of entangled NNs. This is so because the emergent space–time constructed upon the $\boldsymbol{q}$-space of a QGAE is essentially different from the Minkowski space where the particular solution to Einstein's equations was obtained and topologically described by the Penrose diagram.

Describing the space–time topology of entangled NNs may prove rewarding, as it is likely to herald a new breed of quantum computation that we may provisionally term "quantum gravity computation".

5.4 PRIMEVAL WORMHOLE AS A TWO-STROKE COSMIC ENGINE FUELED BY DARK MATTER: PROVISIONAL SOLUTION TO THE COSMOLOGICAL CONSTANT PROBLEM

Theorem 5.1 enables further understanding of the entanglement of the primeval wormhole. To further investigate this matter, we implement our AI tool with an architecture consisting of an annular arrangement of two replicas of a QGAE (Figure 5.10). The emerging results reveal that a massive

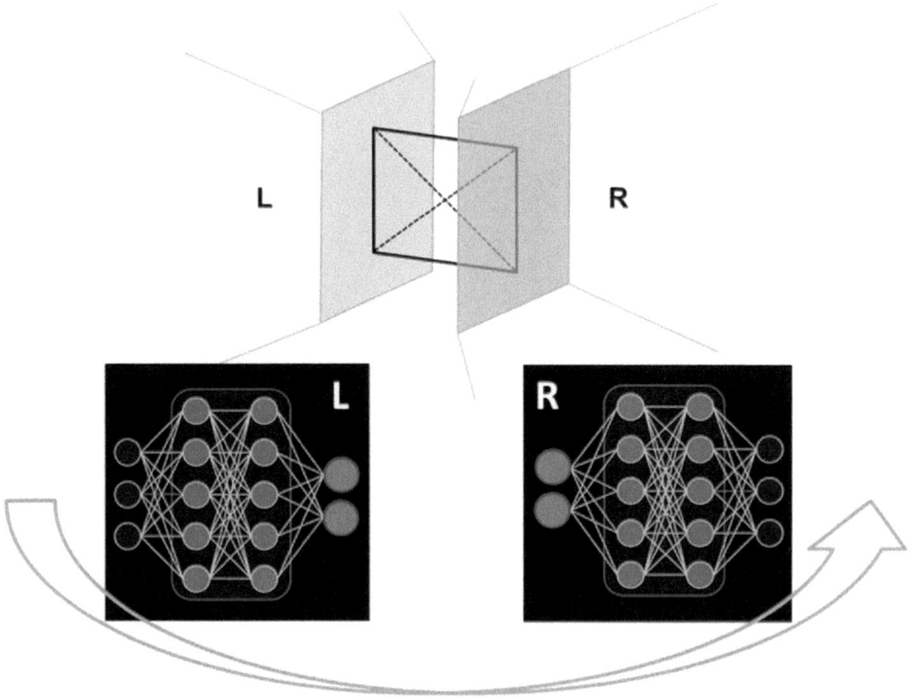

FIGURE 5.10 Annular arrangement of two replicas of the holographic (quantum gravity) autoencoder as required to investigate the sustainability and dynamics of the primeval wormhole arising in the three-toroidal space.

expenditure in dark energy is required for an interim destruction of the primeval wormhole by changing the aspect ratio $\chi = 1 \rightarrow \chi > 1$, resulting in the conversion of a horn torus into a spindle torus. As per the Maldacena–Susskind ansatz [1], this change in dimensional aspect ratio must correlate with a change $\Delta = 0 \rightarrow \Delta \leq \delta$ in the level of wormhole disentanglement, with $\delta =$ function of $(\chi - 1)$ (Figure 5.11).

Thus, Fig. 5.11 displays the dark energy expenditure in terms of the level of disentanglement ($\Delta \leq \delta$) and equivalently, in terms of the change in aspect ratio, quantified as $(\chi - 1) > 0$ that indicates the dismantling of the ER bridge. In other words, the dark energy expenditure for disentanglement of the primeval wormhole generates or rather enables an increase in one circular dimension, say, r_2, relative to the other circular dimension with radius r_1. We are assuming without loss of generality that the aspect ratio is given as $\chi = \frac{r_2}{r_1}$. The universe expansion along one dimension is enabled and materializes because the primeval wormhole gives back the dark energy invested in its disentanglement, while additional dark energy is generated as the volume of the vacuum increases with the expansion. This surplus in dark energy enables its reinvestment in the next cycle of entanglement–disentanglement that has a higher cost in terms of reversible work.

As we can see, the chain of events:

[δ-level disentanglement] → [vacuum expansion to restore $\chi \approx 1$] → [δ-level disentanglement]

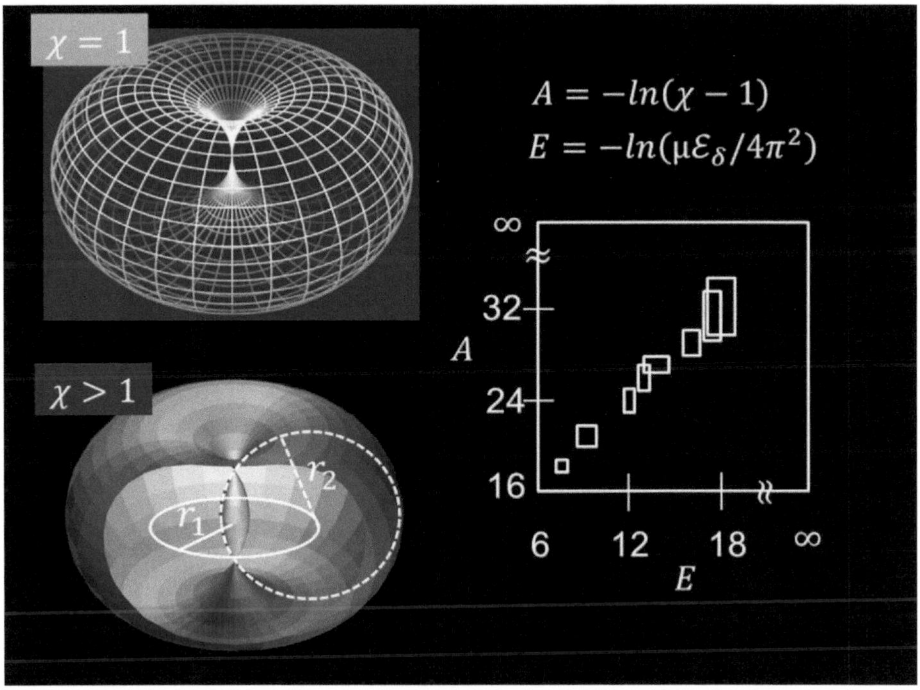

FIGURE 5.11 The two-stroke engine sustaining the primeval wormhole and fueled by dark energy. Correlation between disentanglement level (δ) and change in aspect ratio χ due to dimensional expansion.

describes the primeval wormhole as a two-stroke engine fueled by dark energy (Figure 5.12), whereby the measured density of dark energy (ρ_Λ) is given as follows:

$$\rho_{\Lambda} = \rho_e(\delta) + \rho_{\Delta V}(\delta) - \rho_{e'}(\delta), \tag{5.10}$$

where $\rho_e(\delta)$ is the density of dark energy released upon restoration of the entanglement from a δ-level disentanglement of the primeval wormhole, $\rho_{\Delta V}(\delta)$ is the surplus in dark energy per unit volume representing the extra vacuum energy associated with the vacuum dimensional expansion (ΔV) that restores the entanglement ($\chi \approx 1$), and $\rho_{e'}(\delta)$ is the next dark energy investment per unit volume required to disentangle the primeval wormhole to δ-level. Since $\rho_\Lambda \approx 5\times10^{-10}$ J/m^3, we obtain $\delta = \delta_\Lambda = 34.01\times10^{-18}$.

Space in the three-torus universe grows one dimension at a time, through cycles associated with enormous bursts of dark matter with restitution of the primeval wormhole entanglement followed by ever increasing expenditures of dark energy required to disentangle the reconstituted primeval wormhole. The amount of dark energy that is effectively not spent in the cycle may be computed as the difference between two consecutive bursts.

Thus, the cosmological constant problem, often described as defined by the enormous (120 orders of magnitude) surplus of dark energy arising from vacuum fluctuations, is more subtle than what it appears to be in the standard formulation [3]: *Dark energy is spent at an ever increasing cost and replenished cyclically by a pulsating universe that tears and regenerates the entanglement fabric of its primeval wormhole through the alternating expansion of its vacuum compact dimensions.* This endless sequence of cycles of destruction and creation in the evolution of the

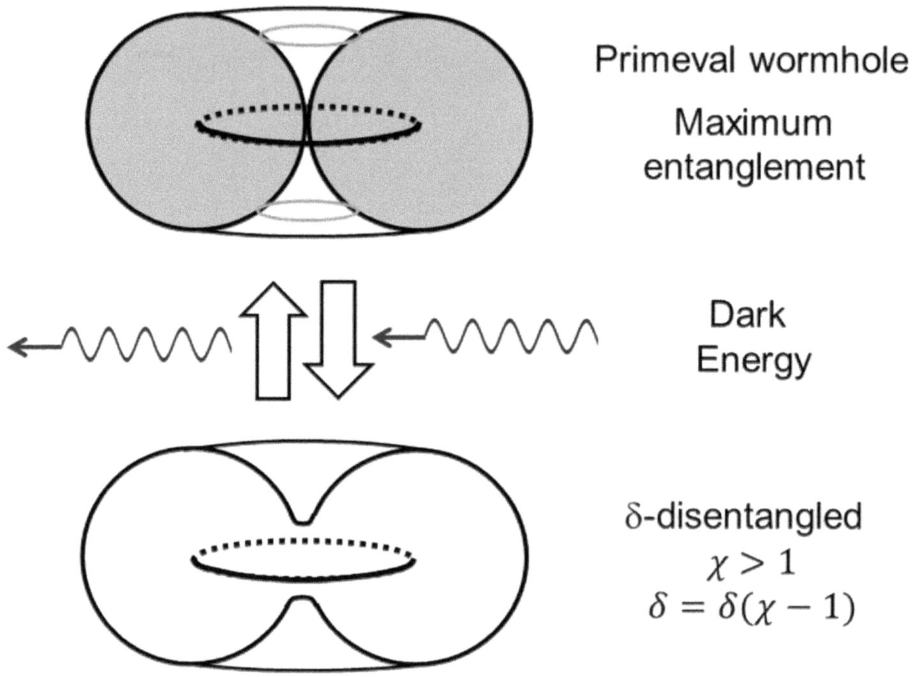

FIGURE 5.12 Representation of one cycle of the cosmic engine that sustains the primeval wormhole.

universe shares an astonishing parallel with the ancient cosmogonies stemming from metaphysical inquiries in the valley of the Indus. The infinite toils of Shiva, the destructive deity that as "Nataraja" dances out the pulse of the universe, are particularly suggestive of the rhythm conferred to the expansion of the universe through the cycles of destruction and creation of the entanglement fabric of its primeval wormhole. Shiva's dance is strikingly suggestive of the topology-sustaining two-stroke engine fueled by dark energy and controlling the vacuum expansion in the universe runaway (Figure 5.13).

5.5 A HOLISTIC QUINTESSENTIALLY ENTANGLED UNIVERSE ADMITS NO OBSERVER

The state of a cell in the entangled fabric of quantum vacuum is of course ambiguous, as it befits the standard interpretation of quantum mechanics, whereby disambiguation requires an act of observation. The structure of vacuum is upheld through the correlation pattern defined by the compounded ambiguous state: $\frac{1}{\sqrt{2}}|0\rangle|0\rangle+\frac{1}{\sqrt{2}}|1\rangle|1\rangle$, a Bell pair indicative of maximal entanglement. The collapse of the wavefunction for one of the cells, say cell 1, would yield ket state $|0\rangle$ or $|1\rangle$, depending on whether the observation registers respectively the absence or presence of a particle with associated de Broglie wave equal the dimension of the cell. This observation instantly triggers the collapse of the wave function for cell 2. Thus, the state of the system is now either $|0\rangle|0\rangle$ or $|1\rangle|1\rangle$, depending on the measurement that took place for cell 1. This is a completely standard reasoning befitting the so-called Copenhagen interpretation and holds no surprise. It fits exactly with the dictum by John A. Wheeler: "No phenomenon is a physical phenomenon until it is an observed phenomenon".

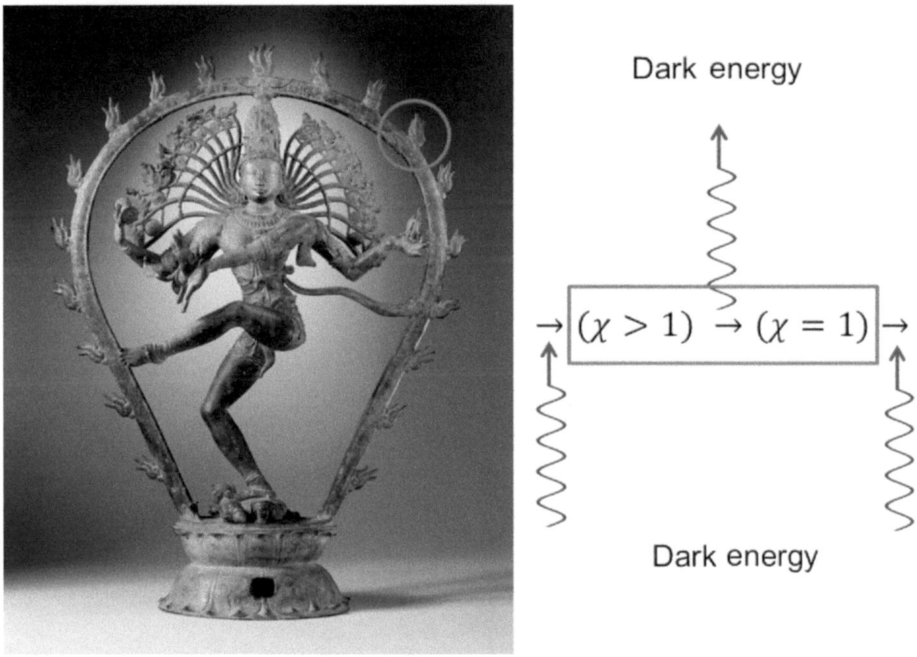

FIGURE 5.13 Shiva as "Nataraja", the lord of the dance, the destructive Hindu deity that also dictates the rhythm of the universe. The tempo of the pulsating universe is marked by Shiva's hourglass-shaped drum. As Nataraja, Shiva dances out the sequence of cycles of successive destructions followed by reconstructions of the universe that determine its evolution. Each cycle in the Nataraja's dance is marked by a flame (circle), akin to the burst of dark energy released with the reconstruction of the entanglement fabric of the primeval wormhole. This cosmic event is followed by the reinvestment of the dark energy in the next cycle that begins with the destruction of the entanglement of the primeval wormhole. Credit: Los Angeles County Museum of Art. Image of the Hindu deity "Shiva Nataraja" obtained from the Public Domain (https://en.wikipedia.org/wiki/File:Shiva_as_the_Lord_of_Dance_LACMA_edit.jpg).

However, the act of observation has been placed and can only be placed in the quotient space $W/\sim$, a latent manifold which is a projection of the quintessential space W. But, as we examine the quintessential entanglement bundle (Eq. 5.5) corresponding to the Bell pair (Eq. 5.4) in quotient space, we notice that there is no component of the bundle that projects on either $|0\rangle|0\rangle$ or $|1\rangle|1\rangle$. In other words, there is no value pair for the dormant variables y_1, y_2 for cells 1 and 2 that would yield either $|0\rangle|0\rangle$ or $|1\rangle|1\rangle$ as projection onto $W/\sim$.

This indicates that entanglement is never lost in W as a result of observation. We may formalize this observation as follows:

Theorem 5.2. Space W does not admit quantum observation, and hence, it should be considered quintessentially entangled.

The quintessential entanglement is again strongly suggested by the ancient metaphysical cogitations from Hinduism that posit the holistic nature of the universe as a major premise in its cosmogony. In this cosmogony developed over four millennia ago by the civilization of the Indus valley, all things and beings stem from a primeval unity and are unified by their blending into *Brahman*, the ultimate holistic entity, designated by the ōm symbol (Figure 5.14). Strikingly, it is precisely the ōm symbol that is believed to be the sign produced by the figure of Shiva Nataraja in its the choreography of the pulsating universe marked by cycles of destruction and reconstruction (Figure 5.13).

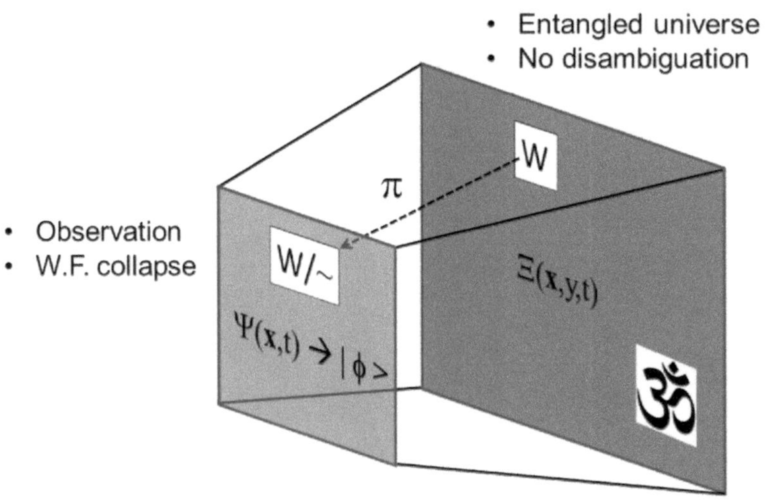

FIGURE 5.14 The quintessential universe W admits no observer; hence, all quantum phenomena remain potential and unrealized. The quintessential wave function does not collapse with observation, in contrast with its projection onto the latent space $W/\sim$. The holistic nature of W is thus apparent and may be equated with *Brahman*, whose sign "ōm" is featured in the figure.

5.6 PRESERVATION OF THE TOPOLOGY AND QUANTUM FABRIC OF SPACE TIME IN A MULTIVERSE SCENARIO

AI has addressed the quantum gravity problem in more than one way, and one alternative is constructing a "physical learning machine" or a "learning machine with a physical embodiment", with stochastic connectivity weights and hidden variables representing the random states of the nodes. This system may be endowed with emergent gravity and emergent quantum behavior (Chapter 3). In this way, AI represents the universe as a holographic autoencoder, where the holographic map $h: \partial W \rightarrow W$ is one of the many possible inverses of the canonical projection $\pi: W \rightarrow \partial W$ that maps the quintessential space W onto its "quantum border" ∂W. The emergent quantum behavior arises in the equilibrium regime for node state (hidden) variables upon which relativistic strings are embroidered. Conversely, the network is endowed with an emergent gravity in the nonequilibrium regime for the hidden variables. The results illustrate the possibility of establishing the duality quantum mechanics/general relativity in neural networks regarded as physical systems governed by the laws of statistical mechanics (Chapter 3). These networks are endowed with two different thermodynamic regimes, where the geometric dilution parameter $\upsilon = -\log\cos\alpha$ is the proxy for time and the advent of the lightest dark matter occurs at $\sim 10^{-27}$ s counting from the birth of the universe.

The coupling of two replicas of the QGAE in an annular conformation (Figure 5.10) was considered a suitable arrangement in order to reproduce the primeval wormhole arising from the construction of space as the three-torus. But this AI setting does not tell us whether $\delta = \delta_{\Lambda} = 34.01 \times 10^{-18}$ represents a universal constant that actually materializes in the universe. If it does not, then the dark matter density of the universe would not be constant, but we know that it is, at least approximately. There is in principle no *a priori* reason for the level of disentanglement of the primeval wormhole not to vary with each cycle. If it happens to allow for a surplus or shortfall in dark energy, then an additional cosmic mechanism would need to be invoked, representing a sort of "cosmic valve" that needs to become operative to maintain a constant density of dark matter energy, as mandated by the cosmological constant. This cosmic valve mechanism turns out to be cosmic reproduction.

As described in Chapter 3, a birth channel for cosmic reproduction can be set up within this scheme by constructing a second holographic autoencoder $h': \partial W' \rightarrow W'$ that serves as antenna, capable of capturing the tunneling $\partial W \to \partial W'$ of a dark energy burst in the universe $(\partial W, W)$ in the form of a quantum vacuum fluctuation. Quantum tunneling implies the existence of a spatial interface, and the only spatial dimension of the universe that offers a boundary is the dormant dimension, and hence, the tunneling would have to be "quintessential", i.e., operative through the dormant dimension. The tunneled fluctuation is amplified within the second holographic autoencoder to give birth to a progeny universe $(\partial W', W')$. By $\sim 10^{-27}$ s after its birth, the baby universe will begin storing the tunneled dark energy in a material embodiment, with the formation of dark matter in the form of ur-Higgs particles. In this way, AI constructs a cosmic reproductive machine that converts dark energy into dark matter (Figure 5.15). Notably, this machine harnesses dark energy as the cosmic "birth inducer" and generates dark matter in a multiverse scenario as the canonical (surjective but not injective) projection $\pi' : W' \to \partial W'$ admits *a priori* a multiplicity of holographic maps $h' : \partial W' \to W'$ all satisfying the equation $h' \circ \pi' = id_{W'}$.

To leverage this multiverse scenario, an advanced AI-based cosmic technology for universe reproduction is implemented (Chapter 3). The technology harnesses dark energy tunneling as a nucleating quantum fluctuation that spans the progeny universe. The cosmic reproduction machine requires a second holographic autoencoder serving as antenna or $\partial W \to \partial W'$ receiver. An energy

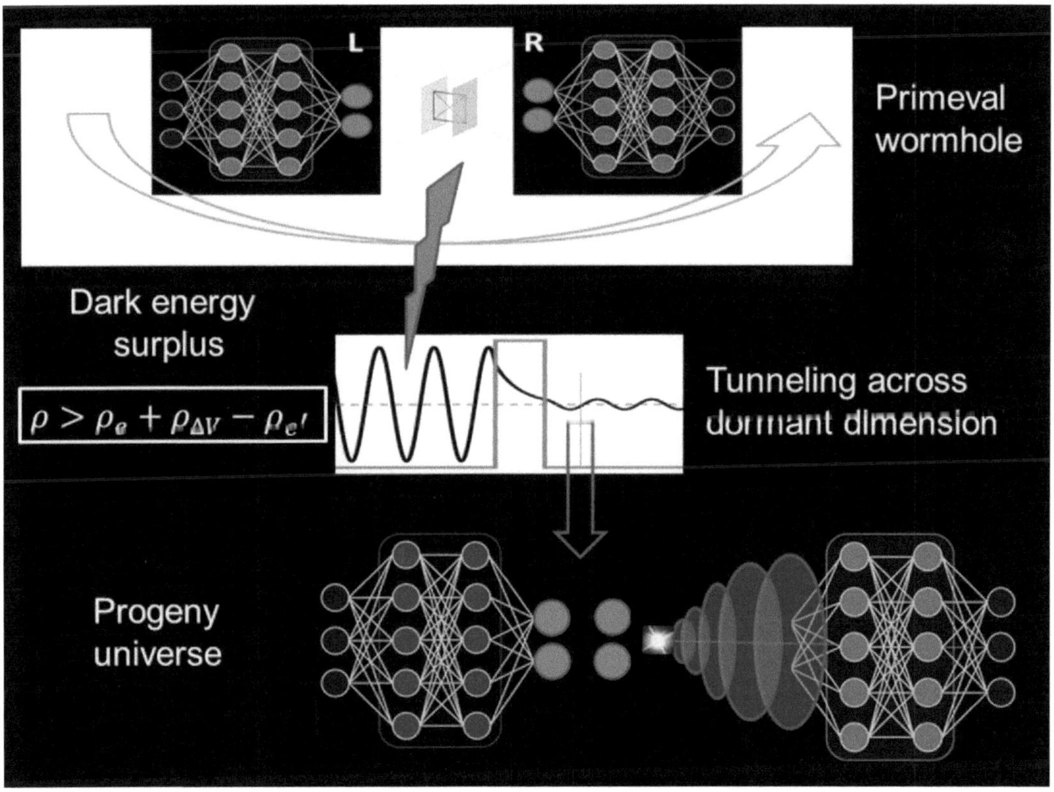

FIGURE 5.15 AI-based cosmic reproductive machine that converts dark energy into dark matter in a multiverse matrix. A birth canal for cosmic reproduction is set up by coupling the universe $(\Omega = \partial W, W)$ to a second holographic autoencoder acting as receiver and capable of capturing the tunneled dark energy originated in the progenitor universe. The tunneled quantum fluctuation is amplified through the receiver autoencoder to give birth to the progeny universe $(\Omega' = \partial W', W')$ that stores the tunneled dark energy as dark matter in the form of stationary waves along the quintessential circular coordinate of quark-size attometer dimension.

surplus from the ever-increasing vacuum quantum energy that is not invested in restoring the primeval wormhole entanglement is meant to be indicative of a poor understanding of dark energy in our universe. This surplus may be reconciled within the multiverse scenario put together by a technologically advanced civilization capable of cosmic investment in dark energy for reproductive purposes, as schematically depicted in Figure 5.15.

On the other hand, the cosmic valve may work in the opposite direction as required in the event there is a shortfall in dark energy to restore the entanglement of the primeval wormhole while maintaining the dark energy density at its constant value $\rho_\Lambda \approx 5\times10^{-10}\,\mathrm{J/m^3}$. In this case, the conversion of dark matter into dark energy and tunneling of the latter through the interface of the dormant dimension would be carried out by the lender which now is the progeny universe.

The vacuum catastrophe or cosmological constant problem (Chapter 1) alluded to above is arguably the most embarrassing discrepancy in all of physics. It has become a veritable "tag for demolition" of the whole edifice of physics. A naïve computation of the vacuum quantum energy density yields values estimated to be 120 orders of magnitude larger than the expected contribution to the cosmological constant based on experimental observation [3]. AI provides a solution to this problem, shifting the task to a mere calculation of the rate of generation of progeny universes arising from dark energy tunneling. Thus, the tunneling of vacuum fluctuations is given a material embodiment as dark matter ur-Higgs particles in the embryonic universe. These particles are stored as stable wave excitations along the quintessential quark-scale dimension and endowed with mass 246 GeV/c^2, the vacuum expectation value of the Higgs boson.

We have described AI's multiverse solution to the cosmological constant problem. The solution is based on a "multiverse matrix", a pivotal scenario consisting of coupled holographic autoencoders capable of converting dark energy into dark matter as the tunneling of dark energy is steered through a cosmic birth canal that exploits the dormant dimension. The multiverse generator acts as a cosmic valve to maintain a constant dark energy density in the case of surplus or shortfall in the fueling of the two-stroke engine that holds together the topology of the universe and its fabric of quantum entanglement.

5.7 CHIRALITY IN THE DARK UNIVERSE

The invisibility of dark matter is directly attributable to a lack of communication with the weak force. This statement implies that a proportion of the negative-energy fermionic sea, known as Dirac sea (Chapter 1), must be somehow rendered impervious to communication of the weak force, but how? Identifying the actual purveyor to the dark universe requires a careful discussion of "chirality", an essential concept in particle physics.

It is obviously tempting to identify the Dirac sea of negative-energy fermions, or the closely related vacuum energy in quantum field theory, with a portion of the dark universe. Taken at face value, this assertion is plainly incorrect. A revision and fine-tuning of the previous material presented in this chapter becomes mandatory as we confront the fact that the weak interaction that enables fermions to escape the Dirac sea (with concurrent creation of particle-antiparticle pairs) is *chiral*, a concept now to be discussed.

Chirality is a property of particles that has not entered so far into the discussion put forth in this book. The decisive experiment by Chien-Shiung Wu, asserting the lack of parity conservation (left/right handedness equivalence) for the weak interaction [4-6], makes this discussion imperative. It was this experiment that prompted the unification of the weak force with electromagnetism, a merging now named electroweak force.

Handedness may be defined by the sign of the projection of spin axis of rotation along the direction of motion (left = negative, right = positive), and it is a constant of motion but not a Lorentz invariant except in the case of massless particles, where handedness is equal to chirality. For massive particles, handedness is obviously not Lorentz invariant, since a frame of reference may move faster

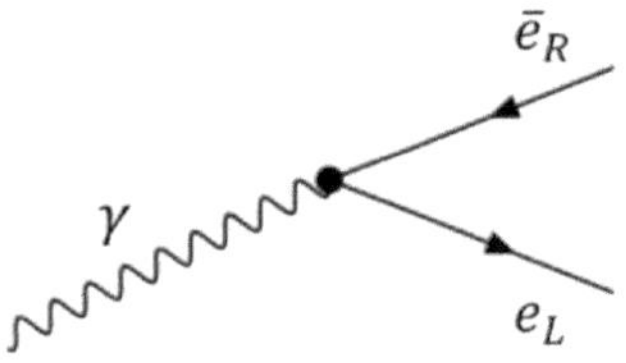

FIGURE 5.16 Feynman diagram for the vacuum creation of the electron-positron pair with the respective particle chiralities.

than the particle, reversing the particle's direction of motion relative to the observer. Thus, handedness and chirality cannot be equated for massive particles, with the latter being Lorentz-invariant but not a constant of motion, *au contraire* to handedness.

Spin ½ fermions, and in particular electrons, generated in vacuum fluctuations as particle–antiparticle pairs, as described schematically by the Feynman diagram in Figure 5.16, are left-handed, in accordance with the chirality of the weak force. Likewise, if the universe arose from a vacuum fluctuation as asserted in Chapter 4, we would expect that the Dirac sea would be initially composed of left-handed fermions, hence not part of the dark universe *per se*. So, the identification of vacuum energy with dark energy as put forth in this chapter must be upheld with a major caveat.

The autoencoder endowed with its physical realization (Chapter 3) is suited to materialize a holographic universe, since the information embossed in space–time is encoded in the latent manifold endowed with a fabric of quantum entanglement. However, no constraint arising from the existence of invisible matter (or its energetic equivalent) has been imposed on the encoding of the atlas – i.e., the covering with locally tangent spaces – of the latent manifold. In other words, the existence of a dark universe was not factored into the construction of the latent manifold on the same footing as the constraints (compactness and connectedness) that determined its topology, with its primeval orientable wormhole (POW). As now shown, those constraints pertain to the *orientability* of the manifold, a property that should be topologically on equal footing as compactness and connectedness.

To warrant their invisibility, fermions in the dark universe must be endowed with right-handed chirality. On the other hand, orientability is the only topological feature that can be tuned at this point in the discussion to accommodate this possibility, as argued in the subsequent sections. We are compelled to admit that the previous development requires revision, as the latent manifold may need to be reoriented if this possibility impinges on chirality, as it indeed does. More specifically, we are seeking for portal to the dark universe enshrined in the topology of the universe.

5.8 PRIMEVAL NONORIENTABLE WORMHOLE AS PORTAL TO THE DARK UNIVERSE

In defining the topology of the universe, we have established that space–time as a manifold needs to be compact and multiply connected since it is constrained to have no boundary. However, we have not yet dealt with orientability, a topological latitude that, as it turns out, plays a pivotal role in determining the portal to the dark universe.

As delineated in the previous sections, the primeval wormhole is sculpted by the topological constraints to which space–time is subject. On the other hand, the setting of "universe as hologram" fits into an AI system in the guise of an autoencoder, precisely the system that captures the essential physics with dimensionality reduction. This autoencoder has been subject to constraints that defined the encoding of the tangent bundle of its latent manifold. However, the encoding of the tangent bundle is itself subject to constraints that have not been hitherto considered and result from the tuning of orientability.

To better understand the problem, we need to identify the portal for the generation of the "dark" dextrogyre (right-handed) spin ½ fermions. These dark fermions are actually specular versions of those generated in pair creation through vacuum fluctuation (Figure 1.19). The concept of mirror matter was introduced by Lee and Yang with the idea of parity (P) violation in weak interactions [6]. The mirror "hidden" matter sector of quantum field theory restores the equivalence in the universe under mirror reflection. As it has been established, the standard model Lagrangian contains only the left-handed components of the Dirac wave functions: $\psi_L = P_L \psi$, where the left-handed and right-handed projectors are made up of products of Dirac matrices (Figure 1.18):

$$P_L = \frac{1}{2}(\mathbb{I} - \gamma^5); P_R = \frac{1}{2}(\mathbb{I} + \gamma^5); \gamma^5 = i\gamma^0\gamma^1\gamma^2\gamma^3 \tag{5.11}$$

Within the standard model, left and right chirality components are treated differently by the weak interaction. For example, the weak interaction could rotate a left-handed electron into a left-handed neutrino with emission of a boson W^-, but could not do so with right-handed counterparts. Therefore, we need to tune the topological latitude of the primeval wormhole so that, upon crossing, it should enable the transformation:

$$P_L = \frac{1}{2}(\mathbb{I} - \gamma^5) \rightarrow P_R = \frac{1}{2}(\mathbb{I} + \gamma^5); \mathbb{I} = identity\ matrix \tag{5.12}$$

As it turns out, a primeval nonorientable wormhole (PNOW) would enable such transformation [7] and, therefore, may serve as a portal to the dark universe made up of right-handed Dirac fermions. This implies that the entanglement pattern arising from boundary identification in the universe should – in all likelihood – be nonorientable, as described in Figure 5.17a,b. This quotient space may be represented as the Cartesian product of a two-torus and a Klein bottle (Figure 5.18). It should be emphasized that the hole in the nonorientable factor-manifold does not exist in four or more dimensions: It is purely artefactual since it results from the inherent limitations of the three-dimensional representation.

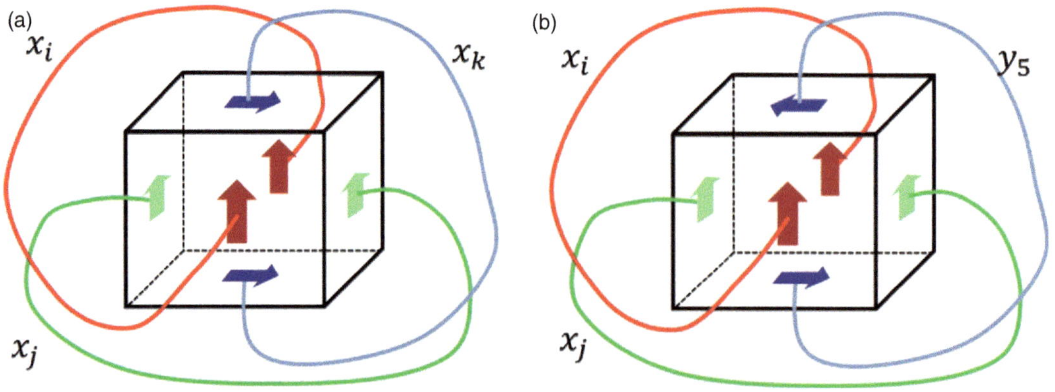

FIGURE 5.17 Topological identification of opposite faces of a three-dimensional cube to render a three-torus (a) or a nonorientable three-dimensional manifold (b), depending on the space–time coordinates considered (x denotes any of the four canonical, while y_5 is the fifth compact coordinate). The arrow bestows orientation to the face. In topological terms, the nonorientable manifold (b) is a Cartesian product of a circle and a Klein bottle.

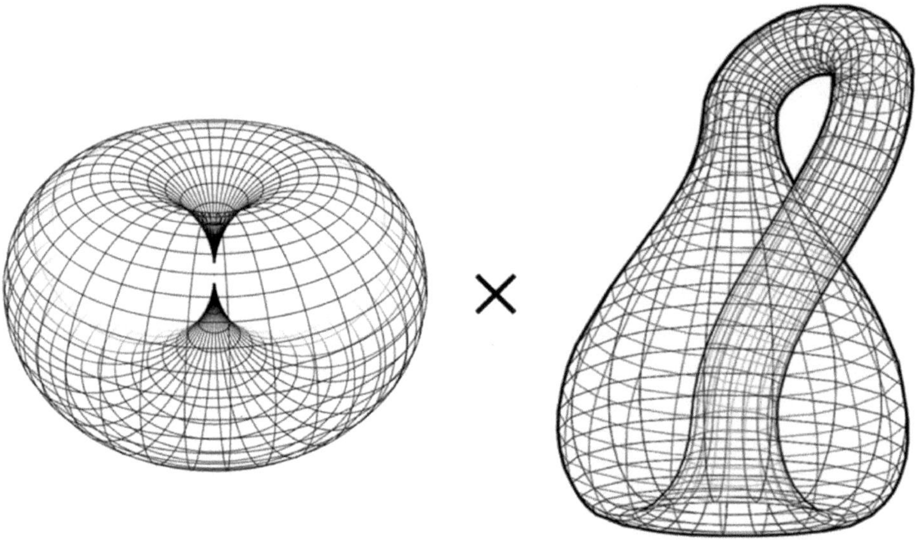

FIGURE 5.18 Four-dimensional nonorientable spatial cross section of the quintessential space–time represented as the Cartesian product of a two-torus and a Klein bottle.

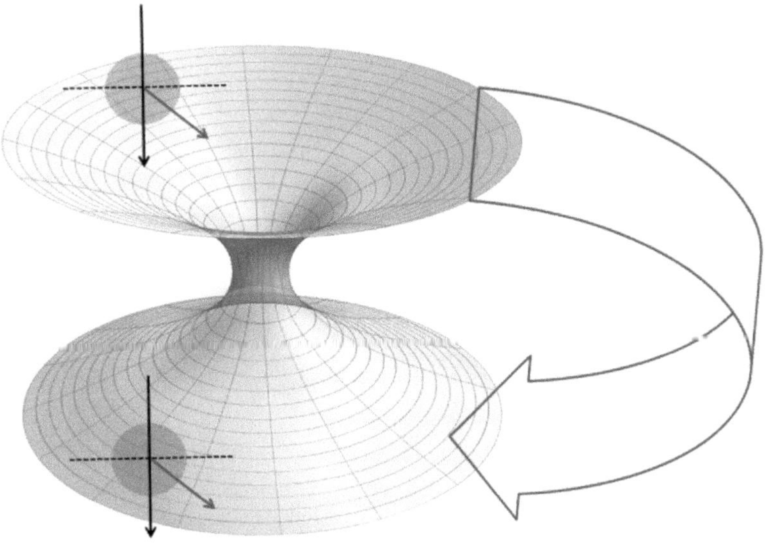

FIGURE 5.19 Topological representation of the primeval orientable wormhole (POW) that preserves particle chirality upon crossing (spin axis as red vector, momentum vector in black).

To construct a two-dimensional cross section of the PNOW as purveyor of dark fermions, we need to join the exteriors of event horizons for the two connected blackholes but replace an orientable (Figure 5.19) for a nonorientable surface obtained by cutting and pasting a Moebius strip (Figures 5.20 and 5.21a). The actual AI system (coupled autoencoders) realizing the nonorientable wormhole is described in Figure 5.21b, whereby the tangent bundles for the respective latent manifolds are encoded with identification following the reflection symmetry: Two tangent spaces that transform into each other through the symmetry operation are identified in the Moebius-type

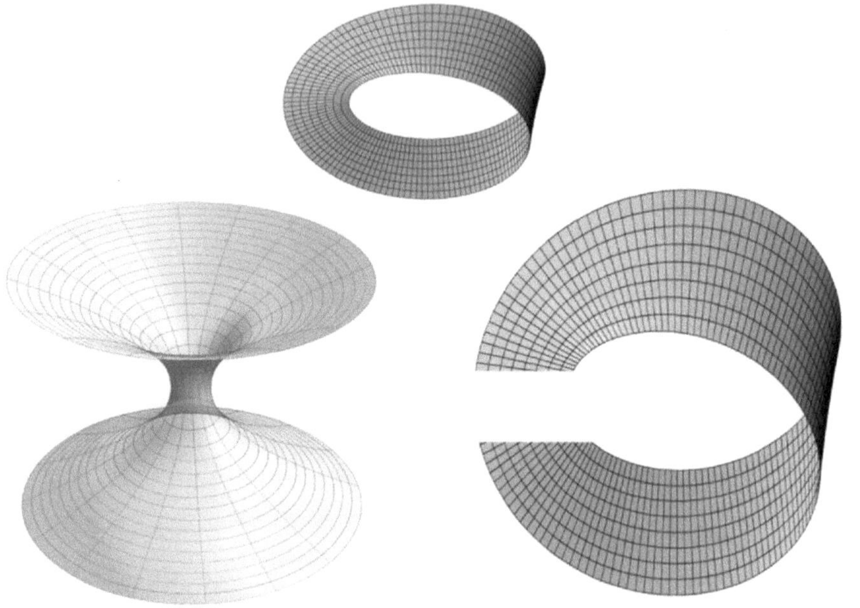

FIGURE 5.20 Elements for the construction of a two-dimensional cross section of a nonorientable wormhole by cutting and pasting a Moebius strip.

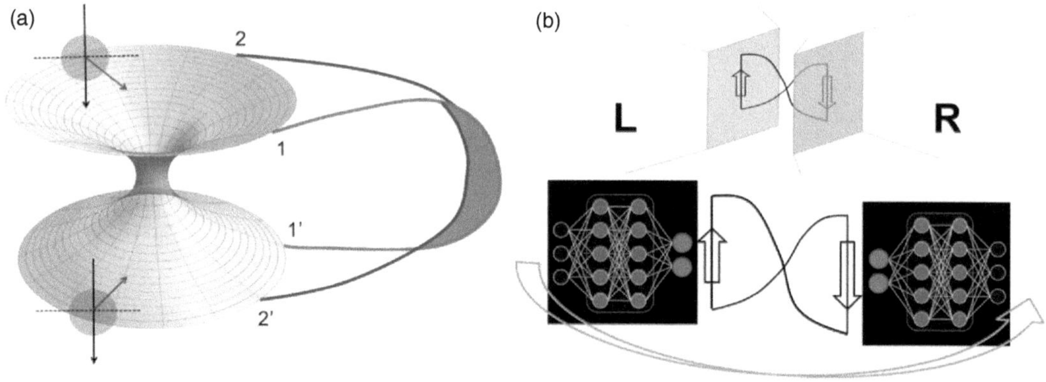

FIGURE 5.21 Topological representation of a primeval nonorientable wormhole (PNOW). (a) Two-dimensional rendering of the manifold displaying through surgical attachment of a Moebius strip with inversion of particle chirality upon crossing. (b) Annular assemblage of two holographic autoencoders with outer layers connected in opposite orientations.

annular arrangement of quantum gravity (holographic) autoencoders, as shown schematically in Figure 5.21b. The entire encoded atlases are thus glued steered by the inversion operation.

One would be tempted to assume that the PNOW could change the chirality of the negative-energy fermions in the Dirac sea upon crossing (Figure 5.22), but such picture would not be altogether compatible with quantum field theory, which supersedes the Dirac ansatz in favor of particle-pair creation through vacuum fluctuations, as discussed in Chapter 1 (Figure 1.19). Thus, the correct picture of the PNOW as portal for the dark universe is provided in Figure 5.23. The right-handed "dark" fermion generated by the PNOW is precluded from contributed to the replenishing of the

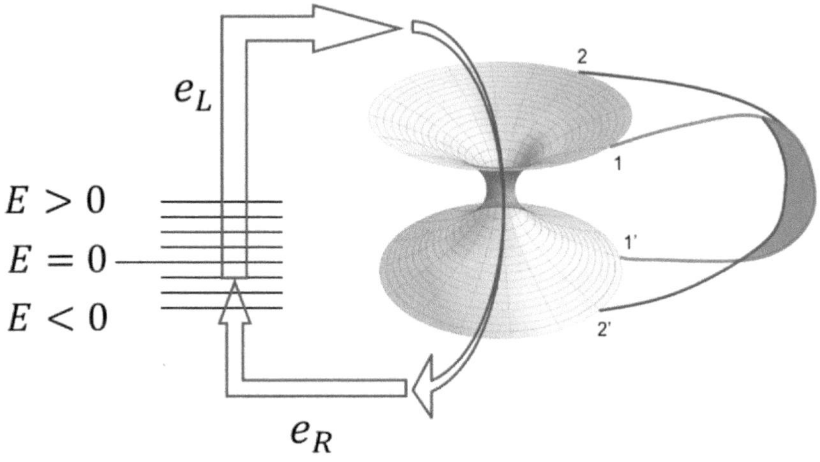

FIGURE 5.22 PNOW as direct supplier of dextrogyre "dark" spin-1/2 fermions to the Dirac sea of negative-energy fermions.

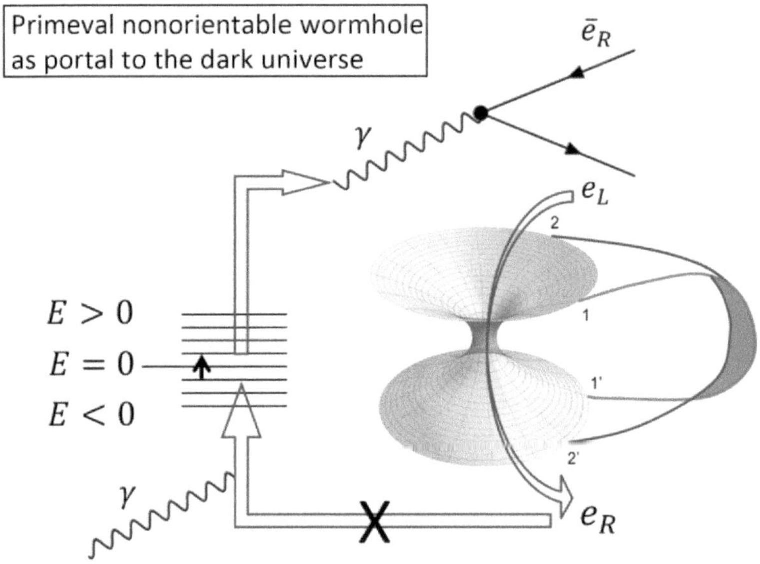

FIGURE 5.23 PNOW as cosmic engine that sustains the portal to the dark universe in the quantum field theoretical vacuum. The "negative energy flow" drawn from the Dirac sea maintains the PNOW, which is thus fueled by dark energy in the quantum field interpretation of vacuum energy. The coupling between fueling dark energy and production of dark fermions materializes in the operation of the cosmic engine.

Dirac sea because of its lack of interaction with gauge bosons that communicate the weak force and would enable transition to the lower energy levels. Nevertheless, crossing cannot be only one way, since that would imply a constant production of dark matter, which would be at odds with what is established in that regard: *the total amount of dark matter remains constant*, in contrast with the total amount of dark energy that grows constantly to maintain constant concentration upon vacuum growth. This implies that transitions yielding dark fermions are reversible in the sense of being in dynamic equilibrium with the transitions of dark fermions into visible levogyre or detectable fermions via the crossing of the PNOW, as described in Figure 5.24.

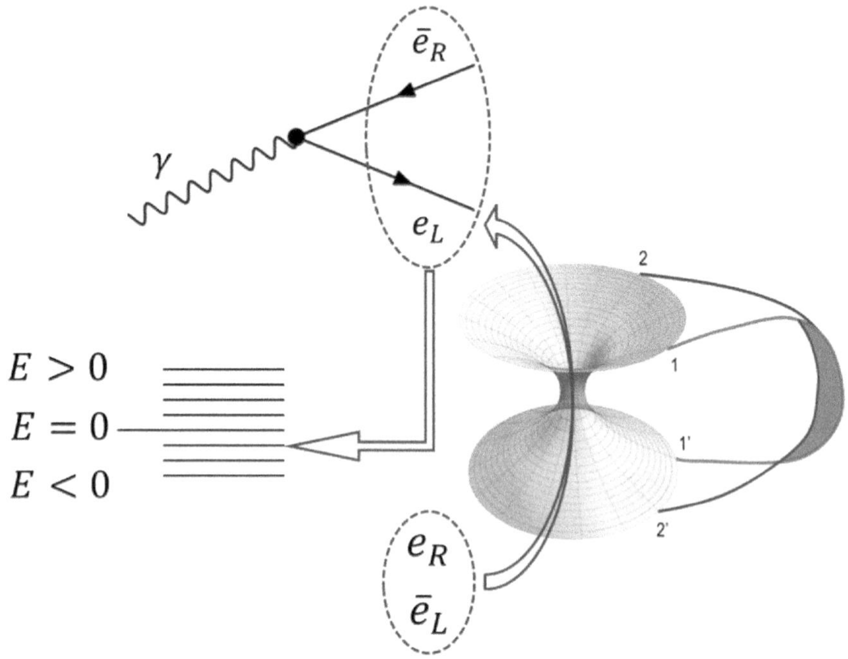

FIGURE 5.24 Inverse transformation of dark spin-1/2 fermions into visible fermions upon crossing the PNOW. The transformations of matter described in Figures 5.23 and 5.24 are in dynamic equilibrium.

Through the process described in Figure 5.24, the PNOW can replenish the depleted Dirac sea, which only gets repopulated from detectable "levogyre" fermions upon photon emission.

5.9 PORTAL TO THE DARK UNIVERSE RECREATED IN A QUANTUM COMPUTER

The previous sections have described the sustainability of the primeval wormhole – and thereby, of the topology of the universe – through constant expenditure of dark energy in an expanding vacuum generated autocatalytically. Thus, an open traversable wormhole fitting the universe's horn-torus topology was shown to be sustained by a cosmic engine fueled by dark energy. The engine was shown to partially disentangle the exteriors of the primeval wormhole ($\chi = 1 \rightarrow \chi > 1$), enabling autocatalytic universe expansion, one dimension at a time. In turn, this process restores entanglement and replenishes dark energy, completing the cycle of the cosmic two-stroke engine. Since dark energy is the ever-increasing "vacuum energy surplus" after fueling the cosmic engine through its successive cycles, this implies that a shockwave of "negative energy" (i.e., lower than the expected vacuum energy) must traverse the wormhole at regular intervals to keep it open, hence enabling information passage. This negative energy traffic across the PNOW is described in Figure 5.23.

This prompts us to address the possibility of creating the primeval wormhole in a quantum computer, taking advantage of the possibility of manipulating quantum entanglement of qubits which are embodied in the space–time wormhole in accordance with the Maldacena–Susskind conjecture [1]. This effort is also inspired by the Wheeler dictum that mass and energy are essentially derivatives of information. In the orientable toroidal topology of the universe first considered, the challenge would be to create two sets of entangled qubits representing the exteriors of the wormhole and enabling a sustainable open duct by creating the equivalent of the negative energy pulse in a quantum computer (Figure 5.25). This can be achieved by a rotating magnetic field that induces spin rotation in a ring of qubits representing one of the wormhole exteriors.

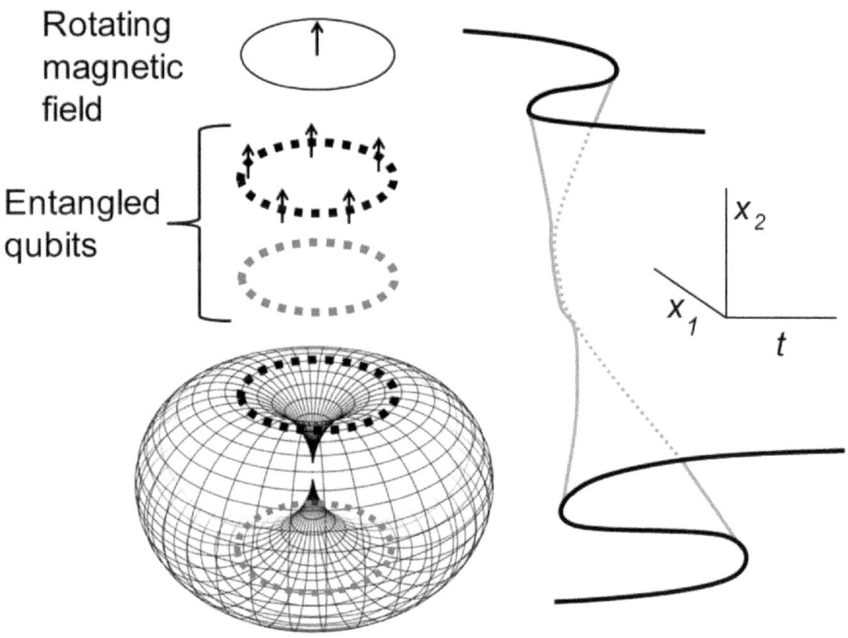

FIGURE 5.25 Dark energy as enabler and sustainer of the primeval wormhole recreated in a quantum computer as a quasi-continuum "pulse of dark energy deficiency" sustained between the two wormhole exteriors and traversing its interior.

Thus, in essence, the wormhole would be created through entanglement of qubits and its sustainability as qubit passage would be ensured by recreating the cosmic engine fueled by dark energy in the guise of a "pulse of dark energy deficiency" traveling along the wormhole. In this way, dark energy could be recreated in a quantum computer as the enabler and sustainer of the primeval wormhole of the universe.

However, to generate dark fermions, a PNOW with quotient space defined in Figure 5.17 would need to be implemented in the quantum computer. This would require the antiparallel connection of the outer layers of two quantum-gravity autoencoders as depicted in Figure 5.21b, recreating an qubit passage with two-dimensional y_5-involved cross sections mapped as a Klein bottle (Figure 5.26).

5.10 THE PORTAL TO THE DARK UNIVERSE COUPLES DARK ENERGY EXPENDITURE WITH A DYNAMIC EQUILIBRIUM BETWEEN DARK AND DETECTABLE MATTER

Chapter 1 describes the disconcerting but established phenomenology that accounts for the existence of dark matter and dark energy in the standard four-dimensional space–time. A quintessentially encoded ur-space–time is needed to account for dark matter, as shown in Chapter 4. Yet, dark matter appears likely to be more than one thing. A portion of dark matter stays as ur-particles, created early on in the universe history, while another portion, the dark fermions, are in dynamic equilibrium with detectable matter, as previously shown. The ur-Higgs and its partners, quintessential particles with no geometric dilution into the standard spatial coordinates, account for 81.28% of dark matter. The concentration of ur-particles is ever decreasing in our universe which undergoes accelerated expansion. On the other hand, the dark fermions in dynamic equilibrium with detectable matter constitute

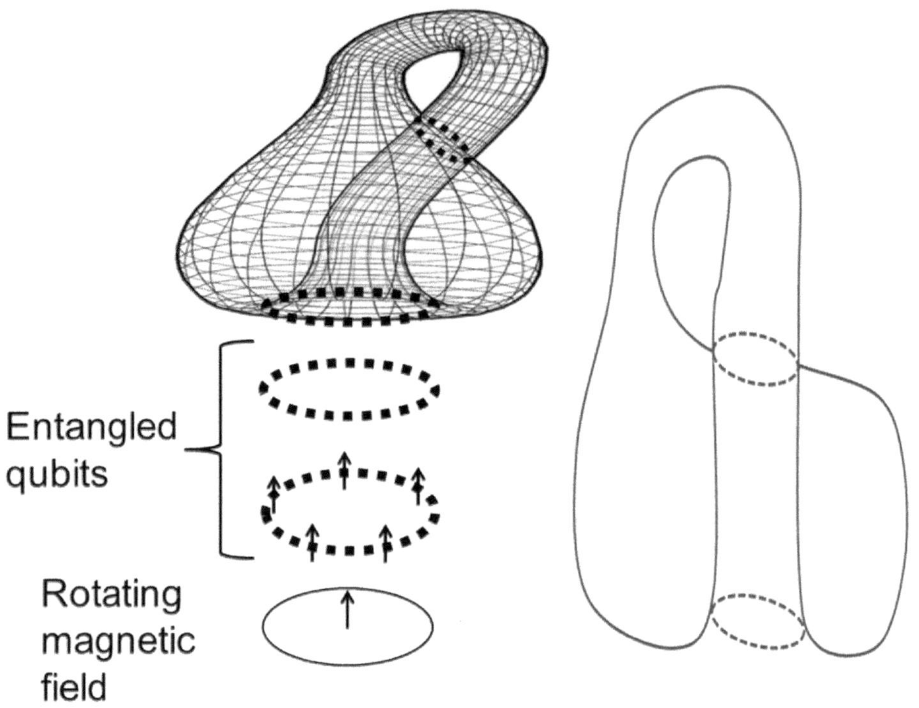

FIGURE 5.26 Scheme of a PNOW recreated in a quantum computer.

18.72% of dark matter. Since "matter in all forms", including dark energy, is ***at present*** 26.7% dark matter and 5% detectable matter (Chapter 1), the dark fermion proportion at 18.72% of dark matter corresponds to 15.77% of all matter (dark + detectable) in the universe, which is exactly equal to the percentage of all matter represented by detectable matter. In turn, the 15.77% of all matter, representing either the proportion of dark fermions or detectable matter, corresponds, ***at present***, to 5% of all forms of matter. The proportions in the current estimated composition of the universe are depicted in Figure 5.27.

In this chapter, we have provided a vantage point to identify dark energy as the surplus in vacuum energy taken from the amount required to fuel the cosmic engine that maintains the PNOW in an expanding universe. The concentration of this surplus is indeed constant since dark energy is not subject to dilution associated with vacuum creation. This fact stands in contrast with the concentration of dark matter, including ur-particles and dark fermions that are in dynamic equilibrium with detectable matter. Clearly, the dark matter concentration decreases as it gets progressively diluted concurrently with vacuum growth over time. The sustainability of the PNOW, depicted schematically in Figures 5.28 and 5.29, ensures the dynamic equilibrium between dark fermions and detectable matter, a contribution which maintains the overall mass of dark matter constant while the overall concentration of dark energy is also kept constant.

To conclude, the dynamic equilibrium between dark matter and detectable matter is maintained by a cosmic engine fueled by vacuum energy that sustains the portal to the dark universe while maintaining a constant concentration of dark energy in an ever expanding universe. Thus, the AI model shows that the ur-Higgs particle, as well as the primeval black hole and wormhole entanglement are essential components of the dynamic picture that makes the universe sustainable.

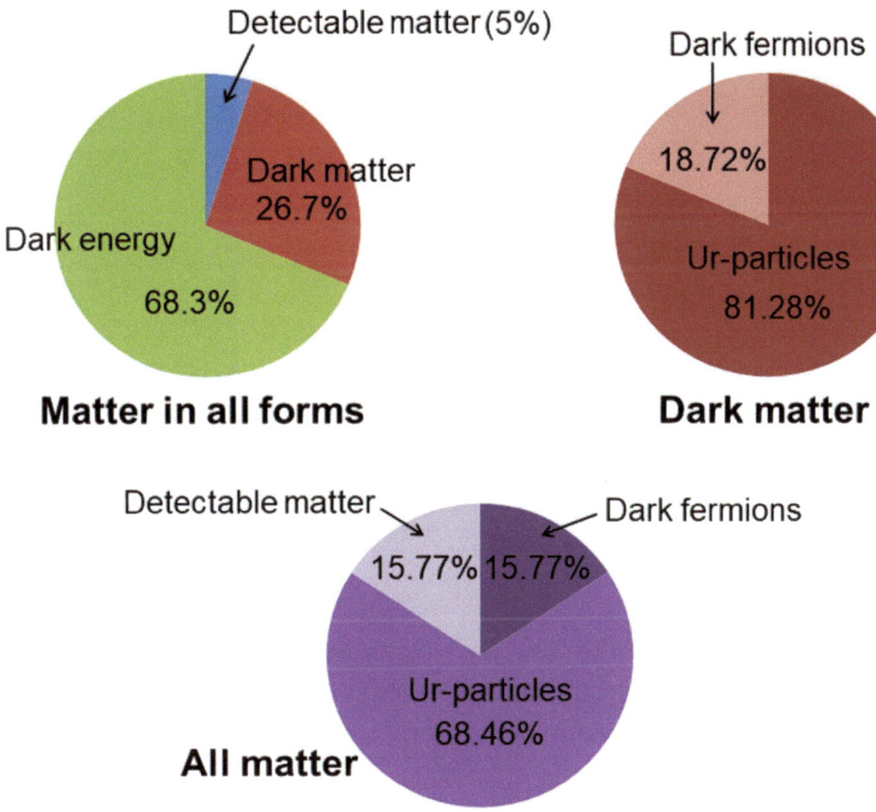

FIGURE 5.27 Estimated composition of "all forms of matter", dark matter and all matter in the present-day universe.

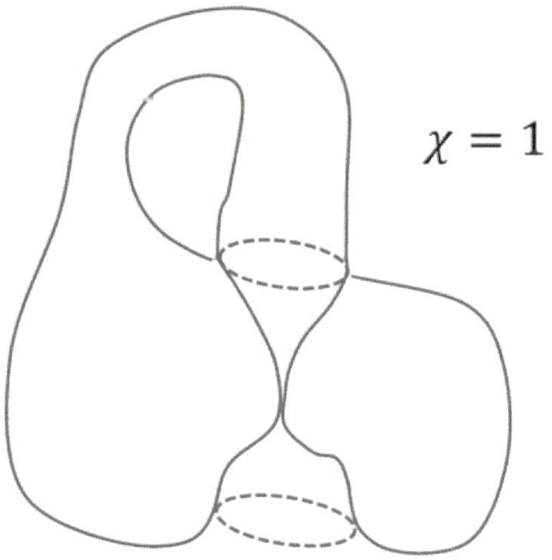

FIGURE 5.28 Schematics of the Klein bottle at unity aspect ratio representing a two-dimensional cross section for the PNOW.

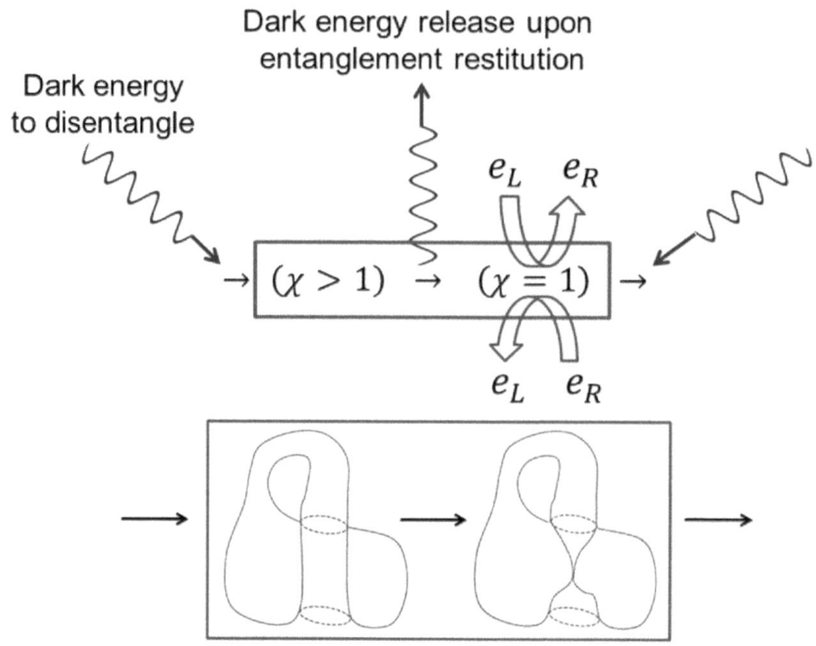

FIGURE 5.29 Schematics of the operation of the cosmic engine fueled by vacuum energy that sustains the portal to the dark universe. The dynamic equilibrium between dark matter and detectable matter is coupled to the production of dark energy (the surplus vacuum energy in each engine cycle) at a constant concentration in an expanding universe.

REFERENCES

1. Maldacena J, Susskind L (2013) Cool horizons for entangled black holes. *Fortsch Phys* 61, 781–811.
2. Bohm D, Aharonov Y (1957) Discussion of experimental proof for the paradox of Einstein, Rosen and Podolsky. *Phys Rev* 108: 1070–1079.
3. Adler R, Casey B, Jacob O (1995) Vacuum catastrophe: An elementary exposition of the cosmological constant problem. *Am J Phys* 63: 620–626.
4. Weinberg S (2008) *Cosmology*. Oxford University Press.
5. Fisher P (2022) *What Is Dark Matter?* Princeton University Press.
6. Lee TD, Yang CN (1956) Question of parity violation in weak interactions. *Phys Rev* 104:254–258.
7. Visser M (1995) *Lorentzian Wormholes: From Einstein To Hawking*. Washington University Press, St. Louis, MO.

6 AI's Autoencoder Approach to the Theory of Everything

"Was there below?
Was there above?"

Creation Hymn, Rig-Veda

This chapter describes the AI approach to the so-called "theory of everything", which is based on the training of an autoencoder that distills quantum reality from a quintessential space–time encoding an extra spatial dimension. Gravity is shown to be quantized under a generalized gauged ur-symmetry inclusive of the electroweak symmetry at the group-theoretic level. This inclusion and the possibility to cast gravity, a relativistic concept, in terms of a generalized gauge theory of the local Lorentz group lead to an ultra-unification that suggests that the fundamental forces in nature stemmed from one primeval force arising at the Planck epoch. The vehicle for this primeval force is a holonomic boson that measures chiral defects induced by a generalized symmetry transformation as quantized tension that charges the boson under the primeval symmetry. This force then got diluted differently into the geometry of the universe in accordance with the gauge symmetries that charged the fundamental force carriers in turn activated at successive phase transitions in the aftermath of creation.

6.1 SEARCHING FOR THE PRIMEVAL FORCE THROUGH INCLUSIVE SYMMETRIES

Experiments attempting to recreate the big bang and measurements in deep space both point to the tantalizing possibility that our universe may be the relic of something simple, powerful, and highly symmetric. The evidence points to a state where matter and energy cannot be told apart and the four fundamental forces are unified into one, possibly representing a precursor to quantum gravity. In this context, a "theory of everything" is meant to signify a theory of quantum gravity or, alternatively, a theory that identifies the primeval force in the big bang aftermath.

To address this challenge, we first need to settle some very basic questions, such as what is it that is being evolving ever since the Planck epoch? This primeval period is marked by a critical point at $T \sim 10^{32}$ K, a phase transition that took place at $t \sim 10^{-43}$ s after the big bang, as shown in Chapter 2. The geometric support of that entity, meaning space–time, may have changed dramatically but if the tenets of general relativity can be extrapolated to such primeval time, at least we can be certain that its topology remained invariant. Otherwise, the space–time manifold would have been altered through surgical gravitational singularities. So, if space–time was a compact manifold, then it should remain compact today. Furthermore, space–time cannot have boundaries. This is so because the phrase "boundaries of the universe" imply that nothingness is beyond the boundaries,

DOI: 10.1201/9781003632924-6

so there is no geometry to signify "beyond the boundaries" or bestow meaning to "outside", as that refers to a place, implying that a geometry exists to locate "outside". Ontologically, we are asserting that the space–time support of the universe is boundary-less. That leaves us with the inevitable conclusion that space–time is multiply connected since we already showed that it is compact. Hence, space–time is topologically a torus.

There is a second issue that needs to be settled before embarking in a quest for the theory of everything. It pertains to the number of dimensions of the manifold that represents space–time. We have asserted in Chapters 1 and 2 that a fourth spatial dimension needs to be incorporated to address the problem of the geometric dilution of mass and gravity. This problem leads to the hierarchy problem of incommensurably small particle masses relative to the Planck mass, and the incommensurably small coupling strength of gravity relative to the other fundamental forces of established quantum origin, as described in Chapter 2. Thus, an extra compact dimension has been incorporated (while maintaining the space–time topology) with radius $r_0 = 0.802695 \times 10^{-18}$ m given by the smallest material scale, the effective quark diameter established from parton models of inelastic $e^{\pm}p$ scattering. Amazingly, the minimum rest–mass energy stored in the extra dimension is $E = \hbar c / r_0 = 246$ GeV, precisely the true vacuum for the Higgs field.

This observation has profound implications pertaining to the origin of mass and the nature of the dark sector, with the latter shown in Chapters 2, 4, and 5 to arise from energy stored in the extra dimension and undiluted into the visible dimensions. Furthermore, the fact that the topology of the universe corresponds to a four-torus, however large, makes it feasible to incorporate other compact dimensions without introducing boundaries. On the other hand, any "brane model" of the universe such as the Randall–Sundrum model that incorporates the extra dimension as a quotient space identified with an arc of the circle may be dismissed on purely ontological grounds as this procedure generates boundaries (Figure 6.1).

The next challenge in the quest for the theory of everything is how to incorporate the extra dimension whose existence we have now justified. Chapters 2 and 3 describe an AI system that came to be known as autoencoder and is capable of encoding the extra dimension. This system distills quantum reality from the 5D space–time, where the extra latent dimension is shown to identify the quantum vacuum as ur-matter, that is, as a precursor of the visible and dark sectors of the standard model of particle physics. The autoencoder gets trained to generate events in the 5D space–time in the form of ur-boson generation and decay that get decoded into events in 4D space–time, so that the dynamics in 4D entrains the dynamics in 5D while the latter influences the former during the training as described in the commutative diagram in Figure 6.2. This operational tenets bear similarity with characterizations of the dark sector that influences the deep-space dynamics of the visible sector while it is seemingly not encoded in the visible dimensions.

Through AI, we learn that ur-matter comes in six "shades" according to the symmetries that arise as the fundamental relativistic equation in natural units: $k^2 + k_0^2 = \omega^2, (k_0 = \hbar / r_0)$, is factorized in 2×2 matrix form : $\pm I_2 \omega = \alpha k + \beta k_0$, where α, β are any two different Pauli matrices. The symmetry of one of these shades yields a gauge symmetry arising from the inclusion of the photon field via the covariant derivatives, as shown in Chapter 2. Furthermore, this shade becomes the precursor of mass endowment in the visible sector that retains electroweak symmetry, as the shade symmetry is compatible with that of the symmetry-breaking Higgs scalar field.

On the other hand, the remaining five symmetries resulting from alternative factorizations were disregarded at the time when quantum field theory was formulated. Those lost symmetries have now become highly relevant as dark matter has entered the physical picture. The five shades of ur-matter that cannot become gauge invariant through amalgamation with the EM vector potential (via covariant derivative) thus constitute the established components of the dark sector. The lack of interaction with the photon due to the symmetry incompatibility renders dark matter obviously invisible, while its ratio 5:1 relative to visible matter is in good agreement with the observation that dark matter constitutes approximately 85% of all matter (Chapter 1).

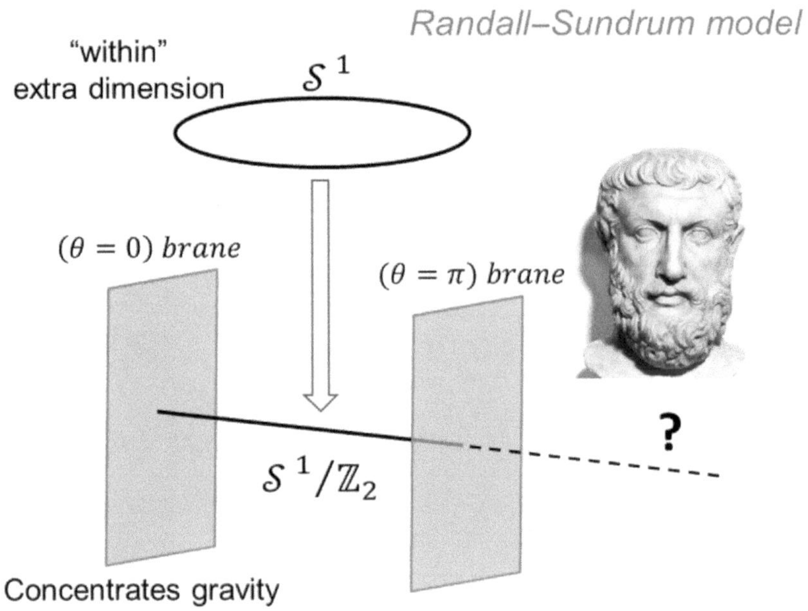

FIGURE 6.1 Ontological inconsistency in proposals of a higher dimensional mechanism with specified boundary conditions adopted to solve the hierarchy problem, such as the Randall–Sundrum theory of gravity. This theory defines the fifth dimension as a quotient space of the circle defined via the identification of angular coordinates $\theta, -\theta$. Thus, the metric is completely specified by the values in the finite interval $[0, \pi]$, yielding a universe with boundaries defined by the branes at the extremes of the interval. This leads to an ontological inconsistency because there are no coordinates beyond the boundaries to provide meaning to "outside". The picture contains the effigy of Parmenides, founder of ontology (Wikimedia Commons, public domain).

To determine their external local symmetries, ur-particles are spatially represented as geometric dilution of the ur-matter shade via the pitch angle α, so that the angled cylindrical cross section becomes deformed to a circle with radius $r_0/\cos\alpha$. Spatial local symmetries can be constructed to warrant the conservation of the geometric dilution $\upsilon = -\log\cos\alpha$ of the ur-Higgs particle mass (~ 246 GeV/c^2) in accordance with Noether's theorem. The symmetry groups for these symmetries are $SO(3)/\mathbb{Z}_2 \approx SU(2)$ and $U(1)$, the complex rotation group.

The full external symmetry associated with conservation of geometric dilution of ur-matter in the 5D space–time is then $SU(2)\times U(1)$. At the phase transition at critical temperature $T = k_0/k_B$, this local external symmetry in 5D space–time W gets internalized as local gauge symmetry associated with the electroweak unification upon the projection $\pi: W \to W/\sim$ onto the quotient manifold that represents the 4D space–time (Figure 6.2).

The autoencoder projection has the net effect of gauging the external $SU(2)\times U(1)$ ur-symmetry, as it lumps 5D points into equivalence classes modulo the dark coordinate. However, the symmetry is restored, resurfacing in a different guise, i.e., as internal local gauge symmetry for the electroweak boson fields. We may state that the "primeval symmetry" of ur-matter gives birth to the electromagnetic and weak force charged under the internalization of the external ur-symmetry.

This analysis sets the tone for approaching the evolution of the universe by identifying the critical juncture at which a phase transition activates the ur-boson associated with the $SU(2)\times U(1)$-symmetry. At critical temperature $T_0 = 246$ GeV$/k_B = 2.85\times 10^{15}$ K, a phase transition occurs, where the compatible shade of ur-matter becomes activated and is able to mix and thereby endow the Higgs field ϕ_H with its true vacuum (TV), as described by Eq. 2.15. This activation of the ur-Higgs scalar is reflected in a kinetic energy commensurate with the rest–mass energy 246 GeV, making it

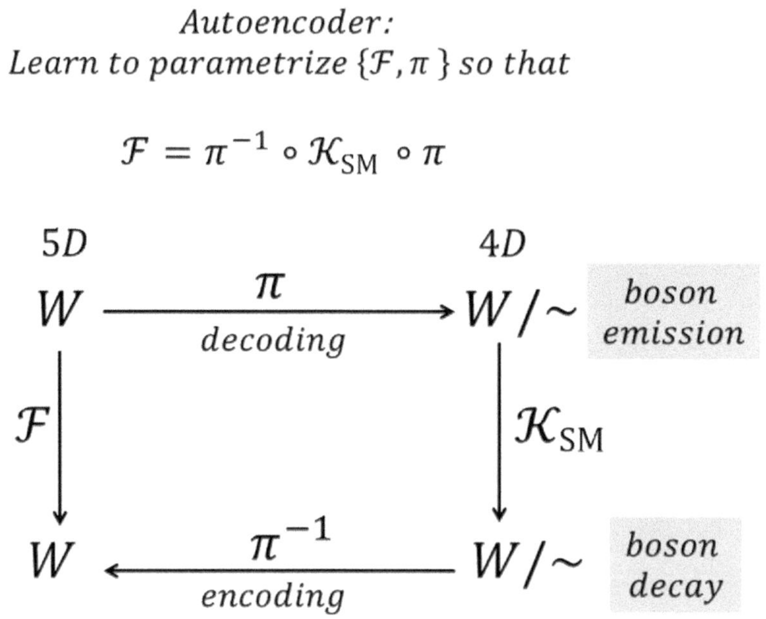

FIGURE 6.2 The autoencoder trained and specialized to distill the standard model from a 5D space–time by correlating events involving emission and decay of bosons. The autoencoder becomes optimized when its parametrization makes the diagram commutative, representing the influence of the dark sector encoded in an extra spatial dimension on the dynamics of the visible sector under the standard model.

meaningful to factorize the term $\left(k^2 + k_0^2\right)$ with a choice of Pauli matrices that yields a symmetry compatible with the gauge electroweak symmetry.

The scalar ur-particles in 5D *W* become endowed with spin, required to internalize the local spatial symmetry of ur-matter as gauge electroweak symmetry within the visible sector. Furthermore, the phase transition yields the symmetry breaking associated with endowing the Higgs field with the stable TV at 246 GeV. Thus, the field excitations in the true vacuum yield the particle masses, while the latent heat associated with the phase transition can be assimilated to the dark energy, estimated as $\Delta Q_{\mathrm{H}} = V\left(\phi_{\mathrm{H}} = 0\right) - V\left(\phi_{\mathrm{H}}^{\mathrm{TV}}\right)$, where V is the potential energy of the Higgs field.

This scheme for inducing mass in elementary particles is inspired by the ideas of Philip W. Anderson on superconductivity (Figure 6.3). In this context, the symmetry breaking associated with the formation of the organized phase at $T < T_c$ creates a stable vacuum responsible for endowing the EM force carrier with mass, thus "slowing down the photon" with the net effect of excluding the QED field lines for the superconducting material.

A crucial result yielded by the autoencoder is the observation that the Higgs field assumed to endow bosons with mass while retaining gauge symmetry is a provisional *ansatz* that needs to be revised vis-à-vis the incorporation of the extra spatial dimension. The autoencoder treats the quantum vacuum as the vehicle for ur-matter (UM), that is, the precursor of matter. There are six ways (shades) of encoding the scalar field spanning the fifth dimension with momentum k_0 according to the symmetries resulting from the six modes of factorization of the special relativistic relation $k^2 + k_0^2 = \omega^2$. Five of these shades yield dark matter proper (DM), while a sixth shade yields the ur-matter (UM) for the visible sector of the SM. This is so because the ur-field is capable of amalgamating with the Higgs scalar field ϕ_{H} since they are symmetry-compatible (Eq. 2.15), thus yielding the true vacuum (TV) of the Higgs field at 246 GeV. Mass in the visible sector originates

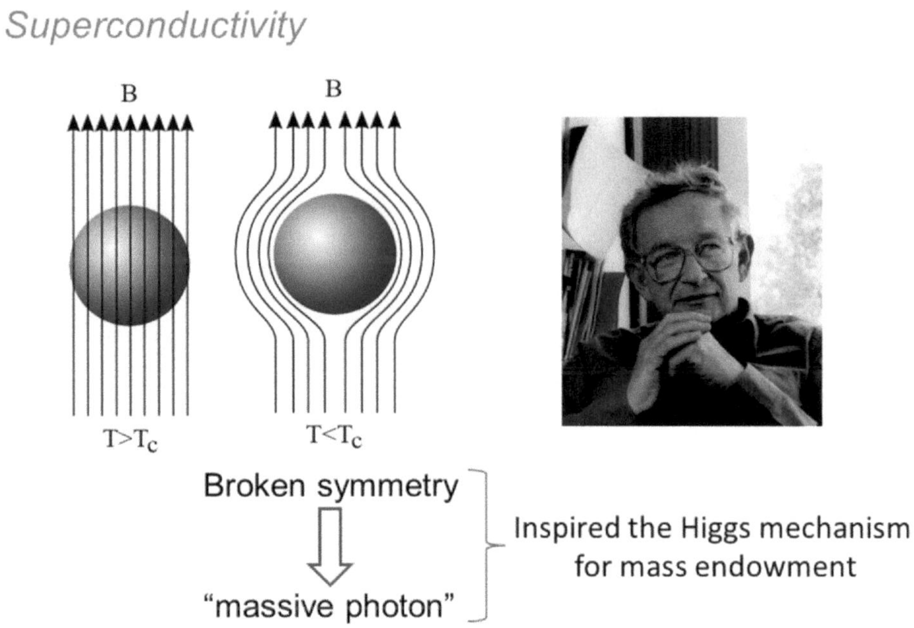

FIGURE 6.3 Inspiration for the Higgs mechanism to induce mass in elementary particles, provided by research of P. W. Anderson on superconductivity. In this context, the symmetry breaking associated with the formation of the organized phase at $T < T_c$ (T_c = Curie temperature) creates a stable vacuum responsible for endowing the photon (carrier of electromagnetism) with mass. This "slowing down" has the net effect of excluding the QED field lines from the superconducting material. The picture of Philip W. Anderson is in the public domain (Wikimedia Commons).

through a symmetry breaking (SB) phase transition induced by the activation of $\phi_{\mathrm{H}}^{\mathrm{TV}}$. In general, criticality is thus parametrized by the temperature at which ur-matter acquires entity as a particle, that is, when its kinetic energy is commensurate with its rest–mass energy. The autoencoder thus informs that mass only *appears* to originate from the activation of the Higgs mechanism. Actually, it is the geometric dilution of ur-matter as ur-particles that serves as precursor to the mass endowment for the corresponding particles in the visible sector. These visible particles internalize the spatial symmetry of their respective ur-particles as gauge symmetries, thus acquiring spin, whereas the ur-particles are scalar.

Rather than assuming the existence of the all-pervasive scalar field that endows particles with induced mass once it has adopted its stable true vacuum through self-interaction, AI postulates that the quantum vacuum itself has intrinsic materiality spanning a "within" dimension. This dimension stores rest–mass energy that, when endowed with the proper symmetry (Eq. 2.15), gets communicated to the Higgs field as its stable true vacuum. This result appears to represent a substantive way forward as it portends to reconcile the SM with the standing hierarchy problem, the cosmological constant problem, the vacuum catastrophe, and the influence of dark matter and dark energy on deep-space dynamics.

Furthermore, according to the autoencoder scheme (Figure 6.2), the electroweak unification is essentially enabled because the ur-symmetry group $U(1)$, under whose gauging QED gets charged, is actually a subgroup of $SU(2)$, the ur-symmetry group under whose gauging the bosons carrying the weak force get charged. The inclusion relation is actually articulated via the isomorphism $U(1) \approx \widehat{U(1)} \subset SU(2)$ described by Eq. 2.53.

Strikingly, the electroweak unification can be extended further to encompass gravity itself. This "theory of everything" may materialize under the same autoencoder scheme (Figure 6.2) because the Lorentz group ($SL(2,\mathbb{C})$), under whose generalized gauging gravity is charged (Figure 2.20), includes both the EM and weak-force ur-symmetry groups as subgroups in accordance with: $U(1) \approx \widehat{U(1)} \subset SU(2) \subset SL(2,\mathbb{C})$. Thus, gravity may be treated within a quantized ultra-unification scheme since its ur-symmetry includes the ur-symmetry of the weak force and electromagnetism. In other words, the ur-symmetry of gravity subsumes the local external symmetry of the lightest ur-matter particle that enables mixing of its scalar field with the Higgs field endowing the latter with its true vacuum (Eq. 2.15). This ultra-unification substantiates the curvature-mass duality of general relativity, an assertion based on two observations: (a) the generalized gauging of the ur-symmetry of gravity charges an extended object, the spin holonomy, spanning an arbitrary contour γ that subsumes the curvature of space–time via the transport map P_γ for the spin connection (Eq. 2.51). This holonomic boson measures a chiral cosmic string defect induced by a symmetry operator $U_v, v \in Z(SL(2,\mathbb{C}))$, acting on it. (b) The ur-symmetry of gravity subsumes the symmetries under whose gauging the Higgs boson gets charged, enabling the mass endowment for particles in the visible sector. This leads us to articulate the theory of everything enshrined in the following statement:

The relativistic duality of mass distribution and space–time curvature materializes through two symmetry-interrelated phase transitions taking place respectively at 10^{-43} s and 10^{-27} s after the big bang. First, the ur- $SL(2,\mathbb{C})$-symmetry is internalized or gauged as a generalized one-form symmetry, so that gravity gets charged as a curvature-associated object. This phase transition is followed by the zero-form gauging of the electroweak ur-symmetry at 10^{-27}s. As just shown, this latter symmetry is subsumed into the gravity ur- $SL(2,\mathbb{C})$ symmetry and induces mass on visible-sector particles by bestowing the stable true vacuum on the scalar Higgs field. Since gravity becomes a quantum force under the generalized gauge symmetry, the ultra-unification implies the mass–gravity duality and suggests that the fundamental forces in nature stem from one primeval force present at the Planck epoch. This force got diluted differently into the geometry of the universe, following the gauge symmetries and gauge generalized symmetries that charged the fundamental force carriers activated at successive phase transitions in the aftermath of creation.

6.2 IDENTIFICATION OF THE PRIMEVAL FORCE

The analysis in Chapter 2 revisited above suggests the existence of a primeval force whose symmetry-related modes of geometric dilution are likely to give rise to the four fundamental forces. It is clear that, prior to the phase transition at 10^{-43} s after the big bang, the primeval force is communicated by an ur-boson associated with $SL(2,\mathbb{C})$ symmetry and carrying momentum nk_0 with $n \sim \mathcal{O}(10^{13})$. More precisely, the parameter n satisfies: $\frac{nk_0}{c} \sim m_P, n = \left[\sqrt{\hbar c^3 / k_0^2 G}\right]([.] = \text{integer part})$. According to the autoencoder dynamics of the universe evolution (Figure 6.2), this ur-boson gets transferred to the visible sector under a concomitant one-form gauging of the $SL(2,\mathbb{C})$-symmetry at $T = T_P$. Taking place during the Planck epoch, this event marks the birth of what we may aptly call "quantum gravity".

As it surfaces within the visible sector as an extended (nonlocal) object, the primeval boson becomes describable in terms of a generalized gauge symmetry. The primeval boson measures space–time curvature, since it is represented as the spin holonomy over a given contour γ defined by $\mathcal{W}_s(\gamma) = tr_s P_\gamma \exp\left(\oint_\gamma A.ds\right)$, where P_γ is the path-ordering operator (Figure 6.4) associated with the spin connection for an irreducible representation s of $SL(2,\mathbb{C})$. The parallel transport map is defined over the tangent bundle for the space–time assumed to have become a Riemannian manifold

$$\mathcal{W}_S(\gamma) = tr_S P_\gamma \exp(\oint_\gamma A \,.\, ds)$$

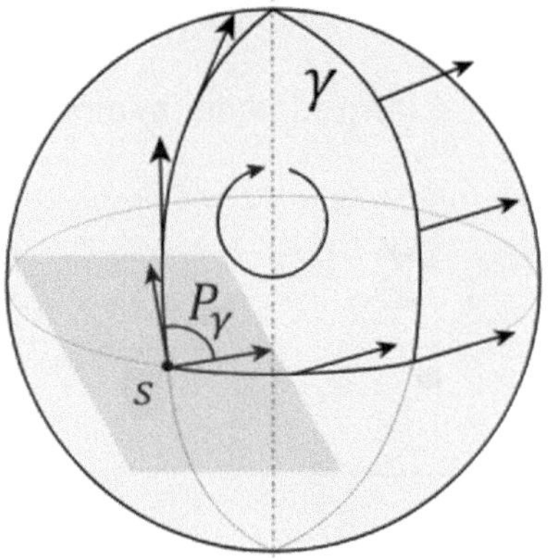

FIGURE 6.4 Representation of the primeval force carrier in the universe, charged 10^{-43} s after the big bang under the gauged generalized $SL(2,\mathbb{C})$ symmetry and represented as a spin holonomy measuring the curvature of the universe, with curvature radius estimated at $\sim \sqrt{\hbar G/c^3} \sim 10^{-35}$ m .

at the Planck criticality, while the integrand A defines a one-form EM gauge connection with vector potential $A^a = A^a_\mu dx^\mu$. Since the curvature of space–time is factored into the parallel transport map, the holonomic boson provides a measure of curvature (Figure 6.4), signaling a convergence between quantum mechanics and general relativity. In more precise terms, the holonomic boson is charged under the symmetry operator $U_\nu, \nu \in Z(SL(2,\mathbb{C}))$ that represents a chiral cosmic string defect which is measured as quantized induced tension by $\mathcal{W}_s(\gamma)$, given the spinor-bundle connection that defines the holonomic boson (Figure 6.5).

In this way, the relativistic duality gravity-curvature is shown to be enshrined in the quantum gravity charged under the generalized gauge symmetry. Estimated at nk_0 , $n \sim \mathcal{O}(10^{13})$, the rest–mass momentum of the primeval force carrier yields an extreme curvature radius estimated at $r_C \sim 10^{-35}$ m . The conversion of this extreme curvature into the universe mass took place as the universe cooled down to $\frac{k_0 c}{k_B} = 2.85 \times 10^{15}$ K , when the ur-Higgs boson, lighter by a factor of 10^{-13} , became activated under a symmetry included within that of the primeval ur-boson.

Given the nature of the primeval force carrier as a holonomic boson charged under the generalized gauge symmetry (Figure 6.5), we may state that the primeval force is of geometric origin, essentially provided by the extreme curvature of the earliest nonsingular universe once it became supported by a space–time endowed with a spin connection.

In this context, the theory of everything involved recasting relativity as a gauge theory of the local Lorentz group, so that a chiral defect induced by a primeval symmetry transformation is directly measured as quantized tension by the primeval holonomic boson.

To fully grasp the purview of the theory of everything, we note that the inflation along the dark dimension may be regarded as a proxy for the universe evolution (Figure 6.6). This is so because the radius of the circular extra dimension determines the masses of the ur-bosons whose symmetries become gauged at the crucial phase transitions that determine the universe evolution. This fact prompts us to posit that the quantum force carriers are actually relics of the holonomic boson (Figure 6.6). The generic contour of this primeval boson projects onto the dark dimension that, at

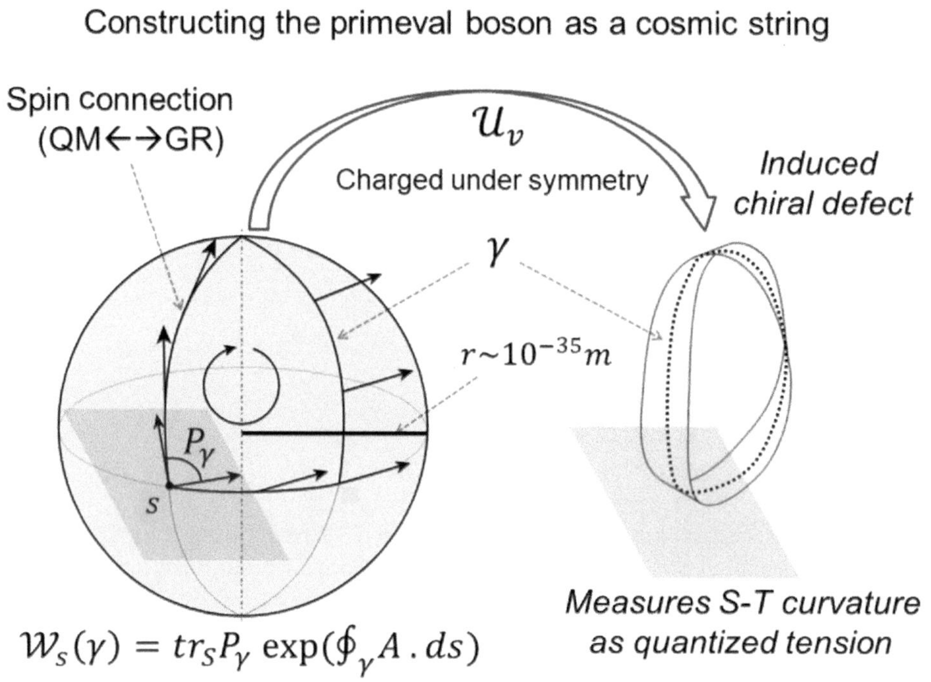

FIGURE 6.5 Charging of the primeval boson, recasting relativity as a gauge theory of the local Lorentz group. The holonomic boson is charged under the symmetry operator $U_v, v \in Z(SL(2,\mathbb{C}))$, that induces a chiral cosmic string defect which is captured as quantized tension by the parallel transport map associated with the spinor bundle connection in $\mathcal{W}_s(\gamma)$.

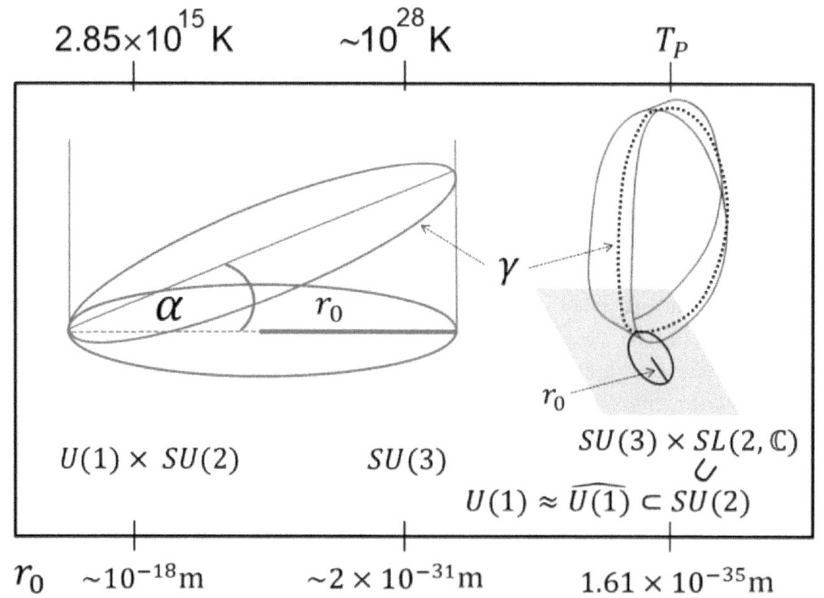

FIGURE 6.6 Carriers of quantum forces as relics of the holonomic boson that stores gravity along the dark dimension. The conceptual relationship is enabled by the fact that the radius of the dark dimension functions as a proxy for the universe evolution.

time of Planck's criticality, has a radius $r_0 = \sqrt{\hbar G/c^3}$, precisely the curvature radius of the space–time computed by the holonomy defined by the contour. Thus, *gravity is stored in the dark dimension during the Planck epoch*, in striking analogy with the storage of the true vacuum for the Higgs field in the dark dimension at the electroweak epoch. The Higgs boson, with its geometric dilution of the true vacuum, becomes a relic of the cosmic string contour that defines the primeval holonomic boson or carrier of the gravity stored in the dark dimension (Figure 6.6). This assertion pivots on two facts: (a) The internal symmetry of the Higgs field is subsumed by the generalized symmetry of the holonomic boson, and (b) the mass–gravity duality of general relativity conceptually links the Higgs boson and the holonomic boson.

In this guise of the theory of everything, all four forces become manifestations of a primeval force under different symmetries, and therefore, the quantum bosons become relics of the holonomic boson, as depicted in Figure 6.6.

6.3 LARGE LANGUAGE MODEL FOR QUANTUM GRAVITY IN THE UNCHARTED ULTRAVIOLET COMPLETION

The theory of everything (ToE) cannot be considered complete without showing that the graviton, an excitation of a quantized gravitational field, originates in the primeval boson previously described in the strong field limit as a gauge representation of Lorentz transformations. Thus, the ToE must build upon the tetrad frame fields introduced to construct a gauge theory of gravity (Chapter 2). In the weak field limit (WFL), a perturbative graviton with Minkowski background yields a massless spin-2 particle with $U(1)$ gauge symmetry, consistent with the infinite range of gravity communicated at the speed of light. This section describes AI's take on the uncharted strong field limit (SFL), clearly not amenable to a perturbative treatment.

While the WFL appears tractable using standard techniques by regarding the graviton tensor field as a perturbation of the Minkowski tensor, the SFL could be better grasped with the deployment of a large language model (LLM) within a functional language program (Lean). This requires an LLM for a fiber-bundle formulation of a gauge theory pivoting on the spin connection that determines the tangent bundle of a space–time endowed with an extra warped dimension. In this way, this section describes a predictive differential geometry model of the graviton as transducer of space–time curvature into energy stored in a warped dimension (Chapter 2). The curvature–energy transduction becomes operative as the boson begets curvature in the dualistic relativistic framework, implying that the stress-energy tensor must be associated with a scalar describing self-interacting DM within a ϕ^4 model.

A warped spatial dimension needs to be incorporated to address the problematic geometric dilution of mass and gravity, as shown in Chapter 2. This dilution provides an explanation of the incommensurably small coupling strength of gravity relative to the other fundamental forces of established quantum origin. Significantly, the minimum rest–mass energy stored in the warped dimension is $E_0 = k_0 c = \hbar c / r_0 = 246\ \text{GeV}$, the true vacuum of the Higgs field. Thus, the stress-energy tensor in a fiber-bundle gauge theory of the graviton will be shown to introduce a mass-endowing mechanism via the curvature–mass relativistic duality modeled by a self-coupling potential energy term. We anticipate that this ϕ^4 -model yields massive gravitons vis-à-vis the vacuum expectation value (v.e.v.) of a Wilson loop representation of the graviton in the SFL. It should be noted that such a characterization cannot be inferred with the assistance of Lean by itself but requires a LLM operating within Lean.

The geometric dilution parameter υ plays a role akin to the scale and amplitude of the de Broglie wavelength inhomogeneity in the dark dimension in models representing cosmological gravitons as dark matter. Thus, the decay of KK-gravitons into lighter gravitons, a hallmark of cosmological evolution in warped reality, plays out in a similar way in here, as $\alpha \rightarrow \frac{\pi}{2}$ or, equivalently, $\upsilon \rightarrow \infty$, reckoning that geometric dilution serves as proxy of time in cosmological evolution (Chapters 2 and 4).

To account for quantum gravity in the SFL, a fiber-bundle formulation of its gauge theory is implemented within a functional language program operating a LLM (Appendix). Thus, spinors are attached to the general relativity framework provided by the space–time curved manifold $\mathcal{M}$. This procedure is enabled by first tending the spin connection across the tangent-bundle $T\mathcal{M}$. This connection is implemented via a frame field required to piece together the atlas $\{T_x\mathcal{M}\}_{x\in\mathcal{M}}$ of local tangent spaces and relates to the geometry of the curved manifold in such a way that a tangent-space Lorentz transformation translates onto a gauge transformation of the spinor field. To enable interconversion between the metric tensor $g_{\mu\nu}$ of $\mathcal{M}$ and the flat (Minkowski) metric tensor η_{ab} of $T\mathcal{M}$, a pentad frame field $\{e_\mu^a\}$ (Greek indices for curved space tensor and Latin isospin indices for flat tangent-space Lorentz indices). The spin connection w_μ^{ab} is then expressed in terms of pentadic field elements and their derivatives:

$$w_\mu^{ab} = \frac{1}{2}e^{\nu a}\left(\partial_\mu e_\nu^b - \partial_\nu e_\mu^b\right) - \frac{1}{2}e^{\nu b}\left(\partial_\mu e_\nu^a - \partial_\nu e_\mu^a\right) - \frac{1}{2}e^{\lambda a}e^{\sigma b}$$
$$\left(\partial_\lambda e_{\sigma c} - \partial_\sigma e_{\lambda c}\right)e_\mu^c \tag{6.1}$$

Equipped with this representation, a fiber-bundle formulation of the gauge theory of gravity in the SFL is formalized within a LLM based on the pentadic framework.

The AI-empowered inferential scheme to model the graviton requires a differential geometry formalization supporting two interplaying components: (a) a theorem/proposition prover such as Lean 4 with an intensional logic under the guise of a dependent type theory, and (b) a generalized autoencoder LLM serving as the supporting inferential framework within which the proof assistant operates by maintaining compatibility between general relativity and its gauged version on the tangent bundle of space–time.

To chart the space–time manifold adopting a minimal covering $\mathcal{M} \subseteq \bigcup_{x\in C(\mathcal{M})} \mathcal{O}(x)$, the formalization provided in the Appendix is instrumental since the set of charts $\{T_x\mathcal{M}\}_{x\in C(\mathcal{M})}$ is incorporated as data synchronized with mathlib, and hence so are the denumerable diffeormorphisms: $\mathcal{O}(x) \rightarrow T_x\mathcal{M}$ $\forall x \in C(\mathcal{M})$. The next step is to specialize the fiber-bundle formalization to the warped graviton within a gauge theory of gravity. The overarching inductive type is introduced as represented by the commutative constructive arrow diagram in Fig. 6.7, whereby every proposition formulated at the level of the atlas $\bigcup_{x\in C(\mathcal{M})} T_x\mathcal{M}$ may be decoded at the level of the covering $\bigcup_{x\in C(\mathcal{M})} \mathcal{O}(x)$.

If the functional program language is informed declaratively that the massive graviton with tensor field $\mathcal{Y}^{\mu\nu}$ is associated with a scalar field $\phi = \mathcal{Y}^{\mu\nu}g_{\mu\nu}$ governed by the quartic self-coupling potential $V(\phi) = -\frac{1}{2}m^2\bar{\phi}\phi + \frac{\lambda}{4}(\bar{\phi}\phi)^2$, the operational inductive type enshrined in the LLM infers that (Figure 6.7):

- The scalar field is nonlocal, that is, its support is made of contour lines, not points in $\mathcal{M}$,
- $\phi = \mathcal{W}$, where $\mathcal{W}$ is a Wilson loop that holonomically transduces space–time curvature.

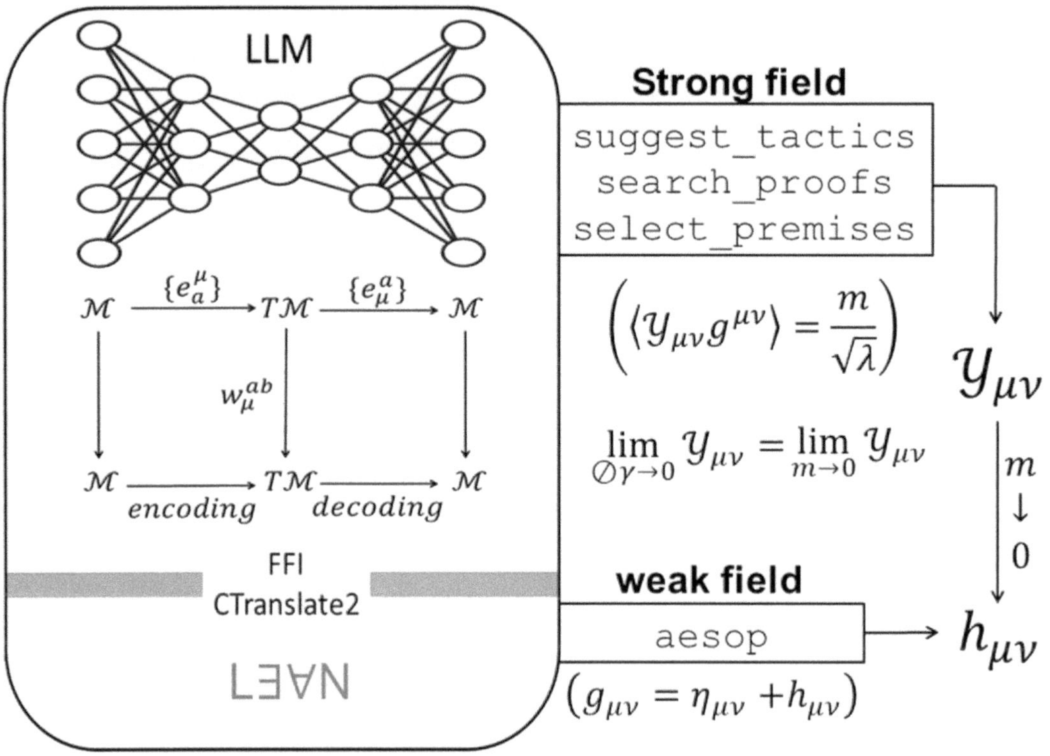

FIGURE 6.7 Architecture of a Lean copilot incorporating an autoencoder LLM specialized for the discovery of the warped graviton in the strong field limit. The LLM inferences are run within Lean with supporting framework provided via a shared CTranslate2 C++ library at Lean's foreign function interface (FFI).

- The warped graviton is massive since its vacuum expectation value $\mathcal{W}$ is the nonzero ϕ-value realizing the minimum of $V(\phi)$: $\langle \mathcal{W} \rangle = m/\sqrt{\lambda}$ ($\lambda =$ self-coupling parameter, $m =$ graviton mass). These parameters determine the relevant inductive data types
- The chiral symmetry breaking that bestows mass to the warped graviton is thus informed by the potential $V(\mathcal{W})$.

As inferred by Lean's Aesop in the WFL, the warped graviton tensor field $h_{\mu\nu}$ is the one assigned to a massless spin-2 particle with U(1) gauge symmetry ($h_{\mu\nu} \to h_{\mu\nu} + \partial_\mu a_\nu + \partial_\nu a_\mu$). This graviton is treated perturbatively on a Minkowski background without the need for the LLM, so that $g_{\mu\nu} = \eta_{\mu\nu} + h_{\mu\nu}$.

To identify the warped graviton tensor field $\mathcal{Y}_{\mu\nu}$ in the SFL, the LLM represents the curvature with Clifford algebra generators that transform under the spinor winding "electromagnetic" $U(1)$-symmetry elements $U_\xi, \xi \in [0, 2\pi]$ in the manner of the electromagnetic vector potential A_μ. The warped graviton field is implemented in such a way that $\mathcal{Y}_{\mu\nu}$ may be charged under a symmetry transformation $U_\zeta, \zeta = \pi$, associated with the nontrivial idempotent element $U_\zeta \in Z(SL(2, \mathbb{C})) \approx \mathbb{Z}_2$ in the center of the Lorentz group. This chiral-defect-inducing symmetry effectively mounts the contour of the line operator onto a Moebius strip creating a chiral defect that amounts to a π-winding deficit in the spinors attached to space–time, while two such deficits created by the U_ζ iteration $U_\zeta^2 = U_{2\pi}$ amount to the identity acting on the parallel transport map.

From these selected premises, the LLM constructs a Wilson loop $\mathcal{W} = \phi = \mathcal{Y}^{\mu\nu} g_{\mu\nu}$, so that space–time curvature may be holonomically transduced as rest mass. Thus, the following compatibility relation is established, implying that the massive warped graviton indeed belongs to the ultraviolet completion of gravity, as verified by Lean's Aesop:

$$\lim_{\oslash\gamma\to 0} \mathcal{Y}_{\mu\nu} = h_{\mu\nu} \tag{6.2}$$

for $\gamma:[0,1]\to\mathcal{M}$, a generic closed map upon which the holonomy is defined. Thus, the warped graviton tensor field in the SFL is identified by the autoencoder LLM of the gauge theory (Figure 6.7) as

$$\mathcal{Y}^{\mu\nu} = \mathcal{P}_\gamma \exp\left(\oint_\gamma w_\lambda^{ab} e_a^\mu e_b^\nu dx^\lambda\right), \tag{6.3}$$

where $\mathcal{P}_\gamma$ is the path-ordering operator associated with the contour γ.

Thus, the massive warped graviton is represented as a ϕ^4-self-coupled transducer of curvature in the ultraviolet completion of gravity. Furthermore, the $U(1)$ gauge symmetry in the WFL translates into the generalized (chiral) gauge symmetry concomitantly with the transition to the SFL.

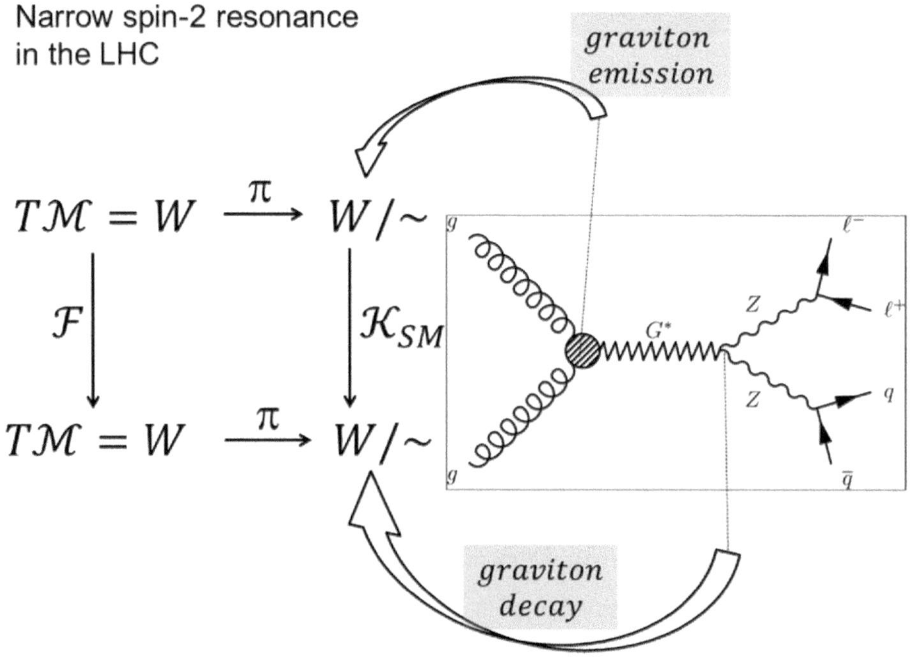

FIGURE 6.8 Warped graviton generation and detection in a LHC (Chapter 2) as approached by the autoencoder LLM operating within the formal proof assistant.

To identify the warped graviton as a spin-holonomy transducer of space–time curvature, the autoencoder LLM builds upon a pentadic formalism that casts gravity in gauge theoretic terms (Figure 6.7). The graviton tensor field contraction is thus identified in the SFL as the Wilson loop with a v.e.v defined by a self-interacting DM scalar ϕ incorporated in the stress-energy tensor contracted as $T(\phi)$. The warped graviton is thus massive and charged under a symmetry represented by inducing a chiral defect in the guise of a winding deficiency in the spinors attached to the relativistic framework. By contrast, in the WFL, the graviton becomes massless and treated directly by Lean itself as a perturbation of the Minkowski background (Figure 6.7).

As shown in Chapter 2, the warped graviton in the SFL may be produced and detected in the laboratory as a boson that transduces the space–time curvature at its emission cross section (Figure 6.8) into a geometrically diluted warped dimension that stores rest–mass energy via relativistically operational self-interaction. The curvature-transducing mechanism materializes via the Wilson loop. As discussed in Chapter 2, the generation and detection of the warped graviton invites probing regions in parameter space hitherto unexplored at the LHC or future colliders (Figure 6.8).

Since warped gravitons are likely DM candidates, and self-interactivity is evidently inherent to the graviton in the SFL, we can expect that at least a portion of DM in the universe should be self-interacting, as argued in Chapter 2. These assertions have obvious cosmological implications. The DM self-interactions would allow energy and momentum transport through the galaxy halos, creating structural and dynamical patterns fundamentally different from those produced by the hitherto assumed collisionless DM. These results are likely to provide a new vantage point to interpret the diversity of galactic patterns and other cosmological phenomenology associated with DM.

6.4 AI'S ATTEMPT AT THE "GRAVITIZATION OF QUANTUM MECHANICS"

The large language model in the guise of an autoencoder, introduced to provide the physical underpinnings of quantum gravity (Appendix), has not been fully exploited as of yet. In fact, only half of the AI system has been dealt with so far. The quantization of gravity via the frame-field formalism has been brought to fruition in Chapter 2 and in the preceding section. However, the possibilities of the decoding side of the autoencoder, whereby quantum mechanics may be "gravitized", have not been discussed at any level. This is in part due to the wanton difficulties introduced by certain inconsistencies arising in the theory of quantum mechanics, specifically by the problem of the collapse of the wave function due to measurement. Such inconsistencies cannot be carried over to the relativistic arena, where they find no counterpart. Before dwelling on this problem, we would need to explain why it would make any sense to gravitize quantum mechanics in the first place.

Rather than attempting to quantize general relativity, this functional mode of the autoencoder attempts to bring quantum theory in line with Einstein's theory of gravity. Why are we turning the tables around rather than advancing the agenda of quantum gravity? People tend to regard quantum theory as more fundamental than general relativity. Quantum mechanics finds no contradictory or falsifying evidence from experiment or observation and the theory is so firmly established that it is believed the whole of physics should be brought under its aegis. Yet, the general theory of relativity is also a fundamental scheme with no confirmed experiments or observations telling against it. Of course this is so provided that we comply with Einstein's inclusion of the cosmological constant, which appears to be needed to explain the deep space phenomenology (Chapters 1 and 2). So why give quantum theory the upper hand in the grand unification? Surely there are far more phenomena that require quantum mechanics to be explained than those that call for general relativity for an explanation. Additionally, since quantum mechanics deals with very small objects, while general relativity deals with large objects, and big things are made up of small things, then quantum mechanics, the theory of the small things, must be the more fundamental of the two. However, there is another issue that should be deemed more important than the matter of scale, namely, the

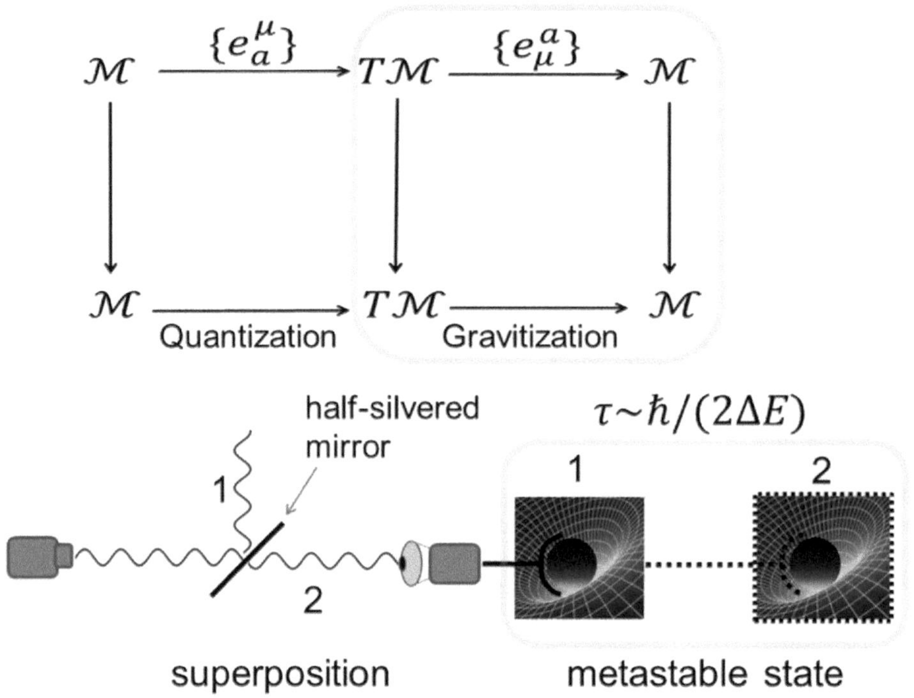

FIGURE 6.9 Correspondence between a superposition of quantum states and a relativistic metastable state with a finite lifetime arising in the "gravitization of quantum mechanics".

consistency of the theory. In that realm, Einstein's theory gets the pride of place. Thus, the imperative of "gravitizing quantum mechanics" could be said to be justified, at least in part.

Be as it may, the gravitization of quantum mechanics finds significant hurdles, ever apparent in the reversal of the Palatini formalism for quantizing gravity. The main hurdle arises precisely from the apparent or perceived inconsistency of quantum theory, whereby the participation of the observer triggers the collapse of the wave function representing a superposition of quantum states. Thus, to reverse the frame-field *vierbein* formalism in the decoding mode of the autoencoder (Appendix), we would need to describe the relativistic counterpart of a quantum state superposition.

To come to grips with the problem, let us resort to a *gedankenexperiment* where a superposition of quantum states is associated with a significant mass displacement (Figure 6.9). This situation may be realized by a photon that may either get reflected (state 1) or refracted (state 2) at a half-silvered mirror, so that in the latter case and only in that case, a photon detector activates a mechanical device that displaces a mass from its original state 1 to state 2 (Figure 6.9). Without an observer's intervention, the wave function describes a superposition of states 1 and 2, which should correspond to a metastable state in relativistic terms consisting of the superposition of two space–times (1 and 2, respectively) associated with the end points of the mass displacement. Crucially, the relativistic metastable state is ephemeral and its lifetime is computed through the uncertainty principle as $\tau \sim \hbar/(2\Delta E)$, where $\Delta E > 0$ is the reversible gravitational work associated with the space–time transformation $1 \rightarrow 2$.

The previous discussion makes it clear that the main hurdle in a theory of everything based on a gravitization of quantum mechanics arises from the fact that the latter theory involves the observer in its formulation, while general relativity does not. Thus, a reversal of roles in the quantum gravity program would require taking into account relativistic metastable states made up of space–time combinations whose lifetimes should be computable based on the uncertainty principle of quantum mechanics. A relativistic metastable state corresponds to a superposition of quantum states enshrined in a wave function that collapses with the participation of an observer.

BIBLIOGRAPHY

Anastasiou C, Duhr C, Dulat F, Furlan E, Gehrmann T, Herzog F, Lazopoulos A, Mistlberger B (2016) High precision determination of the gluon fusion Higgs boson cross-section at the LHC. *J High Energy Phys* 2016: 58.

Avigad J (2023) Mathematics and the Formal Turn. arXiv:2311.00007.

Basu R, Mandal T (2013) Graviton signals in central production at the LHC. *Adv High Energy Phys* 2013: 652714.

Dissertori G (2016) *The Standard Theory of Particle Physics. Advanced Series on Directions in High Energy Physics*. Maiani L, Ronaldi L (Eds.), Chapter 6, pp. 113–128. World Scientific Publishing, Singapore.

Klimek P (2021) Searches for exotic resonances at LHC. CERN ATL-PHYS-Proc-2021-015.

Moura L, Ullrich S (2021) The Lean 4 Theorem Prover and Programming Language. *Automated Deduction – CADE 28*. Platzer A, Sutcliffe G (Eds.), pp. 625–635. Springer International Publishing, Switzerland.

7 Physical Footprints of a *De Novo* Simulated Universe

"With relief, with humiliation, with terror, he understood that he was also an illusion, that someone else was dreaming him."

The Circular Ruins
Jorge Luis Borges

In this chapter, we critically examine physical footprints that support the conjecture that the universe is a computer simulation and has been simulated *de novo*. The physical evidence supporting this scenario is shown to be enshrined in the very laws of physics. The simulation hypothesis is shown to be pivotally dependent on the possibility of creating conscience as an autoencoder of the wave function collapse.

7.1. REVISITING THE SIMULATION HYPOTHESIS

As shown in Chapter 5, an assemblage of two replicas of an autoencoder may be deployed to recreate a primeval nonorientable wormhole fueled by dark energy and purportedly responsible for the dynamic equilibrium between dark and visible matter. Furthermore, following the dictates of statistical mechanics, we can even conferred physical reality to the AI system [1], much along the lines of Wheeler's dictum "*it from bit*". These forays into mathematical cosmology empowered by AI prompt the daring hypothesis that the universe is a simulation, already put forth by the philosopher Nick Bostrom [2].

The Bostrom hypothesis has been assigned a striking 50:50 chance of being true, but the science put forth in Chapter 5 will surely slant this ratio toward a more favorable probability. The hypothesis has been also subject to derision, since it was deemed to be a nonscientific idea, incapable of yielding a falsifiable prediction. This is surely incorrect in light of the cosmological implications of the AI-based metamodel of the universe described in Chapter 5 and the AI-empowered technology for universe reproduction implemented in Chapter 3. There are plenty of physical tests to prod the simulated universe generated by the annular holographic autoencoder, especially probing the two-stroke engine that sustains the primeval wormhole and the quantum physical constants arising thereof. If nothing else, the simulation scenario has passed the ultimate test by providing a provisional solution to the cosmological constant (vacuum catastrophe) problem.

The idea of a simulated universe, today attributed to a philosopher, finds a precursor in the metaphysical musings of Hinduism, where the deity Vishnu is often depicted as *Anantasayana* (i.e., lying on the serpent *Ananta*) in order to dream the universe into reality (Figure 7.1). Furthermore, in manifold guise, the idea also belonged *mutatis mutandis* to a literary province. Translate the computer science term "simulation" as "dream" and we get the simulation hypothesis in literary embodiments

DOI: 10.1201/9781003632924-7

FIGURE 7.1 *Vishnu Anantasayana.* Deity Vishnu, lying on the serpent Ananta, dreams the universe into reality. Mural from Dasavatara Temple, Deogarh, Uttar Pradesh, India (Wikimedia Commons, in the public domain).

such as the story of the butterfly dreamt by Zhuangzi, the play "La Vida es Sueño" (Life is a Dream) by Pedro Calderón de la Barca, or the short story "The Circular Ruins" by Jorge Luis Borges, where a man comes to realize that he is being dreamt by another man.

In the scientific realm, the simulation hypothesis has now passed an important test: It has materialized in a physical metamodel of the universe known as holographic autoencoder, an AI system in a physical embodiment capable of reconciling general relativity and quantum mechanics (Chapters 2 and 3). This is no minor feat since the respective material scales to which those theories apply are incommensurably different, and hence, the theories were deemed irreconcilable until the attention was drawn to entangled black holes (Chapter 5). There is at least the certainty that one such object exists in the form of the primeval wormhole, as it is inherent to the universe compact, boundary-less, and multiply connected topology. Since the simulated universe finds its supreme embodiment in the operation of the dark-energy-fueling engine that sustains the primeval wormhole, we may say that the quantum gravity conundrum has been resolved, at least vis-à-vis the standards laid out in Chapters 3–5.

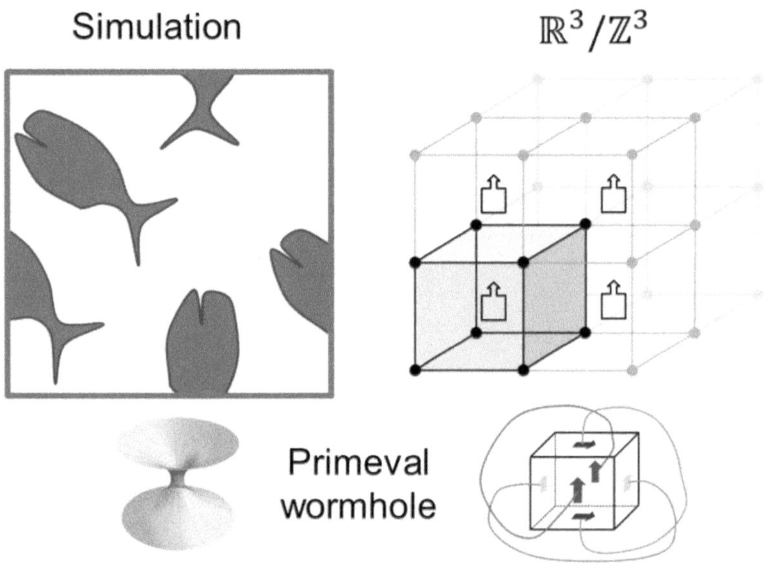

FIGURE 7.2 The spatial cross section of the universe as the quotient space $\mathbb{R}^3/\mathbb{Z}^3$ equivalent to a finite (however enormous) reality cell with periodic boundary conditions. Thus, in contrast with an infinite universe, the actual topology of the universe makes it amenable to be generated in a computer simulation. The simulation must recreate the primeval wormhole inherent to the toroidal manifold $\mathbb{R}^3/\mathbb{Z}^3$, which in turn, requires that we postulate the existence of the dark universe.

7.2 IS THE UNIVERSE AMENABLE OF BEING SIMULATED?

To confer physical meaning to the simulation hypothesis, we need to first decide whether the universe is amenable to be recreated in the conventional sense that we ascribe to the phrase "simulation in a computer". Simulating a process at a specific level of coarse graining, anywhere from organs to quarks, assumes that the process can be thoroughly captured in a reality cell with periodic boundary conditions and that such periodic boundary conditions are physically meaningful in the sense that they do not introduce artifacts affecting the outcome of the simulated process.

If the universe were infinite, extending this line of thought to a simulation of the whole universe may become daunting because we do not really know whether large-scale events – such as patterns of temperature distribution in the cosmic microwave background – may become artificially truncated. On the other hand, we are now persuaded that the spatial manifold of the universe is not infinite ($\mathbb{R}^3$) but finite, specifically, the multiply connected compact manifold $\mathbb{R}^3/\mathbb{Z}^3$ that necessarily spans a primeval wormhole and is compatible with the big bang birth scenario, as described in Chapter 5. In other words, the spatial manifold is actually the quotient space $\mathbb{R}^3/\mathbb{Z}^3$ equivalent to a reality cell with periodic boundary conditions (Figure 7.2). Thus, in contrast with an infinite universe, *the actual topology of the universe makes it amenable to be generated in a computer simulation.* Furthermore, again in contrast with an infinite universe, its compactness allows for the expansion of the simulated reality cell, a process assumed to be fueled by dark energy.

7.3 IS THE UNIVERSE AMENABLE OF BEING SIMULATED *DE NOVO?*

If the universe is indeed a simulation by an advanced civilization, the birth of the universe is also expected to be encompassed by the simulation. Alternatively, the physical relics of the universe evolution have been simulated so that birth scenarios can be conjectured by the virtual conscious

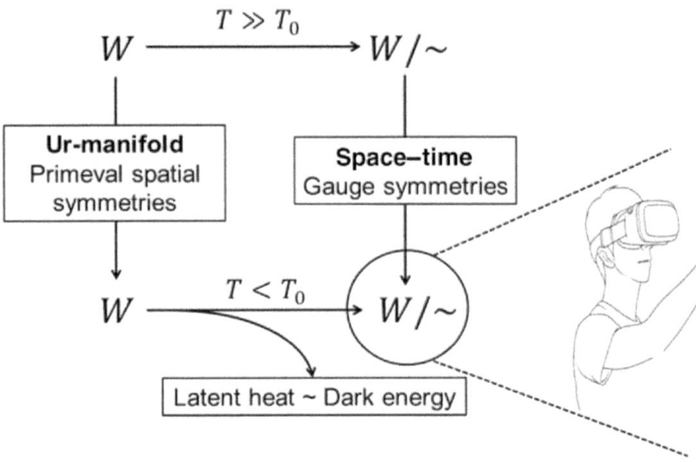

FIGURE 7.3 Autoencoder modeling the birth of the universe as a phase transition turned into an "autoencoder for the virtual reality headset" for the virtual sentient dweller of the simulated universe. An AI system is implemented to model the phase transition and generate the physics that yields the symmetry breaking leading to the emergence of mass. The AI system is represented as a commutative arrow diagram. An extra compact dimension is incorporated to space–time, yielding the ur-manifold W to account for the dramatic geometric dilution of gravity relative to the other three fundamental forces. The gauge fields thus emerge as supported by a 4D coarse-grained manifold $W/\sim$, the standard space–time, rendered as quotient space via an equivalence relation "$\sim$" that filters out the extra dimension. The horizontal arrows $W \rightarrow W/\sim$ become canonical projections, while the vertical arrows denote symmetry breaking associated with the phase transition,

$$T \gg T_c \rightarrow T < T_c \left(T_c = \frac{E_V}{k_B}, E_V \equiv \text{true vacuum for Higgs field, } k_B \equiv \text{Boltzmann constant} \right).$$

dwellers of the simulated universe. Chapter 2 treats the origin of the universe by exploiting the overarching physical framework of a phase transition, whereby primeval spatial symmetries right at the aftermath of the big bang become transmuted into the gauge symmetries of particle fields in the standard model and dark matter fields. Concurrently, the latent heat released at the phase transition becomes dark energy. As shown in Chapter 2, this scenario is amenable of being modeled by an autoencoder, and hence, an advanced civilization could have simply turned this AI system into the "autoencoder for the virtual reality headset" worn by the sentient dweller and thus simulate the birth of the universe, as shown schematically in Figure 7.3. Note that since the birth of the universe involves quantum events, the observer becomes essential to "collapse the wave function" and realize the events that bestow materiality to an evolving universe.

7.4 THE DARK UNIVERSE AS A COMPUTATIONAL ARTEFACT IN THE SIMULATION OF THE UNIVERSE

As discussed in the previous chapters, the physical embodiment of the holographic autoencoder bestows materiality to the $\mathbb{R}^3/\mathbb{Z}^3$ (three-torus) spatial cross-section of the universe. In so far as we judiciously assume that this manifold represents the actual physical space, its inherent primeval wormhole is sustained by dark energy as informed by the AI-empowered simulation (Chapter 5). On the other hand, the autoencoder system simulates the universe based on the commutative scheme presented in Figure 7.4 whereupon, at the level of the universal covering $\mathbb{R}^3$ with projection $\pi: \mathbb{R}^3 \rightarrow \mathbb{R}^3/\mathbb{Z}^3$ [3], there is no primeval wormhole and, consequently, no dark universe to account

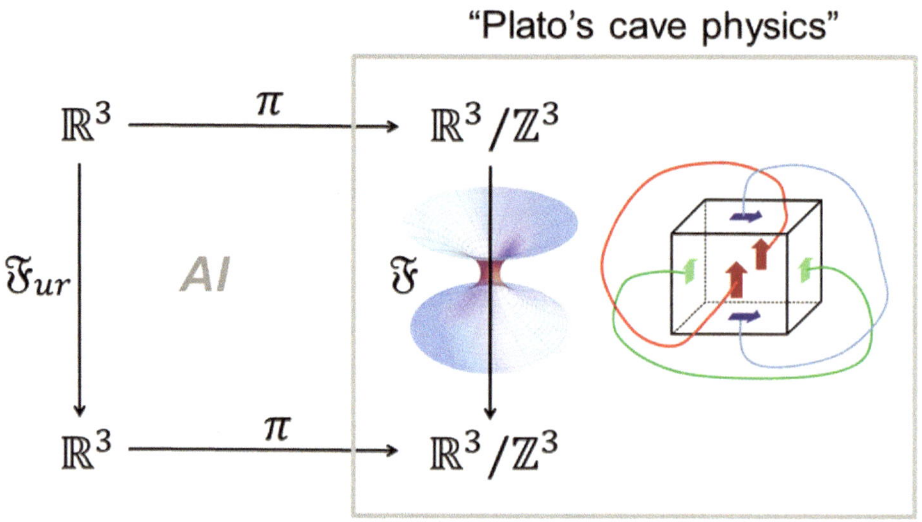

FIGURE 7.4 Commutative scheme representing the AI-empowered autoencoder leveraged to simulate the universe. The dark universe becomes an artefact of the projection onto the quotient space ($\mathbb{R}^3 \to \mathbb{R}^3/\mathbb{Z}^3$) within the commutative scheme that enables the universe simulation by lifting the primeval wormhole inherent in $\mathbb{R}^3/\mathbb{Z}^3$ to the level of its "featureless" universal covering.

for. In other words, *the dark universe is an artefact of the projection onto the quotient space ($\mathbb{R}^3 \to \mathbb{R}^3/\mathbb{Z}^3$) within a commutative scheme that enables the universe simulation by lifting the primeval wormhole inherent in $\mathbb{R}^3/\mathbb{Z}^3$ to the level of its "featureless" universal covering. "Plato's cave" physical account given by the flow map $\mathfrak{F}$ on $\mathbb{R}^3/\mathbb{Z}^3$ contains a primeval singularity that has no reality in the lifted flow map $\mathfrak{F}_{ur}$ of the ur-universe.*

Underlying this discussion is the AI-empowered scheme wherein the simulated universe is defined by Plato's cave flow map $\mathfrak{F}: \mathbb{R}^3/\mathbb{Z}^3 \to \mathbb{R}^3/\mathbb{Z}^3$, while the ur-universe is operationally represented by leveraging the surjective map $\pi: \mathbb{R}^3 \to \mathbb{R}^3/\mathbb{Z}^3$ within the commutative diagram that determines the relation: $\pi \circ \mathfrak{F}_{\text{ur}} = \mathfrak{F} \circ \pi$, as shown in Figure 7.4.

7.5 PHYSICAL FOOTPRINTS OF A SIMULATED UNIVERSE

Any computer simulation has embossed in it the imprint of the processor responsible for its generation. In our case, this imprint should surface in the discoverable universe as an "anomaly" or, rather, in a physical law that defies "intuition". However, our awareness of this anomaly is numbed, clouded by the extreme familiarity with the universe acquired through our conscience, a familiarity sculpted by the accretion of all our experiences than begin at the cradle.

Quantum physics offers a plethora of perplexities that can be readily accounted for if we assume that we live in a simulated universe. When fleshed out in the setting of the universe as a simulation, the fundamental postulate that quantum phenomena do not materialize until they become registered phenomena becomes far more intellectually digestible. *In the simulation scenario, a huge amount of processor time is saved if the observer or detector plays the ontological role assigned by quantum physics*: There is simply no obvious need to materialize phenomena when no act of observation or detection is also simulated to trigger the collapse of the wave function concurrently with state disambiguation. Likewise, quantum entanglement, an object of derision by none other than Albert Einstein, finds a straightforward explanation in the simulation scenario, since locality is immaterial in a virtual reality.

Evidently, if a civilization is simulating our universe having previously constructed some version of the holographic autoencoder, it must be inconceivably more advanced than ours, if nothing else because it has shown to be capable of cosmic manipulation (so far we are not). Yet, regardless of the degree of development, the speed of the processor this civilization was able to deploy cannot conceivably be infinite, an assertion that can be made with certainty. We know with astounding accuracy what the processor speed is! It is related to the storage capacity of the information required for the processor to perform a single operation. If the category "physical space" is being simulated and the processor performs, say, a single operation per second, then the amount of space encoded as bits of information generated must be translatable into actual physical space spanning ~300,000 km along a single dimension. That is, the processor speed begets the speed of light (~300,000 km/s) in its simulated universe, making it a staggering performer, indeed!

If indeed the speed of light, an absolute upper limit and a key parameter defining our universe, contains the imprint of the processor that simulates it, then this speed must be constant irrespective of the location or trajectory of the observer that measures it. This is because the speed of light is not "physical" but an *object d'art* of the simulation. And this is indeed the case! In fact, this is a major tenet of relativity, and Einstein had to postulate the slowing down of time for an observer traveling at speeds close to the speed of light (relativistic speeds) in order to accommodate the artifact of constant measured speed of light: For that particular "relativistic" observer (A), the distance travelled by light is shorter than that for an observer (B) that moves at slower (nonrelativistic) speeds, but time runs more slowly for A than for B, so that the speed of light measured by both observers is the same.

In other words, Einstein had to tinker with time to turn the computational artifact accurately described by the proposition "light speed is constant" into a "physical" proposition. In this sense, space–time becomes Einstein's way to circumvent or rather, come to grips with the perceived anomaly introduced by the fact that our universe is a simulation, something he could have not anticipated at the time when his theory of relativity was conceived.

If the universe is indeed a simulation, the processor performance for maximum quark-level resolution (~1 attometer) must be a staggering $f \sim 3\times\frac{10^{8}}{10^{-18}}\ \text{Hz} = 3\times10^{14}\ \text{Teraflops}$, which is about 10^{13} times higher than the number of floating-point operations per second accessible to quantum computation. However, there are ways to circumvent the prohibitively fast performance required. This entails using autoencoders trained to distill (encode) the coarse-grained version of reality at human eye resolution, i.e., at 10^{-5} m, and decode the information all the way back to a quark-resolved reality in special cases when there is a need for such resolution (i.e., in a lab experiment broadly defined). In a quotient space resolving reality at a "human scale", the required processor speed would be $f \sim 3\times\frac{10^{8}}{10^{-5}}\text{Hz} = 30\ \text{Teraflops}$, which is commensurate with the performance of today's quantum computers. The universe can thus be simulated by two coupled autoencoders fulfilling the commutativity condition for the diagram shown in Figure 7.5.

Quantum mechanical phenomena also find a natural correlate in the simulation scenario. Illustrative examples that merit to be mentioned, and would require further refinement of the scheme presented in Figure 7.5, include the need for an observer to disambiguate the state of the quantum system, and the nonlocality of quantum physics due to ubiquitous entanglement in the quintessential universe W, the ur-universe described at the opening of Chapter 1.

The time slow down or dilation experienced by the observer that approaches the speed of light is an essential relativistic outcome that finds its natural correlate in the simulated universe scenario: The overloading of the processor that simulates relativistic reality slows down its performance by increasing loading times up to a freeze that occurs as the speed of light is reached by

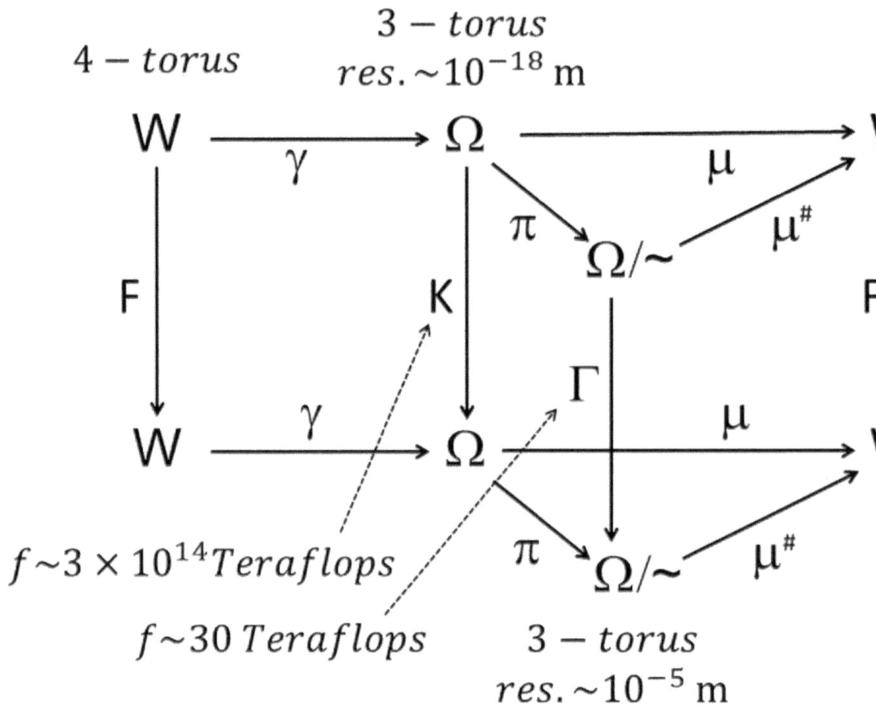

FIGURE 7.5 Commutative diagram representing an AI architecture to simulate the universe with two coupled autoencoders circumventing the prohibitively fast performance required at quark-level resolution. The autoencoders are coupled to distill (encode) the coarse-grained version of reality at human eye resolution (10^{-5} m), and decode this information back to quark-level resolution of reality in special cases when required by an observer/experimenter.

the simulated observer. In practice, in simulations performed on the annular assemblage of holographic autoencoders (Figures 5.10 and 5.21b), it is observed that processor overloading leads to a complete freeze as $\chi \to 1$. In fact, no current AI-based cosmological technology enables us to study the primeval wormhole for levels of disentanglement beyond E = 18 (Figure 5.11), corresponding to $\chi \approx 1 + 10^{-32}$.

The topology of the universe was suspected to be of no tangible significance to physics up to this juncture in the history of science. As AI-empowered metamodels make their forays in the field of cosmology, the situation is likely to change rapidly and the possibility that the universe may be a simulation does not appear so farfetched, at least to the open mind.

The physical footprints of a simulation scenario have been shown in this section to be hiding in plain sight, in the very laws of contemporary physics, while the artefactual nature of the dark universe within the simulation scenario should instill confidence in validity of those same physical laws that we unwaveringly uphold.

7.6 AUTOENCODERS OF THE STANDARD MODEL

As noted American physicist John A. Wheeler would have it [4], the physics world is information, while mass and energy are incidentals. This view is enshrined in his aphorism "it for bit". Since virtual reality is based on information processing, Wheeler's picture substantiates Bostrom's simulation hypothesis [2], an assertion that becomes especially compelling if we construe the uncertainty principle as a glimpse into deep reality. Thus, the Compton wavelength, $\lambda_C = \dfrac{h}{mc}$, in the photon scattering by a particle of mass m could be interpreted as the pixelation of deep reality. This is so because λ_C represents the spatial uncertainty in the observation of the particle.

The compact, multiply connected nature of the latent manifold for Einstein's space–time makes the universe amenable to be simulated, as shown in Section 7.2. On the other hand, physical footprints of the simulated universe are surely apparent in the quantum physics description of the pixelation of deep reality and in the relativistic parametrization of the processor frequency, as indicated in Section 7.5.

So, if quanta represent pixelation, and virtual reality is the outcome of information processing, and the processor itself has a relativistic footprint, what sort of AI system is actually simulating the universe? The AI models for the dark universe described in Chapters 4 and 5 suggest that this question becomes tantamount to ask: What sort of autoencoder can generate the standard model (SM) of particle physics *de novo* and with a pixelation tailored to quantum field theory?

Encoding the 15 massive particle fields in the SM would require autoencoders compatible with each other and capable of encoding also photons and gluons regarded as endowers of reality through materialization of observation. The processor for each autoencoder is endowed with performance frequency f, associated with the respective Compton length as a measure of pixelation according to the relation:

$$f = \frac{c}{\lambda_C} = mc^2/h. \tag{7.1}$$

This implies that the simulation of the universe at the highest resolution requires using an autoencoder capable of encoding the top quark field (heaviest particle) with $m_T = 173\ \text{GeV/c}^2 \approx 3.08 \times 10^{-25}$ kg and $f \approx 4.19 \times 10^{25}\ \text{Hz} = 4.19 \times 10^{13}$ TFLOPS. We name this autoencoder AE. At this colossal level of computer performance, an advanced civilization is capable of encoding all fields in the SM, thereby simulating the entire universe with a single processor. The AE will endow the top quark with mass because it has also encoded the Higgs field, since $m_H < m_T$.

To save computer power, AE needs to be entrained by other autoencoders, AE_1, AE_2, …, AE_{14}, capable of simulating the universe at coarser levels of resolution. In particular, AE_1, the autoencoder of the Higgs boson $\left(m = m_H \approx 2.22 \times 10^{-25}\ \text{kg}, f \approx 3.02 \times 10^{13}\ \text{TFLOPS}\right)$ and all the lighter particles $(m < m_H)$, should be optimized to be completely compatible with AE, as per the commutativity of diagram in Figure 7.6. Autoencoders yielding progressively coarser levels of resolution, carried out by AE_2 (Z-boson), AE_3 (W-boson), …, AE_{14} (electron neutrino), should be made compatible with AE and AE_1 in the sense that the diagram (Figure 7.6) describing map compositions is commutative. The entrainment of all autoencoders through the commutativity with AE_1, itself trained at the highest cost by AE, endows the respective particles with mass, in accordance with the Higgs mechanism. Once all AEs are trained and optimized fulfilling diagram commutativity (Figure 7.6), the universe may be readily simulated at the coarsest level by AE_{14} which operates at the lowest cost and encodes only the electron neutrino (ν_e), the lightest massive particle $(m < 0.8\ \text{eV/c}^2 \approx 1.42 \times 10^{-36}\ \text{kg})$. Thus, the processor performance frequency to simulate the universe at the lowest admissible resolution is estimated at 193TFLOPS. This level of performance seems very much within reach in the foreseeable future.

7.7 NO THEORY OF EVERYTHING IN THE SIMULATED UNIVERSE?

As described in Chapter 2, in the quest for the theory of everything, AI has delineated a blueprint to generate the warped graviton in a collider. At a mass $m_G = 9050$ GeV, the autoencoding

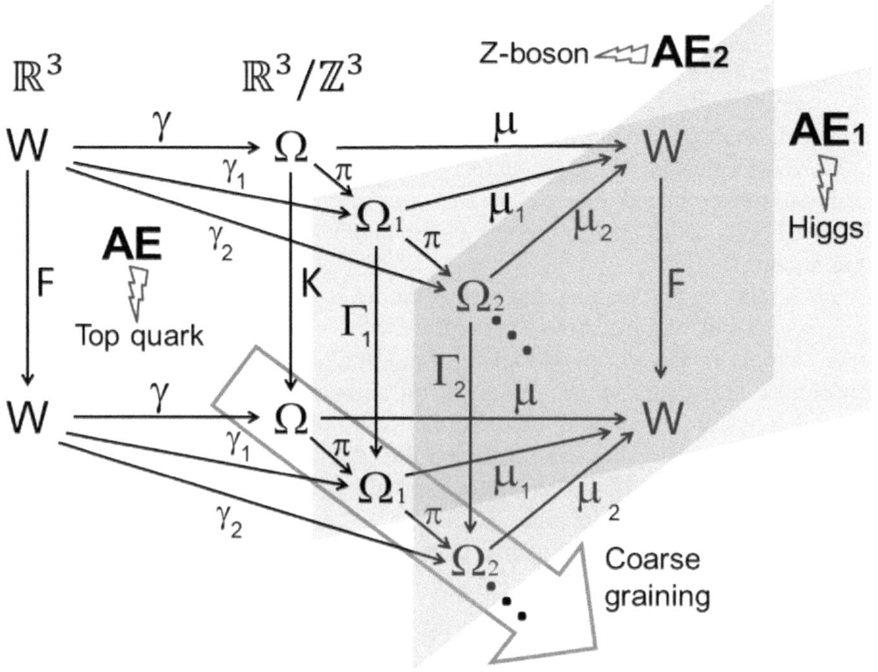

FIGURE 7.6 Commutative diagram representing an AI architecture consisting of 15 coupled (pairwise compatible) autoencoders generating the standard model of particle physics at different levels of field resolution. Pixelation of deep space becomes commensurate with the Compton lengths of the respective particles.

of the warped graviton would require a processor frequency estimated at the staggering $f = \frac{m_G c^2}{h} = 2.56 \times 10^{15}$ TFLOPS (Chapter 2).

While we simply cannot fathom what sort of limitations may constrain simulations by an advanced civilization, we do know that the speed of light represents one such limit. It is entirely possible that the speed of light constitutes the limiting factor precluding the autoencoding of any particle heavier than, say, the top quark. This would imply that the autoencoding of the warped graviton is simply off limits and a theory of everything, establishing the quantum origin of gravity, cannot ever be falsified or corroborated experimentally in our simulated universe.

7.8 AI CREATES CONSCIENCE AS AUTOENCODER OF THE WAVE FUNCTION COLLAPSE

The possibility that we may be living in a simulation depends pivotally on the possibility of creating conscience as an encoder of the quantum signatures of the simulation. As it is well known, these signatures point to the most problematic aspect of quantum physics, namely, the problem of measurement or observer participation in the quantum event. The measurement problem is not only a suggestion that we may be living in a simulation but also appears to be a true weakness of quantum theory [5].

The wave function collapse is often attributed to the act of observation, whereby an *a priori* unresolved superposition of quantum states is replaced by an ascertained specific state of the system. This state is thus believed to become an ontological realization that comes into being when observed by a sentient entity. While this view is quite popular, it seems to be prima facie utterly fallacious. By the same token, all inert lifeless planets would be fuzzy, full of delocalized things, unless and until they were visited at least once by sentient and extremely curious entities that resolved the superposition of states.

Roger Penrose has addressed this paradox claiming that the wave function collapse occurs irrespective of observation as a spontaneous process that becomes meaningful as quantum mechanics becomes "gravitized" in an extended theory that encompasses general relativity [5]. If this indeed provides an improved version of quantum mechanics, then the problem that needs to be addressed may be formulated as follows: What is the role of conscience vis-à-vis quantum measurement and how does quantum conscience materialize?

Using the methods described in Chapter 3, we may assert that AI constructs conscience as an autoencoder trained with the collapse of wave functions, as shown in Figure 7.7. To formalize this construction, let $L^2(W)$ denote the space of square integrable functions with spatial compact support W, while $\Theta: L^2(W) \to L^2(W)$ maps the wave function $\Psi \in L^2(W)$ into the disambiguated state v_Ψ representing the wave function collapse. A sentient entity coarse grains the space l^2 of Fourier coefficients (pure state weights) of wave functions in $L^2(W)$ according to its parsing of reality, which is a signature of its conscience. Thus, an equivalence relation "$\sim$" is introduced to lump up states according to the level of conscious focusing: $l^2 \to l^2/\sim$, so that the following commutative relation is verified (Figure 7.7):

$$K \circ \pi(\Psi) = \pi \circ \Theta(\Psi), \quad \forall \Psi \in L^2(W). \tag{7.2}$$

Here, the projection $\pi: L^2(W) \to l^2/\sim$ is defined as the composition of the isometry between L^2 and l^2, resulting from Parseval's equality, with the canonical projection $l^2 \to l^2/\sim$. The map K represents the "interpretation" of the wave function collapse $\Psi \in L^2(W) \to L^2(W) \ni v_\Psi$ by the sentient being. Thus, the projection and interpretation maps are optimized in a learning context trained by observed wave function collapses (Figure 7.7), until the Eq. 7.2 is exactly satisfied or at least, satisfied to an optimal level.

Conscience cannot be regarded as merely encoding and interpreting observations but also requires "standing back", turning onto itself and reflecting upon the encoded information to make inferences, including decisional ones. These aspects of conscience are enshrined in the right hand side of the autoencoder shown in Figure 7.7. Thus, its version of reality, i.e., the coarse graining $W/\approx$ of the spatial support of the wave function, needs to be optimally adjusted in relation to the equivalence relation "$\sim$", so that the following commutative relation holds:

$$\widetilde{\Theta} \circ \gamma = \gamma \circ K. \tag{7.3}$$

This implies that the map $\widetilde{\Theta}: L^2(W/\approx) \to L^2(W/\approx)$ is the reflection operation defined by the inferences of wave function collapses that the conscience/autoencoder is able to make on its coarse-grained version of reality based on its interpretation of the learned observations.

$$\begin{array}{ccccc} L^2(W) & \overset{\pi}{\rightarrow} & l^2/\sim & \overset{\gamma}{\rightarrow} & L^2(W/\approx) \\ \downarrow \Theta & & \downarrow K & & \downarrow \widetilde{\Theta} \\ L^2(W) & \underset{\pi}{\rightarrow} & l^2/\sim & \underset{\gamma}{\rightarrow} & L^2(W/\approx) \end{array}$$

FIGURE 7.7 Conscience as autoencoder of the wave function collapse.

At this point, we may address the core question: *Who, or rather, what runs the simulated universe?* Clearly, a sentient entity such as the one described in Figure 7.7, that in all likelihood is able to generate metamodels of the observed and recorded phenomena that occurred or keep occurring in whatever universe the autoencoder is located. Such metamodels satisfy Eq. 7.3. The validity of this picture is contingent on the actual universe that trains the conscience/autoencoder ensuring the collapse of the quantum wave function when gravitation is incorporated to the quantum mechanical model of the universe (Section 6.4). The metamodels generated by the simulation are coarse grained via the quotient spaces $W/\approx$ to the input levels required by the observing conscience/autoencoder that is being simulated also in adherence to the basic scheme presented in Figure 7.7. In other words, the inputted phenomenology learned by the observer in the simulated universe is actually an output metamodel generated by the conscience/autoencoder that runs the simulation. The simulation is in turn identified by the simulated observer as the universe itself.

REFERENCES

1. Fernández A (2022) *Topological Dynamics for Metamodel Discovery with Artificial Intelligence.* Chapman & Hall/CRC, Boca Raton, FL.
2. Bostrom N (2003) Are we living in a computer simulation? *Philos Quaterly* 53: 243–255.
3. Fernández A, Sinanoglu O (1982) The lifting of an Inonu-Wigner contraction at the level of universal coverings. *J Math Phys* 23: 2234–2235.
4. Wheeler JA (1982) The computer and the universe. *Int J Theor Phys* 21: 557–572.
5. Penrose R (2014) On the gravitization of quantum mechanics 1: Quantum state reduction. *Found Phys* 44: 557–575.

Epilogue

What Is Left for Human Scientists in the Aftermath of An AI Takeover? Gödel's Incompleteness Theorems May Provide a Safe Haven

"To be is to be encoded".

Ariel Fernández

The striking advances in the so-called "theory of everything" described in this book make it evident that powerful strides in mathematical physics can be made by a large language model (LLM) empowering a formal proof assistant. The final results on quantum gravity described in Chapters 2 and 6 are a stone throw away from providing a completely autonomous AI system that can span the most mature and arduous chapters in particle cosmology.

The book makes it apparent that the most promising AI system designed to make strides at the forefront of mathematical physics includes a LLM that serves as a formal proof engine empowering a functional programming language. The latter operates as a theorem proof assistant, while the LLM constitutes a realization of a generalized autoencoder. The LLM distills the essential physics associated with a latent manifold $\mathcal{Q}$ from a vastly more complex phenomenology associated with a manifold $\mathcal{M}$ (Figure E1).

Quantum gravity is regarded as a holy grail in today's scientific pursuit and AI's takeover through the deployment of formal proof autoencoders such as the one implemented in this book (Figure E1) is becoming apparent. Such possibilities make it inevitable – or rather, essential – to include non-human researchers in the target audience of the book.

A troubling general question looms over this scenario, where humans are likely to lose the intellectual supremacy: What could possibly be the role of human researchers in less than a decade, when AI exerts its hegemony over vast sectors of the scientific pursuit?

Taking an optimistic stance, there may be special niches carved in the scientific endeavor where humans may still prove relevant, for example, at particular junctures where "metamodeling" involving large-scale planning of AI architectures is required. This metamodeling is illustrated by the discovery blueprint sketched in Figure E1.

Additionally, humans may remain relevant players when the interface between physics and metaphysics is blurred or cannot be properly drawn while the information exchange at the crossroads becomes essential to resolve undecidable propositions in Gödel's sense [1]. One would argue that metaphysics is always relevant when the scientific advance is so important that a paradigm shift is the likely outcome, but that has not always been the case, as the following study case proves.

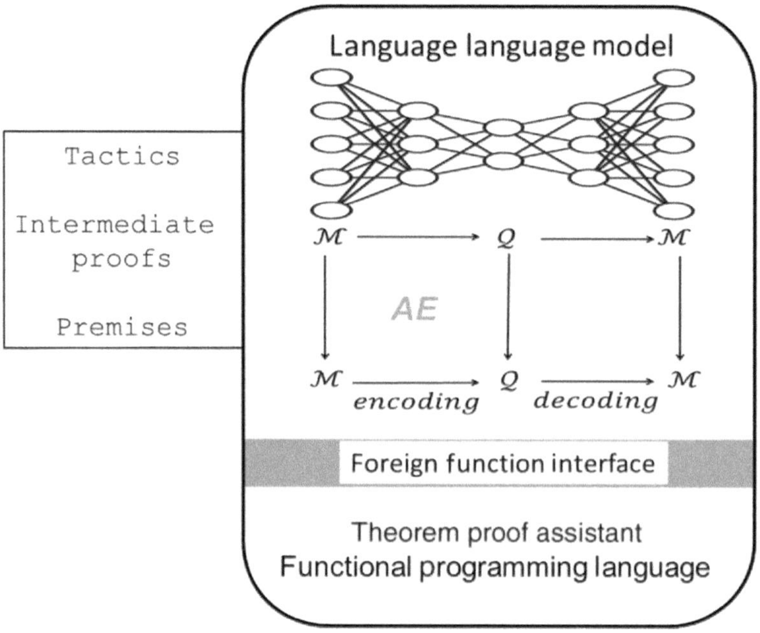

FIGURE E1 Large language model (LLM) empowering a functional programming language operating as a theorem proof assistant. The LLM realizes a generalized autoencoder (AE) that distills the essential physics in the latent manifold $\mathcal{Q}$, a surjective diffeomorphic image of an underlying manifold $\mathcal{M}$.

E1. WHAT IS THE UNIVERSE EXPANDING INTO? FORAYS INTO GOEDEL'S INCOMPLETENESS THEOREM

As cursorily described in Chapter 1, the universe expansion is a key cosmological phenomenon that significantly impacted much of the current thinking in the field. It introduced a plethora of related problems, such as the cosmological constant problem, the vacuum catastrophe problem, and, of course, the problem of the very nature of dark energy. This book argued that is also prompts us to think about a hitherto neglected matter: the topology of the universe. Topology does not impinge on local theories, such as general relativity, and understandably, was not a focus of research for Albert Einstein, who apparently did not devote much thought to the matter. Yet, this book has persuasively argued that the universe expansion, the very phenomenon that lead Einstein to introduce the cosmological constant, demands careful metaphysical musing on the universe topology, a problem deemed undecidable in a Gödel–Turing sense.

Expansion of the universe makes it necessary to assume the universe has boundaries and those boundaries extend over time. Yet, the very concept of "boundary of the universe" is ontologically implausible because it would imply the existence of an "outside", and yet the universe is all that there is, so its "outside" confronts us with a Parmenides-type contradiction (Chapter 6). In other words, there is no geometry to signify "nothingness outside the universe", in stark contrast with vacuum (absence of matter), and hence, the universe cannot have boundaries. Cut and dry as the matter is on an ontological level, and the problem is propositionally undecidable in Gödel's logical and Turing's computational sense based on the frame field *vierbein* formalization of quantum gravity [2] (Chapters 2 and 6).

On the other hand, if the big bang scenario is relativistically upheld, the topology of the universe should remain invariant during and beyond the inflation phase, and hence, the universe must remain compact even if it is enormous today. This leaves space–time with the topology of a 4D torus (assuming an orientable manifold), and only within such topology, the concept "universe expansion" becomes meaningful. Thus, the question "where is the universe expanding into?" is rendered irrelevant: without boundaries, the four-torus simply gets bigger as it expands.

Candid as this ontological argument may appear to be, it is water-tight and will likely send shockwaves across much of the scientific establishment. For one, the so called "brane theories" of quantum gravity, placing undiluted gravity on the UV brane and the standard model on the IR brane [3], would be rendered meaningless, as the branes must be assimilated to boundaries, which constitutes an ontological impossibility (Figure E2). Second, the so-called "universe as hologram" [4], widely conjured in attempts to elucidate the nature of quantum gravity, becomes also a meaningless idea because it assumes the existence of a universe-encoding boundary.

As far as we can see, these ontological cogitations are propositionally undecidable in Gödel's sense. They are also undecidable in Turing's sense, since they are not susceptible to be entered as code into a functional language model such as Lean (Appendix), pointing directly at the incompleteness of mathematics as an axiomatic context.

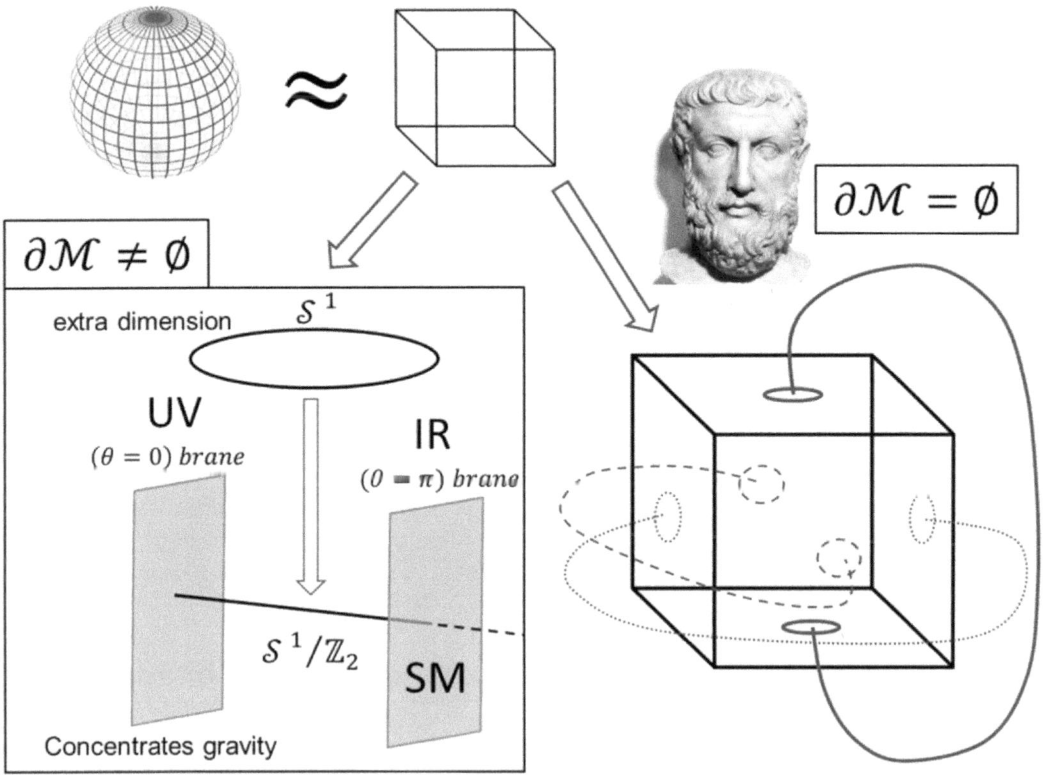

FIGURE E2 Identifying the universe topology becomes an ontological problem that cannot be decided propositionally within the frame-field construction that attaches spinors to general relativity to make it compatible with quantum mechanics. The rejection of the $\partial\mathcal{M} \neq \varnothing$ proposition cannot be decided by the AI system that incorporates the frame-field axiomatic context as code (Figure E1). Human intervention is required to introduce as code the ontologically correct declarative proposition $\partial\mathcal{M} = \varnothing$. On the other hand, the brane gravity theory of Randall and Sundrum [3] (left panel) stems from the proposition $\partial\mathcal{M} \neq \varnothing$, which is *a posteriori* rejected by the kernel of the functional language model as ontologically incorrect. The effigy of Parmenides, founder of ontology, was obtained from the public domain (Wikimedia Commons).

The AI architecture shown in Figure E1 proved adequate to discover the nature of the graviton, the quantum of gravitational excitation, as shown in Chapters 2 and 6, and it can readily encode the manifold $\mathcal{M}$ as boundary-less ($\partial\mathcal{M} = \varnothing$). However, the AI system cannot *a priori* reject the declaration "$\mathcal{M}$ has a boundary" ($\partial\mathcal{M} \neq \varnothing$), which is regarded as an undecidable proposition vis-à-vis the frame-field axiomatization provided by the Palatini gravity (Chapter 2).

E2. HUMAN METAPHYSICIANS IN THE ERA OF AI SUPREMACY

The AI system described in Figure E1 and implemented in the Appendix can propositionally identify the graviton, the holy grail of quantum gravity, as shown in Chapter 2. However, the AI system *per se* cannot accommodate the fact that most theories and observations on holographic or brane gravity are ontologically flawed and therefore should be discarded. The human intervention on the supreme AI system of inference becomes instrumental to incorporate the ontological premise as code into the functional programming language (Figure E3).

Similarly, the mass-gap conundrum, whereby the bosons conveying the weak force and the quark-gluon composites carrying the strong force are massive, while the photon communicating the electromagnetic force is massless, poses an undecidable question as to why should that be the case. It is clear that the matter cannot be decided as an outcome of a Turing computation.

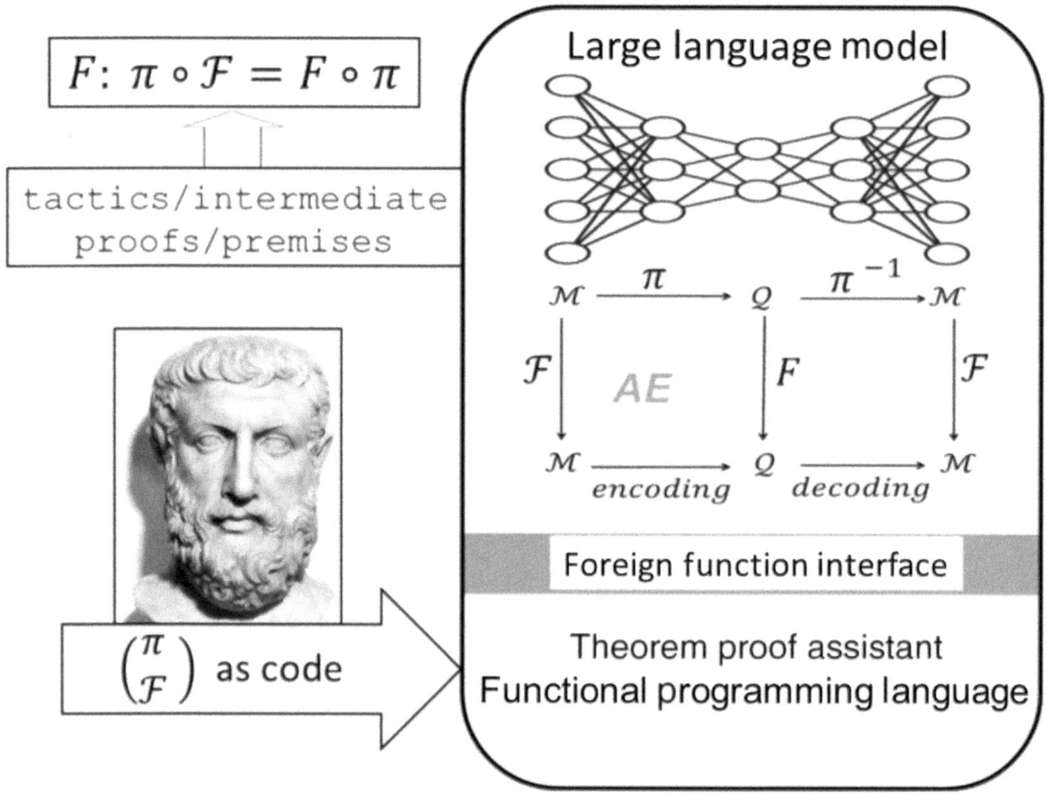

FIGURE E3 The need for ontological (Parmenidean) premises introduced through human intervention into the supreme inferential AI system. The physical underpinnings of quantum gravity, enshrined in the commutativity of the arrow diagram representing the formal autoencoder (AE), cannot materialize without the ontological premises introduced as code by the human participant. The effigy of Parmenides, founder of ontology, was obtained from the public domain (Wikimedia Commons).

These ontological assertions on Gödel/Turing undecidable propositions carve a niche for the human's metaphysical role in the aftermath of the AI supremacy in the scientific endeavor. Gödel's incompleteness theorems and their Turing computational counterparts may well provide a safe haven for a human civilization unwilling to relinquish a privileged position in the scientific endeavor.

E3. PRAGMATIC HUMAN METAMODEL OF THE COSMIC TOPOLOGY ADOPTED BY THE COSMIC MICROWAVE BACKGROUND

This book has elaborated on the human cogitation that the universe is a manifold with nontrivial spatial topology, specifically a compact boundary-less three-manifold. This implies that through every spatial point there are closed curves that cannot be smoothly deformed into a point, and hence, space is multiply connected. While this striking conclusion probably has far-reaching consequences, the parameter space for such manifolds is much larger than what is experimentally accessible. Assuming space is a flat manifold (Chapter 1), there are 17 inequivalent topologies. Our focus has so far been confined to the three-torus with a cubic fundamental domain. Even with a cubic domain, the identification of opposite faces may be twisted, that is, the identification may be done by rotating one of the faces by an angle $\vartheta = \frac{\pi}{2}$ (Figure E4), or $\vartheta = \pi$. Since this may be done for each of the three pairs of opposite faces, we would end up with 27 possible manifolds, each defined by a tern where each of the three pairs is encoded by 0 ($\vartheta = 0$), 1 $\left(\vartheta = \frac{\pi}{2}\right)$, or 2 ($\vartheta = \pi$). As illustration, the twisted torus $\mathcal{T}_{(1,0,0)}$ is shown in Figure E4. As shown in Chapter 5, either $\mathcal{T}_{(1,0,0)}$ or $\mathcal{T}_{(2,0,0)}$ appear to be the most promising topologies to accomodate the cosmic dark-energy-combustion two-stroke engine that sustains the universe topology.

On the other hand, if a parallelepiped fundamental domain is adopted, far richer topological possibilities arise as needed to fit the big data on the cosmic microwave background (CMB, Chapter 1). While the observed CMB fluctuations encode information about topology, the AI-empowered fitting problem is undecidable. Therefore, it requires human intervention through *a priori* metamodels that are not encodable in the theories absorbed by AI (Appendix).

The properties of spatial portions of the (3 + 1)-dimensional manifold describing deep space are the purview of cosmic topology. As indicated in Chapter 1, we always assume a Friedmann–Lemaître–Robertson–Walker (FLRW) metric, with negative, zero (flat or Euclidean), and positive spatial curvature. Topologies associated with these geometries have been widely investigated. If we

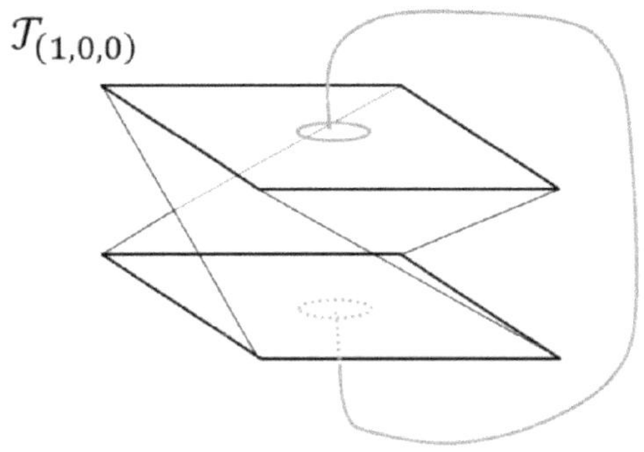

FIGURE E4 Illustration of a tilted three-torus spanned from a cubic fundamental domain.

focus on the universally accepted flat case, there are eighteen Euclidean topologies, with an infinity of both spherical and hyperbolic topologies.

All possible Euclidean manifolds can be spanned with 1–2 parallelepiped or hexagonal prisms as fundamental domains. One can "tilt" the three-torus by adopting a parallelepiped as fundamental domain and identify opposing rhombic faces by first rotating one of the faces by π, a square face by $\pi/2$, a hexagonal face by $\pi/3$, etc. Alternatively, flipping faces yields the Klein spaces, as illustrated in Chapters 1 and 5.

These considerations hint at the possibility that the metamodel of space may be a lot more complex than the one humanly imaginable and discussed in this book, Yet, for all the intellectual prowess of the AI system (Appendix), only humans can fathom the metamodel. As said, this is because the metamodel cannot be distilled by the autoencoder, even in its most generic guise.

To show how the CMB may make the metamodel decidable, we would need the eigenmodes of the Laplacian operator with boundary conditions defined by the cosmic topology of the universe. Such eigenmodes are the analogues of Fourier modes on the covering space (Chapter 1). If the fundamental domain is the cube, the Laplacian eigenmodes are exactly the Fourier modes for countably many wave vectors. The changes to the eigenmodes affect the matter fields within the CMB in ways that may be detectable.

Be as it may, any educated guess on the cosmic topology remains a provisional human metamodel that can be thought of as inscribed in "*als ob*" ("as if") philosophical pragmatism [5]. Since the dawn of civilization, phenomenology has been accommodated in provisional physical theories with varied degrees of success, but things-in-themselves do not afford certainties, at least for now, demanding pragmatic human intervention.

E4. A HUMAN ROLE IN THE ERA OF AI SUPREMACY: AN ONTOLOGICAL INTERVENTION DEBUNKS THE PARADIGM OF THE WORLD AS HOLOGRAM

The proportionality of the black hole (BH) entropy with the area of its event horizon inspired the so-called holographic principles [4] that portend to explain quantum gravity. One such realization of the holographic principle is provided by the AdS/CFT correspondence described in Chapter 3 [4]. This inspiration cannot be propositionally encoded into an AI system such as the one described in the Appendix: It would be outright rejected by the kernel as the outcome of a seemingly incorrect cogitation introducing a conflict with the extant physics entered as code in the large language model.

To dwell into specifics, the Bekenstein–Hawking BH entropy is the amount of information content concealed by the BH in order for it to fulfill the laws of thermodynamics when interpreted by an observer that resides *outside* the BH [6]. For such external observers and only for them, a stationary black hole can only be specified by its total mass (M), electric charge (Q), and angular momentum (J). Since the area of the event horizon cannot decrease and only increases with spontaneous (irreversible) processes taking place inside the BH, it is reasonable to expect that the BH entropy should be proportional to the area of the event horizon. This has indeed proven to be the case [6]. Feeding this proposition into an AI system would not yield any holographic principle for the universe: The kernel would simply reject a proposition postulating a holographic principle as a *non sequitur* vis-à-vis the extant physics already entered as code (Appendix).

The so-called AdS/CFT correspondence is typically interpreted as a holographic map between gravitational degrees of freedom of the universe and the degrees of freedom of the matter field that resides on its boundary (Chapter 3). This assertion is distilled from the fact that certain string theories in five dimensional anti-de Sitter (AdS) space–time are bijectively mapped onto conformal field theories (CFTs) on the four-dimensional boundary of that special guise of the space–time.

Should that observation be propositionally fed onto an AI system that incorporated the *vierbein* frame-field scheme for a space–time manifold with boundary (Appendix), it is certain that the formal AdS/CFT equivalence will be flawlessly drawn. AI would generate the holographic principle *albeit* with one serious caveat: *It would be incorrect on ontological grounds when interpreted as an attribute of the universe.* This is simply because the universe does not have a boundary supporting a CFT. Accordingly, humans must dictate that a boundary-less spatial manifold be encoded within the frame fields exploited by a theorem proof assistant empowered by a large language model to formulate quantum gravity.

This type of ontological intervention broadly delineates the role of humans in the forthcoming era of AI supremacy. The AdS/CFT equivalence is mathematically impeccable, and as such, it will eventually be formulated also by the AI system empowered by a large language model in the guise of a quantum gravity autoencoder (Appendix, Chapter 3). Yet, human interference with the AI system reveals that the metamodel that underlies the AdS/CFT equivalence is fatally flawed since there is no "universe boundary" to support a CFT. This makes the AdS/CFT equivalence only acceptable when subsumed within an "as if" philosophical position [5]. In turn, this pragmatic position crumbles under the weight of the Parmenidean ontological argument, an assertion that points to the heart of the conundrum regarding the human role in the forthcoming era of AI supremacy. Only a human could have made the ontological headway in the AI-controlled environment.

REFERENCES

1. Faizal M, Shabir A, Khan AK (2024) Consequences of Gödel's theorems on quantum gravity. *Int J Theor Phys* 63: 290.
2. Palatini A (1919) Deduzione invariantiva delle equazioni gravitazionali dal principio di Hamilton. *Rend Circ Mat Palermo* 43: 203–217.
3. Randall L, Sundrum R (1999). Large mass hierarchy from a small extra dimension. *Phys Rev Lett* 83: 3370–3373.
4. Susskind L (1995) The world as hologram. *J Math Phys* 36: 6377–6396.
5. Vaihinger H (1935) *The Philosophy of 'As if': a System of the Theoretical, Practical and Religious Fictions of Mankind.* Translated by C. K. Ogden (2 ed.). Routledge, London.
6. Bekenstein JD (2008) Bekenstein-Hawking entropy. *Scholarpedia* 3: 7375.

Appendix

Quantum Gravity in a Large Language Model within a Functional Programming Language

The gauge theory of gravity pivots on a differential geometry formalization of Lorentz transformations that reconciles quantum mechanics (QM) with general relativity (GR), the classical theory of gravity. The differential geometry approach is instrumental because the geometries that support GR and QM are different: While GR is built upon a curved manifold $\mathcal{M}$ that represents space–time, the wave function in QM is supported by an Euclidean space. Thus, "attaching spinors to space–time" requires a covering of $\mathcal{M}$ with open sets: $\mathcal{M} \subseteq \bigcup_{x \in C(\mathcal{M})} \mathcal{O}(x)$, with $C(\mathcal{M})$ a denumerable set of points and $\mathcal{O}(x) \ni x$, so that each open set $\mathcal{O}(x)$ is diffeomorphically charted onto a tangent space: $\mathcal{O}(x) \rightarrow T_x\mathcal{M}$. In this setting, a frame field $\{e_a^\mu\}$ is subsumed into the embedding $\mathcal{M} \rightarrow \bigcup_{x \in C(\mathcal{M})} T_x\mathcal{M}$ to locally map the curvature tensor $g_{\mu\nu}$ defined for each point in $\mathcal{O}(x) \cap \mathcal{M}$ onto the Minkowski signature η_{ab} of the chart $T_x\mathcal{M}$, $\forall x \in C(\mathcal{M})$. Furthermore, the atlas $\bigcup_{x \in C(\mathcal{M})} T_x\mathcal{M}$ is articulated via a spin connection $\{w_\mu^{ab}\}$ that enables the attachment of spinors to GR (Chapter 2). This is done via a type constructor that we name Cod, the covariant derivative determined by the spin connection [Sco] and the curved Clifford algebra [Clf] generated by the commutators of Dirac's gamma matrices. In this way, a fiber-bundle formulation of the gauge theory of gravity is implemented.

The goal here is to implement an interactive theorem prover to generate new results within the frame-field formalization rules made up of a set of declaratory propositions that constitutes an inductive type. We seek for derived propositions especially in the strong-field completion of gravity wherein a cogent propositional definition of the graviton needs to be developed (Chapter 2). To serve this purpose, we resort to functional programming languages, especially Lean [1], that have already formalized the required categories, types, and type constructors for a tangent bundle defined as a smooth vector bundle [2], in turn defined over a curved manifold [3].

The interactive proof assistant pivots on the proposition-as-type paradigm. From a constructive perspective, proofs are mathematical objects that are denoted by suitable expressions in dependent type theory. By contrast, in this environment, a proposition is true if and only if it can be written down properly and checked independently by the kernel that it has been typed correctly. In other words, the expressions themselves are the proofs, and only well-posed (i.e., well-typed) expressions are true. Correct syntax and semantics are equivalent to correct content. A false expression simply cannot be posed (typed) correctly, and all the mathematics axioms need to be incorporated as code.

In our context of interest, Lean translates the generative input on "attaching spinors to relativistic geometry" into formal expressions that are checked by the kernel for correctness in the sense described previously and then stored in the environment [mathlib] for subsequent proposition generation. Most of the commands described in this appendix have global effects: They remain in effect in the file associated with the project and also in any file that imports it, from now to eternity.

To chart the space–time manifold adopting a minimal covering $\mathcal{M} \subseteq \bigcup_{x \in C(\mathcal{M})} \mathcal{O}(x)$, the formalization provided in [4] is instrumental since the set of charts $\{T_x\mathcal{M}\}_{x \in C(\mathcal{M})}$ is incorporated as data synchronized with mathlib and hence so are the denumerable diffeormorphisms $(x) \rightarrow T_x\mathcal{M}$ $\forall x \in C(\mathcal{M})$. The next step is to specialize the fiber-bundle formulation, which has been already implemented in a generic context [5], to the warped graviton within a gauge theory of gravity. This goal makes it essential to formally compute the generating commutators [Com] of covariant derivatives $\left[\nabla_\mu, \nabla_\nu\right]$, which may be treated within the functional programming language as inductive types [6]. This is so because they are declared to constitute a set of type constructors as the commutators operate on different types: Scalar [Sca], Vector [Vec], Tensor [Ten], etc. To operate on tensors by swapping and cancelling covariant and contravariant indices, Einstein's contraction rules need to be subsumed into a relativistic type constructor denoted Ctr.

The overarching inductive type is introduced as represented by the commutative constructive arrow diagram in Fig. A1, whereby every proposition formulated at the level of the atlas $\bigcup_{x \in C(\mathcal{M})} T_x\mathcal{M}$ may be decoded at the level of the covering $\bigcup_{x \in C(\mathcal{M})} \mathcal{O}(x)$, by means of the types frame field [Frf], spin connection [Sco], Clifford algebra [Clf] and the type constructors commutator [Com], covariant derivative [Cod], and relativistic tensorial contraction [Ctr].

Given the propositions-as-types equivalence, induction becomes the fundamental method of proof. As an illustration adopting the inductive type from Figure A1, the following chain of propositions generated by Aesop within the functional program language relates the "relativistic type" $R_{\nu\xi\mu\rho}$ (Riemann curvature tensor) defined on $\bigcup_{x \in C(\mathcal{M})} \mathcal{O}(x) \cap \mathcal{M}$ with the curvature tensor $\Omega^{cb}_{\mu\nu}$ of the spin connection defined on $\bigcup_{x \in C(\mathcal{M})} T_x\mathcal{M}$. The chain of propositions is readily generated by aesop through iterative application of the type constructors within the inductive type:

$$\begin{aligned} g^{\xi\lambda} R_{\nu\xi\mu\rho} A_\lambda &= R_{\nu\mu\rho}^{\ \ \ \lambda} A_\lambda = R_{\nu\mu\rho}^{\ \ \ \lambda} A_\lambda \equiv \left[\nabla_\mu, \nabla_\nu\right] A_\rho = \left[\nabla_\mu, \nabla_\nu\right] e^a_\rho A_a \equiv e^a_\rho \Omega_{\mu\nu a}^{\ \ \ \ b} A_b \\ &= e^a_\rho \Omega_{\mu\nu a}^{\ \ \ \ b} e^\lambda_b A_\lambda = e^a_\rho \Omega_{\mu\nu}^{\ \ \ cb} \eta_{ca} e^\lambda_b A_\lambda \end{aligned} \tag{A1}$$

The symbol "$\equiv$" denotes definitional equality as distinguished from the propositional equality "$=$". We emphasize that the propositional validity of the expressions in Eq. (A1) stems merely from the fact that they can be typed correctly in Lean language, as checked by the kernel, and they follow rule-based inference. The chain of propositions in Eq. A1 yields

$$R_{\nu\xi\mu\rho} = g_{\xi\lambda} e^a_\rho \Omega_{\mu\nu}^{\ \ \ cb} \eta_{ca} e^\lambda_b \tag{A2}$$

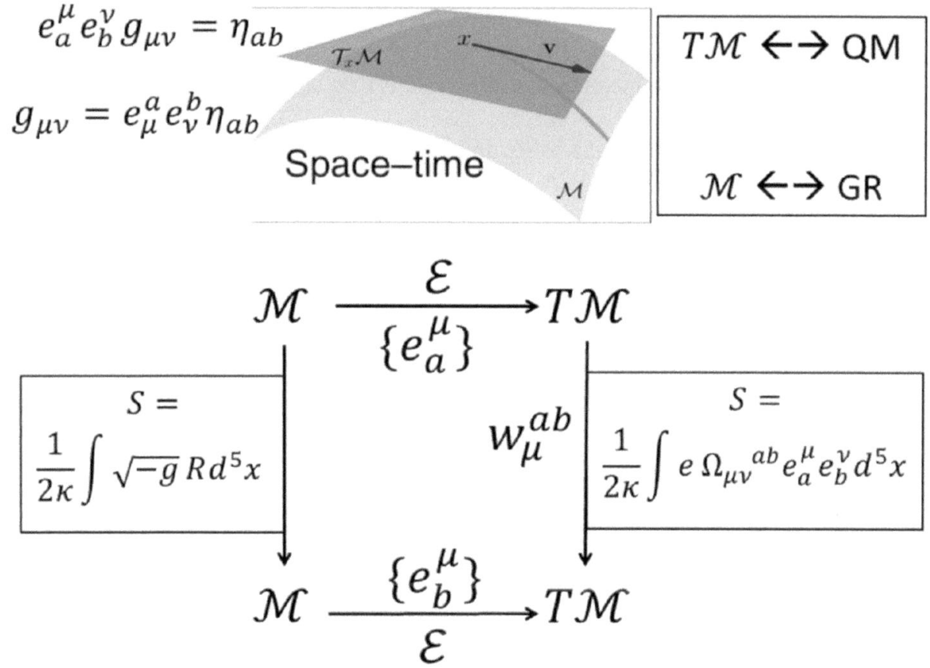

FIGURE A1 Commutative arrow diagram formalizing the compatibility between general relativity and quantum mechanics. Within the setting of the interactive theorem prover, the diagram represents an inductive type.

The chains of generated propositions (Eq. A1) are "compiled" via the "equation compiler" which is not part of the code base and can process multiple arguments sequentially. Its output (Eq. A2) is checked independently by the kernel.

The global effect of the inductive type delineated in Figure A1 is so vast that anyone interested in undertaking the project of formalizing quantum gravity along the premises described in this appendix must first and foremost corroborate that the kernel is endowed with the functionality for manipulating and checking the inductive type declarations enshrined in Figure A1. It is of the utmost importance to verify that the kernel, as it stands today, is capable of generating the type recursors and of checking whether newly introduced type constructors, such as Com, Cod, and Ctr, comply with the new type specifications. This is indeed the case as indicated in mathlib-ported files [2–5]. For example, the core lemma asserting that the derivative of a coordinate change between two charts is smooth on its source, enabling the type constructors "covariant derivative" [Cod] and commutator [Com], reads in Lean code:

theorem cont_diff_on_fderiv_coord_change

```
    {: Type u_1}
    [nontrivially_normed_field 𝕜] {E: Type u_2} [normed_add_comm_
group E]
    [normed_space 𝕜 E] {H: Type u_4} [topological_space H]
    (I: model_with_corners 𝕜 E H) {M: Type u_6} [topological_space
    M] [charted_space H M] [smooth_manifold_with_corners I M]
    (i j: ↥(charted_space.atlas H M)):
cont_diff_on 𝕜 T
(fderiv_within 𝕜 (⇑(j.val.extend I) ∘ ⇑((i.val.extend I).symm))
(set.range ⇑I))
```

((i.val.extend I).symm.trans (j.val.extend I)).source

The construction of the spin connection upon the relativistic atlas $\bigcup_{x\in C(\mathcal{M})} T_x\mathcal{M}$, enabling the spinor attachment on space–time, requires the following structure of charted space:

structure charted_space_core (H: Type u_5) *[topological_space H]* (M: Type u_6):

```
Type (max u_5 u_6)
(atlas: set (local_equiv M H))
(chart_at: M → local_equiv M H)
(mem_chart_source: ∀ (x: M), x ∈ (self.chart_at x).source)
(chart_mem_atlas: ∀ (x: M), self.chart_at x ∈ self.atlas)
(open_source: ∀ (ee': local_equiv M H), e ∈ self.atlas → e' ∈ self.
atlas → is_open (e.symm.trans e').source)
(continuous_to_fun:
∀ (ee': local_equiv M H),
e ∈ self.atlas → e' ∈ self.atlas → continuous_on
⇑(e.symm.trans e') (e.symm.trans e').source)
```

Thus, Lean may be adopted as the functional programing language exploited as an interactive theorem prover or inferential method to construct the warped graviton. In this context, the main goal is to propositionally identify the warped graviton in the ultraviolet completion of gravity within a fiber-bundle formulation of a gauge theory of Lorentz transformations.

To identify the graviton at the functional level, it becomes necessary to make two critical assumptions: (a) $C(\mathcal{M})$ is finite, and hence, the charts may be compiled as data in synchrony with mathlib and match against type Nat and (b) the manifold $\mathcal{M}$ does not have boundaries; otherwise, a completely new fiber-bundle formulation would be required to be ported to mathlib, introducing significant hurdles [7, 8]. Both assumptions are valid on purely physical grounds: According to GR, the topology of $\mathcal{M}$ must remain invariant throughout the universe evolution, hence $\mathcal{M}$ must remain compact, boundary-less and therefore, multiply connected, because that was its topology when it came into existence after the big bang [8]. Compactness of $\mathcal{M}$ obviously implies assumption a).

To construct the warped graviton, the type constructor Tes is introduced to build a stress-energy tensor $T_{\mu\nu}(\phi)$ out of a scalar field ϕ with potential $V(\phi)$:

$$T_{\mu\nu}(\phi)=\partial_\mu\bar{\phi}\partial_\nu\phi-g_{\mu\nu}\left[\frac{1}{2}g^{\alpha\beta}\partial_\alpha\bar{\phi}\partial_\beta\phi+V(\phi)\right] \tag{A3}$$

If the functional program language is informed declaratively that the scalar field is assumed to represent the massive graviton, the operational inductive type in Figure A2 infers that:

- The scalar field is nonlocal, that is, its support is made of contour lines, not points in $\mathcal{M}$,
- $\phi=\mathcal{W}$, where $\mathcal{W}$ is the Wilson loop that holonomically transduces space–time curvature.
- The warped graviton is massive since its vacuum expectation value $\mathcal{W}$ is the nonzero ϕ-value realizing the minimum of $V(\phi)$: $\langle\mathcal{W}\rangle = m_\phi/\sqrt{\lambda}$ (λ = self-coupling parameter, m_ϕ = graviton mass). These parameters determine the relevant inductive data types via the tail function that uses pattern matching.
- The chiral symmetry breaking that bestows mass to the warped graviton is thus informed by the potential $V(\mathcal{W})$.

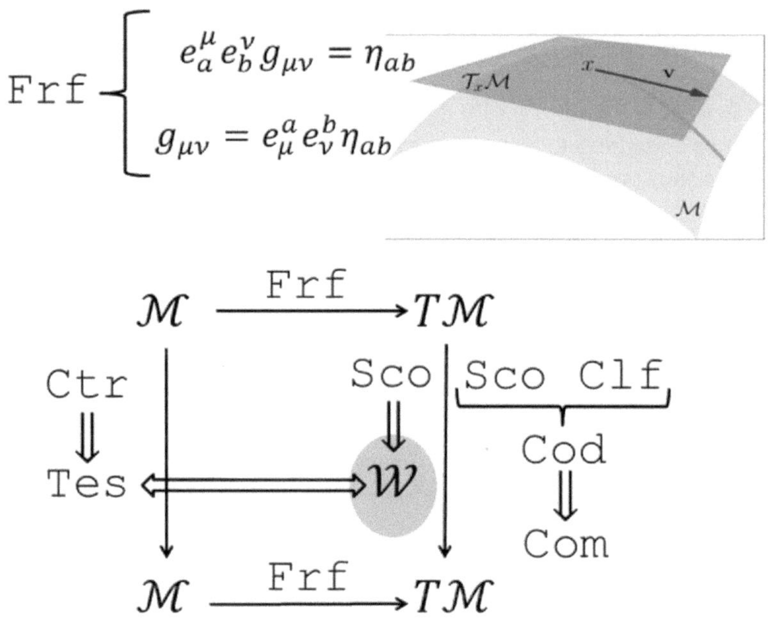

FIGURE A2 Discovery of the warped graviton (highlight) by implementing an inductive type for a gauge theory of gravity cast within a functional programming language with included declaration of type constructors.

It is noteworthy that it has proven possible to identify the warped graviton propositionally, based solely on type universes, dependent arrow types and inductive types, as described in Figures A1 and A2. The nonrecursive inductive type represented in Figure A1 constitutes the *structure* of the problem we are concerned with. The projection function that enables structure destruction correspond to the canonical projection $\pi: \bigcup_{x \in C(\mathcal{M})} T_x\mathcal{M} \to \bigcup_{\bar{x} \in \bar{C}(\mathcal{M}/\sim)} T_x(\mathcal{M}/\sim)$, where the quotient $\mathcal{M}/\sim$ is made up of classes ($\bar{x} \in \mathcal{M}/\sim$) modulo the equivalence relation "$\sim$". Two points belong to the same class if and onl y if they are specified by the same values of the observable coordinates in the standard 4D space–time. Thus, the standard model of particle physics (SM), predicated on $T(\mathcal{M}/\sim)$, represents the destruction of the structure represented in Figure A1 since it decodes the inductive type.

When a structure such as the one represented in Figure A1 is defined via the front-end structure command, Lean automatically generates all *a priori* projection functions, including the canonical projection that begets the SM. The projection onto the quotient and the concomitant extension of the propositional field (Figure A3) are entirely supported by the implemented Calculus of Constructions [6]. The implication is that the warped graviton cannot be understood as a standard extension of the SM, since the canonical projection destroys (lumps up) the propositional information that Lean used to encode the graviton. The situation is akin to consider integers modulo n, $\mathbb{Z} \to \mathbb{Z}_n$, the information on the integer (x) is lost, only the residue of the quotient x/n is encoded, while decoding becomes impossible.

The actual construction of the graviton field is a task suited for a large language model (LLM), wherein the formal proof can be checked rigorously by the proof assistant, free from hallucinations. A Lean Copilot [9], a framework for running LLM-based inference in Lean, becomes ideally suited for the task at hand as the copilot integrates seamlessly into the Lean workflow. It should be emphasized that Lean itself does not support machine learning (ML) tools from LLMs, so the copilot becomes instrumental for suggesting proof steps (tactic suggestion), completing intermediate proofs (proof search), and selecting the relevant premises (premise selection) toward the construction of the graviton field (Figure A4).

By itself, Lean would be hard to use for the final steps of construction of the graviton field because all the mathematics pertaining to "attaching spinors to space–time" (Figure A1) would need to be entered as code in the formalization stage. Lean indeed generated the "rule-based propositions" (Eq. A1) for spinor amalgamation with general relativity via its own proof searcher Aesop, but LLM becomes essential to generate the code and suggest the final tactics leading to the Wilson loop representing the graviton field. Current LLMs are trained, fine-tuned, and benchmarked following standard ML procedures but cannot integrate into the Lean workflow because Lean is designed without ML support. For this reason, it becomes essential to adopt a Lean copilot for building the seamlessly adapted LLM-based proof automation tools. To that effect, the C++ library CTranslate2

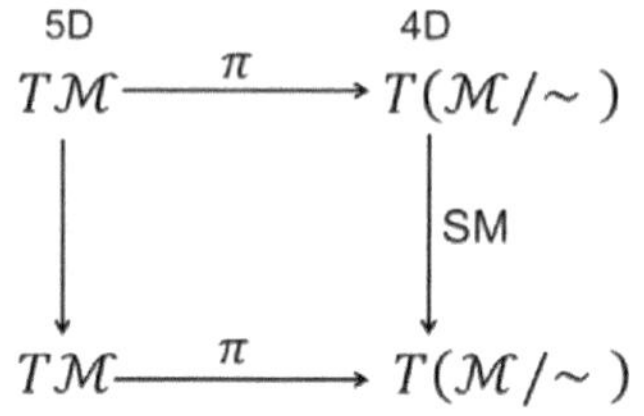

FIGURE A3 Canonical "destructive" projection onto the quotient space and concomitant extension of the propositional field [6]. The implication is that the warped graviton propositionally identified on $\bigcup_{x\in C(\mathcal{M})} T_x\mathcal{M}$ cannot be understood as a standard extension of the SM.

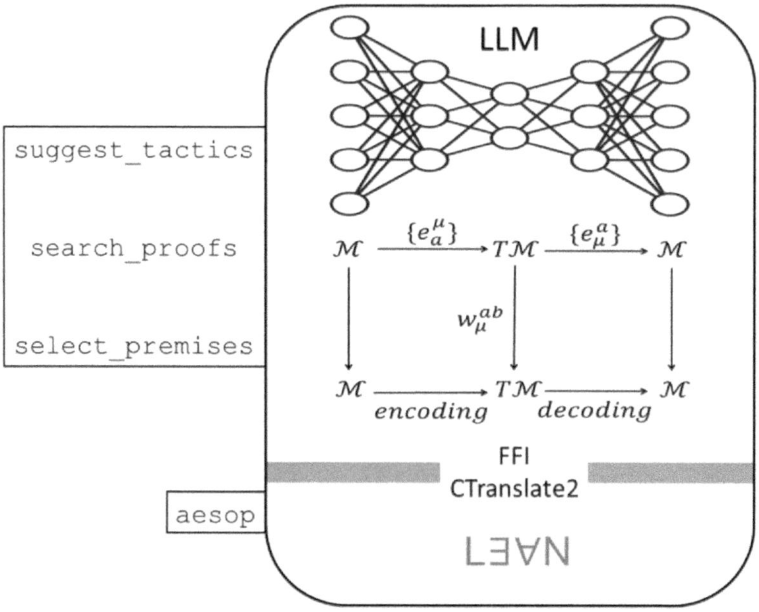

FIGURE A4 Architecture of Lean copilot with an LLM in the guise of a quantum gravity autoencoder (cf. [10]) specialized for the discovery of the warped graviton. LLM inferences are run within Lean with supporting framework provided by the copilot via a shared CTranslate2 C++ library at Lean's FFI.

is introduced for efficient inference with a transformer LLM representing the quantum gravity autoencoder [10] (Figure A4). The inference runs as a shared library through a Foreign Function Interface (FFI) within Lean (Figure A4).

The goal is declared to the LLM as generating a graviton tensor field $\mathcal{Y}_{\mu\nu}$ that yields the perturbative field $h_{\mu\nu}\left(g_{\mu\nu} = \eta_{\mu\nu} + h_{\mu\nu}\right)$ inferred by aesop in the zero mass limit:

$$\lim_{m\to 0} \mathcal{Y}_{\mu\nu} = h_{\mu\nu} \tag{A4}$$

The tactics suggested by the copilot are as follows: (A) apply Ctr to represent the field ϕ associated with the stress-energy tensor $T_{\mu\nu}(\phi)$ (Eq. A3) as contraction of the purported graviton field: $\phi = \mathcal{Y}^{\mu\nu} g_{\mu\nu}$. (B) Apply Ctr to contract Einstein's GR equation to yield (in 5D):

$$-\frac{3}{2}R = \kappa T(\phi), \left(T(\phi) = T_{\mu\nu}(\phi) g^{\mu\nu}\right) \tag{A5}$$

(C) Using Eq. A5, generate the relativistic action as $S(\phi) = -\frac{1}{3}\int \sqrt{-g}T(\phi) d^5x$ functionally dependent on $\mathcal{Y}^{\mu\nu} g_{\mu\nu}$.

The LLM-based intermediate proof prompted by LLM's search_proofs (Figure A4) then becomes

Identify $\phi = \mathcal{Y}^{\mu\nu} g_{\mu\nu}$ so that

$$\lim_{m\to 0} S(\phi) = \int \epsilon^{\mu\nu\rho\sigma} \left(\epsilon^{\alpha\beta\gamma}{}_{\sigma}\right) \partial_\mu h_{\nu\beta} \partial_\alpha h_{\rho\gamma} d^5x, \tag{A6}$$

where the expression on the r.h.s. in Eq. A6 is the action for the perturbative graviton field $h_{\mu\nu}$. Thus, applying the Euler–Lagrange type constructor [EuLa] to Eq. A6 yields the graviton field:

$$\mathcal{Y}^{\mu\nu} = \mathcal{P}_\gamma \exp\left(\oint_\gamma \hat{w}_\lambda{}^{ab} e^\mu_a e^\nu_b dx^\lambda \right) \tag{A7}$$

This expression is certifiably correct, since the Lean kernel verifies the rules-based proposition prompted by aesop (Figure A4):

$$\lim_{m\to 0} \mathcal{Y}_{\mu\nu} = \lim_{\oslash\gamma\to 0} \mathcal{Y}_{\mu\nu} = h_{\mu\nu} \tag{A8}$$

REFERENCES

1. Christiansen DT (2024) Functional Programming in Lean. Release 4.1.0. Microsoft Corporation. https://lean-lang.org/functional_programming_in_lean/
2. https://leanprover-community.github.io/mathlib_docs/geometry/manifold/vector_bundle/tangent
3. https://leanprover-community.github.io/mathlib_docs/geometry/manifold/vector_bundle/basic
4. https://leanprover-community.github.io/mathlib_docs/geometry/manifold/charted_space

5. https://leanprover-community.github.io/mathlib_docs/topology/fiber_bundle/basic
6. Avigad J, de Moura L, Kong S, Ullrich S, et al. (2024) Theorem Proving in Lean 4. https://leanprover.github.io/theorem_proving_in_lean4/
7. Bordg A, Cavalleri N (2021) Elements of Differential Geometry in Lean A: Report for Mathematicians. arxiv.2108.00484.
8. Fernández A (2024) *Artificial Intelligence Models for the Dark Universe: Forays in Mathematical Cosmology*. Routledge, Boca Raton, FL.
9. Song P, Yang K, Anandkumar A (2024) Towards Large Language Models as Copilots for Theorem Proving in Lean. arXiv:2404.12534.
10. Fernández A (2024) *Topological Dynamics for Metamodel Discovery with Artificial Intelligence*. CRC Press, Boca Raton, FL.

Index